Breaking Waves

International Union of Theoretical and Applied Mechanics

M. L. Banner, R. H. J. Grimshaw (Eds.)

Breaking Waves

IUTAM Symposium Sydney, Australia 1991

Springer-Verlag
Berlin Heidelberg New York
London Paris Tokyo
Hong Kong Barcelona Budapest

Prof. Michael L. Banner
Prof. Roger H. J. Grimshaw

School of Mathematics
University of New South Wales
Kensington, New South Wales
Australia

ISBN-13:978-3-642-84849-0 e-ISBN-13:978-3-642-84847-6
DOI: 10.1007/978-3-642-84847-6

Library of Congress Cataloging-in-publication Data
Breaking waves : IUTAM symposium. Sydney, Australia, 1991 /
M. L. Banner, R. H. J. Grimshaw (eds.).
At head of title: International Union of Theoretical and Applied Mechanics.
Papers from the Symposium on Breaking Waves, held at the University of New South Wales.
July 15-19, 1991.
ISBN-13:978-3-642-84849-0
1. Wave-motion, Theory of--Congresses.
I. Banner, M. L. (Michael L.). II. Grimshaw, R. H. J. (Roger H. J.). III. International Union of Theoretical and Applied Mechanics.
QA927.B74 1992 532'.0593--dc20 92-30806

Softcover reprint of the hardcover 1st edition 1992

Typesetting: Camera ready by authors

61/3020-5 4 3 2 1 0 – Printed on acid-free paper.

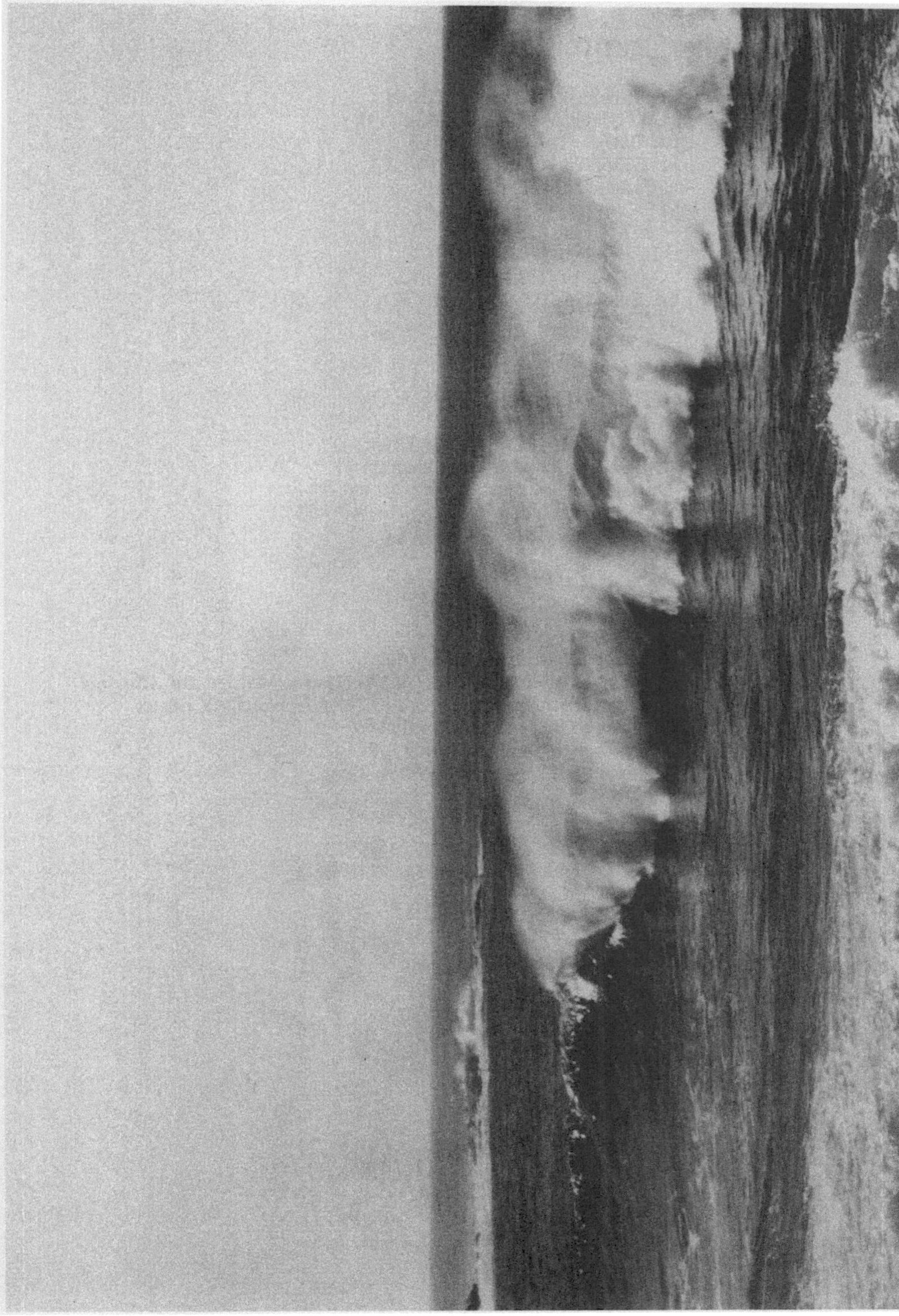

Scientific Committee

R. Grimshaw (Australia) (Chair)
M.L. Banner (Australia) (Secretary)
J.A. Battjes (The Netherlands)
M.A. Donelan (Canada)
Sir James Lighthill (UK)
M.S. Longuet-Higgins (USA)
W.K. Melville (USA)
D.H. Peregrine (UK)
O.M. Phillips (USA)
I.A. Svendsen (USA)
Y. Toba (Japan)

Local Organizing Committee

Professor Roger Grimshaw (Chair)
Professor Michael Banner (Secretary)
Dr Peter Blennerhassett (Treasurer)
Dr Ian Jones
Professor David Wilkinson

Sponsors

Australian Mathematical Society
Applied Mathematics Division, Australian Mathematical Society
International Union of Theoretical and Applied Mechanics
Institute of Engineers Australia
U.S. Office of Naval Research
University of New South Wales

Preface

Wave breaking is a commonly occurring phenomena associated with wave motion in fluids, often inducing significant effects which are of fundamental and technological importance, A familiar illustration is provided with white-capping and microbreaking of the wind-driven ocean surface waves, which is believed to play an important part in the transfers of momentum, mass and heat across the air-sea interface, as well as in the production of underwater ambient noise and augmented microwave backscatter. The enhanced hydrodynamic forces associated with the breaking of the more energetic ocean wave components constitute a significant challenge in ocean engineering, coastal engineering and naval architecture. Other less conspicuous but equally important manifestations are the breaking of internal waves and the filamentation of vorticity interfaces.

Despite recent theoretical and observational progress towards a more complete understanding of wave breaking, mathematical descriptions of its onset and consequences are presently lacking. The aim of this Symposium was to bring together theoretical and observational expertise, with the goal of determining the current state of knowledge of wave breaking and providing a stimulus to future research.

The Symposium focused on water waves of all scales from capillary waves to ocean swell, but also considered internal waves and the filamentation of vorticity interfaces. Specific topics included were:

- Fundamental theoretical studies; wave instabilities; routes to breaking.
- Models of wave breaking.
- Field observations, including statistical information.
- Laboratory studies.
- Shoaling waves, breaking waves on currents, breaking induced by the motion of a ship.
- Consequences of wave breaking for oceanic studies, microwave remote sensing, role of breaking in air-sea fluxes, wind-wave prediction models; acoustics of wave breaking; bubble formation.
- Wave breaking in industrial applications, wave forces on structures.

The Symposium on Breaking Waves was proposed to the General Assembly of the International Union of Theoretical and Applied Mechanics (IUTAM) at Grenoble in 1988, and it was decided it would be held in 1991 in Sydney. The Symposium was held on the campus of the University of New South Wales from July 15-19, 1991. There were 63 participants from 15 countries, and 45 contributed papers, covering fundamental theoretical studies, wave breaking models, field and laboratory observations, and industrial and scientific consequences. Through these contributions and the extensive informal discussions, we believe the Symposium achieved its objectives of establishing the current state of knowledge of wave breaking, and providing a stimulus for future research.

The scientific programme was formulated by the Scientific Committee, and we thank its members for their invaluable advice. All local arrangements were carried out by the Local Organising Committee and we thank them, and many others in the School of Mathematics at the University of New South Wales who contributed to the smooth running of the Symposium. Financial support was generously provided by IUTAM, the U.S. Office of Naval Research, the Institution of Engineers Australia (National Committee on Coastal and Ocean Engineering), and the Australian Mathematical Society.

Michael L. Banner
Roger H.J. Grimshaw

School of Mathematics
University of New South Wales
Kensington, NSW 2033
Australia

2nd March, 1992

Participants

Prof. M.L. Banner, School of Mathematics, U.N.S.W., P.O. Box 1, Kensington N.S.W. 2033, Australia

Dr. J.M. Becker, School of Mathematics, U.N.S.W., P.O. Box 1, Kensington N.S.W. 2033, Australia

Dr. Eugene Benilov, School of Mathematics, U.N.S.W., P.O. Box 1, Kensington N.S.W. 2033, Australia

Dr. P.J. Blennerhassett, School of Mathematics, U.N.S.W., P.O. Box 1, Kensington N.S.W. Australia

Dr. Steve Buchan, Steedman Science and Engineering, 31 Bishop Street, Jolimont W.A. 6014, Australia

Emeritus Professor V.T. Buchwald, School of Mathematics, U.N.S.W., P.O. Box 1, Kensington N.S.W. 2033, Australia

Dr. L. Cavaleri, Istituto Per Lo Studio Della, Dinamica Delle Grandi Masse, ISDGM-CN3, 1364 San Polo, 30125 Venice, Italy

Prof. S. Dai, Shanghai Inst. Appl. Math. & Mech, Shanghai, P.R. of China

Prof. Rolf Deigaard, Inst. of Hydrodynamics & Hydr. Eng., Technical University of Denmark, ISVA - Bldg. 115, DK-2800 Lyngby, Denmark

Dr. R.E. Dewar, Dept. of Physics, Australian National University, G.P.O. Box 4, Canberra A.C.T. 2601, Australia

Dr. Igor N. Didenkulov, Inst. of Applied Physics, USSR Academy of Sciences, 46 Uljanov Str., Nizhny Novgorod, 603600, U.S.S.R.

Dr. W. Drennan, National Water Research Institute, Canada Centre for Inland Waters, P.O. Box 5050, Burlington, Ontario, Canada L7R 4A6

Prof. Kristian B. Dysthe, IMR, University of Tromsö, 9000 Tromsö, Norway

Dr. Naoto Ebuchi, Dept. of Geophys., Tohoku University, Aoba, Sendai 980, Japan

Dr. David M. Farmer, Inst. of Ocean Sciences, P.O. Box 6000, Sidney, BC, Canada V8L 4B2

Prof. R.H.J. Grimshaw, School of Mathematics, U.N.S.W., P.O. Box 1, Kensington N.S.W. 2033, Australia

Dr. Peter Holloway, Dept. Geography and Oceanography, University College, Australian Defence Force Academy, Canberra A.C.T. 2600, Australia

Dr. Hua Feng, First Institute of Oceanography, State Oceanic Administration, P.O. Box 98, Qingdao – Shandong 266003, P.R. China

Prof. Herbert E. Huppert, School of Mathematics, U.N.S.W, P.O. Box 1, Kensington N.S.W. 2033, Australia

Dr. Paul Hutchinson, Centre for Env. Mechanics, C.S.I.R.O., G.P.O. Box 821, Canberra A.C.T. 2601, Australia

Dr. Peter Janssen, K.N.M.I., P.O. Box 201, 3730 AE De Bilt, The Netherlands

Dr. Ian S.F. Jones, Ocean Technology Group, Sydney University, N.S.W. 2006, Australia

Dr. Kimmo Kahma, Merentutkimuslaitos, Finnish Inst. of Marine Res., P.O. Box 33, SF-00931 Helsinki, Finland

Dr. Chang-Gu Kang, Korea Res. Inst. of Ships/Ocean Eng., P.O. Box 1, Daeduk Science Town, Daejeon 305-606, Korea

Prof. Kristina B. Katsaros, Dept. Atmospheric Sc. AK-40, University of Washington, Seattle, W.A. 98195, U.S.A.

Dr. Bernard C. Kenney, Natl. Hydrology Res. Institute, 11 Innovation Boulevard, Saskatoon, Saskatchewan, Canada S7N 3H5

Prof. Sergei A. Kitaigorodskii, Dept. Earth & Planetary Sci., The Johns Hopkins University, Olin Hall, Baltimore, Maryland 21218, U.S.A.

Dr. Gert Klopman, Waterloopk Lab de Voorst, Delft Hydraulics, P.O. Box 152, 8300 Ad. Emmeloord, The Netherlands

Dr. Seung-Joon Lee, Dept. Naval Arch. & Ocean Eng., Chungnam Natl. University, Daejeon 305-764, South Korea

Professor J.C. Li, Institute of Mechanics, Chinese Academy of Sciences, No. 15 Zhong Guan Cun Rd., Beijing 100080, P.R. China

Dr. Jiyue Li, Ocean Engineering Lab., Univ. of California (at Santa B.), Santa Barbara, CA 93106, U.S.A.

Dr. Li Luping, First Institute of Oceanography, State Oceanic Administration, P.O. Box 98, Qingdao – Shandong 266003, P.R. China

Prof. Sung C. Ling, Dept. of Mech. Engr., The Catholic Uni. of America, Washington, D.C. 20064, U.S.A.

Prof. M.S. Longuet-Higgins, Inst. of Nonlinear Science, U.C.S.D., 9500 Gilman Drive, La Jolla CA 92093-0402, U.S.A.

Dr. Stanislaw R. Massel, Austr. Inst. of Marine Science, PMB 3, Townsville, M.C. - QLD. 4810, Australia

Dr. W.D. McKee, School of Mathematics, U.N.S.W., P.O. Box 1, Kensington N.S.W. 2033, Australia

Prof. W. Kendall Melville, Dept. of Civil Eng. BLDG 48-331, M.I.T., Cambridge, Massachusetts 02139, U.S.A.

Dr. G.N. Mercer, Dept. of Applied Mathematics, The University of Adelaide, G.P.O. Box 498, Adelaide S.A. 5001, Australia

Dr. Mark Merrifield, School of Mathematics, U.N.S.W., P.O. Box 1, Kensington N.S.W. 2033, Australia

Prof. Jason H.F. Middleton, School of Mathematics, U.N.S.W., P.O. Box 1, Kensington N.S.W. 2033, Australia

Prof. Hisashi Mitsuyasu, Res.Inst. for Applied Mechanics, Kyushu University 87, Kasuga Koen 6-1, Kasuga, 816 Japan

U Sann Myint, School of Mathematics, U.N.S.W., P.O. Box 1, Kensington N.S.W. 2033, Australia

Prof. D.H. Peregrine, School of Mathematics, University of Bristol, Bristol BS8 1TW, England

Dr. John S. Reid, Div. of Oceanography, C.S.I.R.O., G.P.O. Box 1538, Hobart TAS. 7001, Australia

Dr. L. Shemer, Dept. of Fluid Mechanics, Faculty of Engineering, Tel-Aviv University, Ramat Aviv 69978, Israel

Prof. Victor I. Shrira, Shirshov Inst. of Oceanology, USSR Academy of Sciences, Krasikova 23, Moscow 117218, U.S.S.R.

Dr. Ming-Yang Su, Naval Oceanographic and Atmospheric Research Laboratory, Stennis Space Center, Mississippi 39529, U.S.A.

Prof. Alexander M. Sutin, Inst. of Applied Physics, Academy of Sciences of the USSR, 46 Uljanov Street, 603600 Gorky, U.S.S.R.

Prof. Ib A. Svendsen, College of Engineering, Dept. of Civil Engineering, University of Delaware, 137 Dupont Hall, Newark, Delaware 19716, U.S.A.

Dr. Mitsuhiro Tanaka, Dept. of Appl.Maths - Fac. Eng., Gifu University, 1-1 Yanagido, 501-11 Gifu, Japan

Ms. Yongming Tang, School of Mathematics, U.N.S.W., P.O. Box 1, Kensington N.S.W. 2033, Australia

Dr. A.F. Teles da Silva, Prog. Eng. Mec./Coppe, Caixa Postal 68501, CEP 21945 - Univ. Fed. Do, Rio de Janeiro - RJ, Brazil

Prof. Yoshiaki Toba, Dept. of Geophysics, Tohoku University, Aoba, Sendai 980, Japan

Dr. Hiroshi Tomita, Ship Research Inst., 6-38-1 Shinkawa, Mitaka, Tokyo 181, Japan

Prof. E.O. Tuck, Dept. of Mathematics, University of Adelaide, G.P.O. Box 498, Adelaide S.A. 5001, Australia

Prof. Marshall P. Tulin, Ocean Engineering Lab., Univ. of California (at Santa B.), Santa Barbara, CA 93106, U.S.A.

Dr. Frank Viera, School of Mathematics, U.N.S.W., P.O. Box 1, Kensington N.S.W. 2033, Australia

Prof. David L. Wilkinson, School of Civil Engineering, U.N.S.W., P.O. Box 1, Kensington N.S.W. 2033, Australia

Dr. Takashi Yasuda, Dept. of Civil Engineering, Gifu University, Yanagido 1-1, Gifu 501-11, Japan

Dr. Harry Yeh, Dept. of Civil Engineering 4020, FX-10, University of Washington, Seattle WA 98195, U.S.A.

Prof. Yi Zengxin, NRCMEF, No. 8 Da Hui Si, Haidian Division, Beijing, 100081 P.R. China

Dr. Ian R. Young, Dept. of Civil and Maritime Eng., Aust. Defence Force Academy, Northcott Drive, Canberra A.C.T. 2600, Australia

Prof. V.E. Zakharov, Landau Institute for Theoretical Physics, Moscow, U.S.S.R.

Contents

1. Keynote Lectures

2. Experiments and Observations

3. Theoretical and Numerical Results

5. Wave Breaking in other Contexts

Keynote Lectures

The Physical Consequences of Wave Breaking in Deep Water

M.L. BANNER and M.A. DONELAN

School of Mathematics
University of NSW
Sydney, 2033, Australia
and
National Water Research Institute
Canada Centre for Inland Waters
Burlington, Ontario L7R 4A6, Canada

Summary

The evolving technology of remote sensing and the increasing interest in the distribution of anthropogenic gases have, in recent years, focused attention on the ocean surface. In particular, the complex process of wave breaking has a wide range of consequences for remote sensing, momentum, heat and mass transfer and oceanic acoustics. In this paper we consider some of the more important effects of wave breaking that have recently been subjected to intensive observational scrutiny. These include: changes in microwave reflectivity; stress and gas transfer enhancement; formation of bubble clouds and sound generation; dissipation of surface waves.

Introduction

It is well recognized that at moderate and high winds the sea surface is aerodynamically rough. What is less certain is the relative contribution of various scales (wavelengths) to the roughness. The success of Charnock's formula in describing the roughness length in open ocean conditions suggests that the small waves (several times the peak wavenumber) are the principal stress receptors. This indeed appears to be so when the sea state is well-developed (wave age ~ 1). Underdeveloped waves, on the other hand, produce relatively high drag coefficients and these appear to be related to the relative speed of the peak waves with respect to the wind speed (Kitaigorodskii, 1970; Donelan, 1990). In fact, at short fetch, the waves at the spectral peak are strongly forced and a significant part of the total momentum transfer is absorbed by the waves near the peak (Figure 1).

In fully rough flow the momentum transfer across the interface is due to form drag, i.e., differential pressure on upwind versus downwind slopes. All the momentum transferred in this way enters the wave field initially, but only a fraction remains as wave momentum — at full development the fraction is identically zero. Deductions from fetch-limited

M. L. Banner, R. H. J. Grimshaw (Eds.)
Breaking Waves
IUTAM Symposium Sydney/Australia 1991

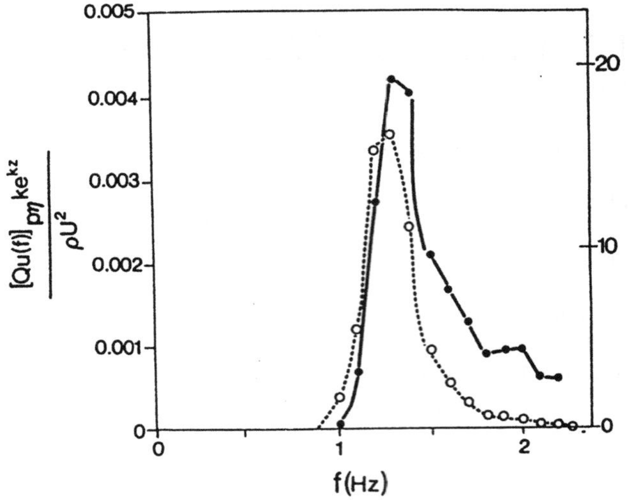

Fig. 1. A laboratory wave spectrum $S_{\eta\eta}(f)$ (broken line) and associated pressure-elevation quadrature spectrum (solid line). The quadrature spectrum has been adjusted by the exponential decay to yield the momentum transfer to the waves. Further normalization by ρU^2 yields the spectrum of contributions to the drag coefficient.

growth curves suggest that developing waves retain about 5% of the total momentum flux (Hasselmann et al., 1973). Direct measurements of differential growth and surface stress (Donelan, 1978) show that as much as 25% may be retained by the waves at very short fetch, in laboratory tanks, but 0 to 8% is more typical of field conditions (Figure 2) and the atmospheric stability has a pronounced influence on it. Presumably, the effect of stability is to alter the wavenumber distribution of waves contributing to the stress transfer. The larger turbulent scales associated with unstable conditions may bias the transfer distribution toward longer waves. Since the short waves are believed to be quasi-saturated, any momentum transferred to them is quickly handed over to currents via enhanced breaking.

Are the short waves in the equilibrium range, indeed, nearly saturated and is the limiting process correctly identified as wave-breaking? These seem to be unresolved issues although a good deal of recent work has attempted to clarify the spectral balance above the peak (Kitaigorodskii, 1983; Phillips, 1985; Donelan and Pierson, 1987; Banner, 1990c). Figure 3 shows two frequency spectra obtained in very different winds (1.2 m/s and 17.4 m/s). In the former case the dominant waves are recent swell and, while the

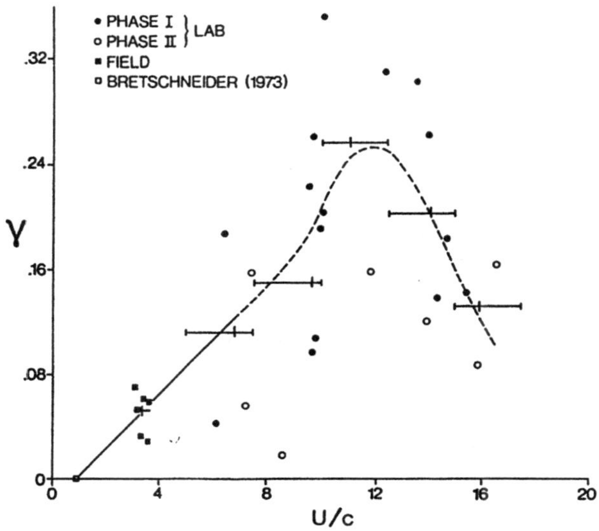

Fig. 2. The fraction of wind momentum retained by waves Υ versus the inverse wave age U/c. For the laboratory data the larger crosses indicate the averages of the data assembled in groups of width 2.5 in U/c. The vertical bar indicates the average value of U/c in the group. The cross representing the field data has been placed at the value of Υ corresponding to neutral atmospheric stability. The point of full development ($U/c = 0.83$, $\Upsilon = 0$) according to Bretschneider (1973) provides another estimate of Υ. The straight line (solid) at lower values of U/c provides convenient access to Υ for wave prediction purposes. The extension of the line to higher values of U/c (dashed) serves only to illustrate the trend in Υ.

wind is well above the minimum for sustaining short wave growth (Kahma and Donelan, 1987), it is very light indeed and it is difficult to believe that the same limiting process maintains the same shape of the high frequency spectra, i.e., ω^{-4}. The ratio of spectral levels is roughly the same as the square root of the wind speed ratio. At high wind speeds there is little doubt that wave breaking is the principal limiting process.

Despite the widespread occurrence of wave breaking in the ocean, our present knowledge of this process is incomplete. However, existing studies indicate that it appears to contribute significantly to basic geophysical air-sea interfacial transfer processes including momentum, mass and heat transfers, as well as in related areas such as underwater ambient noise generation and microwave backscatter from the sea surface. This contribution provides a brief overview of present knowledge, highlighting areas where further research is needed in order to refine our present physical understanding and predictive modelling capabilities.

Breaking Types

Wave breaking is a familiar feature of the wind-driven sea surface, occurring most visibly

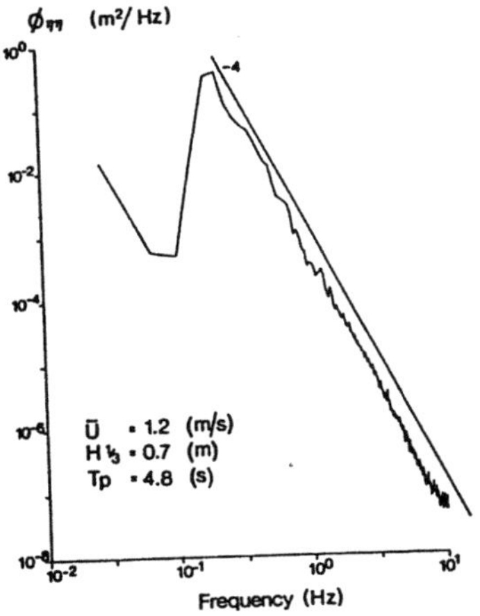

RUN 85111

Fig. 3 a. The spectrum of surface elevation for the case of recent swell. The −4 power law fitted to Fig. 3 b is also shown here.

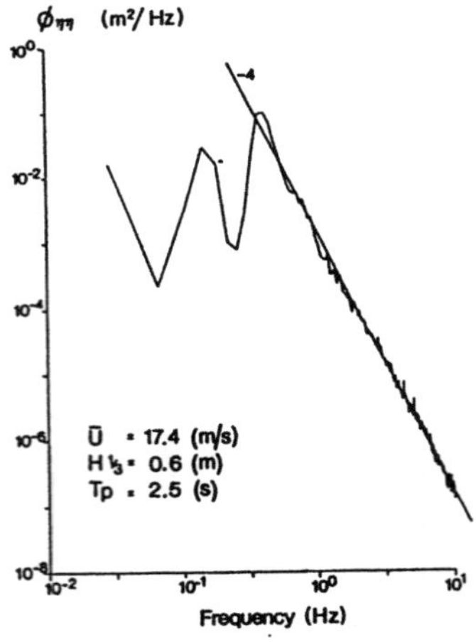

RUN 85159

Fig. 3 b. The spectrum of surface elevation for the strongly wind-generated case. Above the peak, the spectrum conforms to a −4 power law (Donelan et al, 1985).

in the form of whitecaps of various scales. It also occurs on a more widespread basis as *microscale* breaking, but in this form it is less conspicuous because of its lack of air entrainment and smaller scale, involving wavelengths less than about 30 cm. The most prevalent form of deep water wave breaking occurs as spilling breakers, where the breaking crest water surges gradually forward to spill down the forward face of the breaker. The more violent form known as the plunging breaker, though more typical of shallow water breaking, can also be observed. This form is characterised by the breaking crest water jetting forward, and engulfing a pocket of air as it collapses on to the forward face of the breaker.

Momentum and energy transfer

Enhanced Wind Input

Several authors have reported significant influence of wave breaking on the interfacial stress at the onset of breaking. Using a quasi-steady breaking wave realization, Banner and Melville (1976) identified the presence of locally separated flows in the air flow associated with wave breaking, and noted a consequent significant enhancement of the local wind stress. The association of strong air flow separation effects with breaking

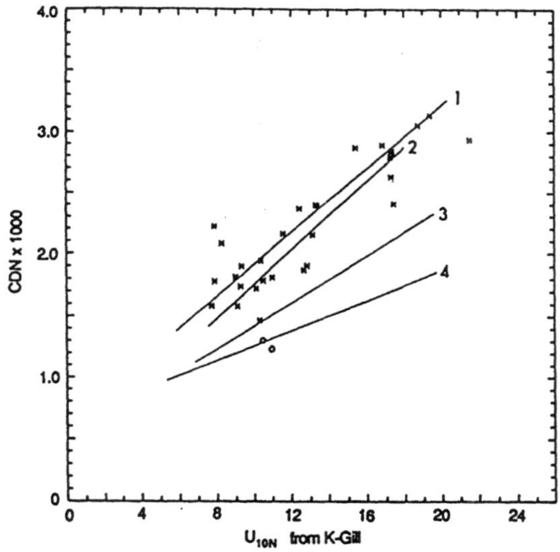

Fig. 4. Neutral drag coefficient, CDN, vs. ten-meter height wind speed for HEXMAX data. Asterisks indicate measurements in slightly unstable conditions, small circles for slightly stable conditions.
Line 1 is best fit to data, Line 2 from Donelan (1982), Line 3 from Geernaert, et al. (1986), Line 4 from Smith (1980) represents open ocean conditions. From De Cosmo et al. (1988).

wind waves was explored in greater depth in detailed flow visualisation studies over propagating wind waves by Kawai (1981, 1982). Strong local enhancements in the wind stress were also found. However, investigations seeking to identify relative contributions to the stress enhancement as normal or tangential have produced somewhat conflicting results. Detailed hydrogen bubble observations in the immediately underlying water by Okuda *et al.* (1977), Okuda (1982) suggest that strong wave-coherent tangential stresses dominate, peaking on the windward face of the wave form. However, measuring near-surface pressure distributions, Banner (1990a) reported a near-doubling of the wind stress when comparing incipient and actively breaking wind-driven waves, with (separation-induced) phase shifts in the wave-coherent pressure producing wave drag increases of about the same size as the total stress enhancement, indicating a lesser influence of enhanced tangential stress. Therefore basic questions still remain on the relative importance of normal and tangential stresses in the actual mechanism of enhanced momentum flux from the wind associated with breaking wind waves.

Such local considerations are germane to the heightened recent interest in the sea state dependence of the wind stress, or equivalently, in the aerodynamic drag coefficient C_d of the sea surface. The trends of recent observational studies shown in Figure 4 indicate a significant sea state dependence in the wind stress [e.g. Donelan (1982), Geernaert, Katsaros and Richter (1986), Smith and Anderson (1988), DeCosmo, Katsaros and Lind (1988)]. These findings are of basic importance in the parameterisation of C_d , a primary parameter in modelling wind waves, ocean circulation, as well as weather and climate prediction, where C_d parameterisations in current use contain *no* sea state dependence.

Theoretical models to account for these observed trends have been proposed [e.g. Janssen (1989), Toba *et al.* (1990), Banner (1990b), Plant (1991)]. However, both observationally and theoretically, there is still a lack of consensus on this problem. The model proposed by Toba *et al.* (1990) predicts that C_d increases with wave age, while most other model studies predict the opposite trend. On the one hand there is a notable lack of wind stress and wave data for strong winds and long fetches: the wind stress is difficult to measure reliably, as are the desired complementary wave measurements (e.g. directional wave spectra, breaking wave probabilities). On the other hand, the proposed spectral models are very sensitive to the form of the wave spectrum and its assumed variation with wave development. Model studies suggest that although the momentum flux to the energetic wind wave components is locally large, the contribution to the wind stress from wave drag is significantly greater from the much larger spectral range embracing wind wave components shorter than about one tenth of the dominant wavelength. It appears that a reliable improvement in the C_d parameterisation will require considerable further research effort, and will need to investigate closely the contributions from the short wind wave scales, including microscale breaking waves.

Breaking influence on spectral transfers

Modulation of very short wind waves by longer waves is an important mechanism influencing the microwave backscattering process from the sea surface, and underlies the data interpretation from a potentially very useful class of active microwave remote sensing sea state and marine meteorological instruments. In this context, an intriguing feature of a wind wave field concerns its modulation by an underlying large scale wave field propagating in the wind direction. In laboratory studies, it has been found by a number of investigators that the short wave spectrum attenuates markedly as the modulating wave steepens [e.g. Mitsuyasu (1966), Phillips and Banner (1974), Donelan (1987)]. Phillips and Banner (1974) attempted to explain this effect on the basis of enhanced breaking due to enhanced wind drift layer influence, but the influence of this mechanism was questioned by Wright (1976). Donelan (1987) observed that the input source term was largely unaffected by the influence of the modulation and suggested that modification

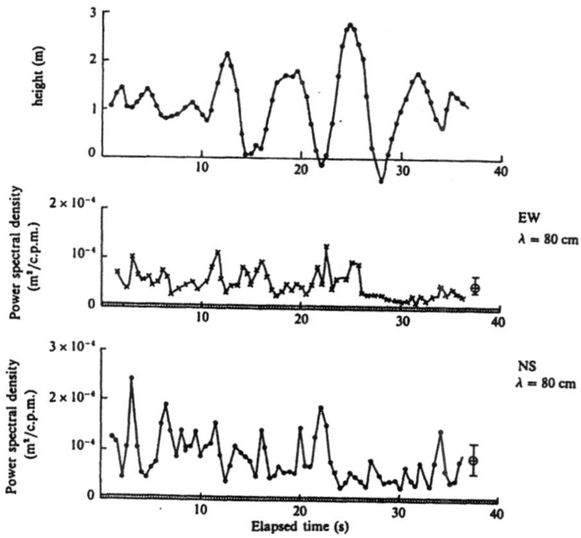

Fig. 5. Time series of power spectral density of wavenumber components during an interval embracing a whitecapping event, uncorrected for noise. Wavelength = 0.80 m. Experiment 2. Water height is with respect to an arbitrary origin. Noise levels are shown for each wavenumber component as a shaded band. 95% confidence limits about the mean level are shown at the right.

of the nonlinear wave-wave interactions was responsible for the observed behaviour. Also, Longuet-Higgins (1987) proposed a two-scale model which considered randomness in both long and short waves and examined the effects of breaking of the short waves under conditions where the short waves were re-generated by the wind. His predictions were in qualitative agreement with the observations. With its potential to elucidate short wave modulation in a continuous spectrum, further research effort is needed to understand the underlying physical mechanisms in such two-scale problems. When the large scale wind wave is itself involved in breaking, it produces a marked local attenuation of the entire short wind wave spectrum in its wake [Banner *et al.* (1989), Figure 5]. This highly nonlinear behaviour has potentially important implications for air-sea interaction and remote sensing and warrants further study, particularly for strong wind forcing conditions with extensive whitecapping.

Momentum flux to currents

Duncan (1981), (1983) studied quasi-steady wave breaking associated with a towed subsurface hydrofoil. He reported a series of interesting geometrical and dynamical relations for these quasi-steady breakers, including the useful result that the stress per unit width of breaking crest was about $0.06c^4/g$, where c is the phase speed, indicating the effect of wave scale. As part of a very detailed laboratory study of ***transient*** breaking, Rapp and Melville (1990) reported systematic results for the breaking-induced momentum flux to the underlying flow due to wave ***packets*** involved in varying degrees of wave breaking embracing incipient, spilling and plunging breaking. Their results showed somewhat lower mean breaking wave stress levels than the quasi-steady breaking wave stress levels reported by Duncan. Similar investigations involving near-monochromatic wave ***trains*** have provided very useful quantitative information on the very interesting three-dimensional instability properties associated with the approach to breaking [e.g. Su and Green (1986)] and significant frequency downshifting which occurs following breaking [e.g. Lake and Yuen (1978)]. Studies exploring these nonlinear aspects are continuing, both theoretically and observationally [e.g. Trulsen and Dysthe (1990)] .

The application of the results of such laboratory studies of the hydrodynamic consequences of breaking waves to broader wind wave spectral situations has not yet been widespread. However, applying these laboratory findings in conjunction with spectral breaking distributions can provide useful guidance in the field: Phillips (1985) invoked Duncan's findings as a basis for modeling the spectral wave energy dissipation rate and the spectral momentum flux from the wave field to the underlying currents.

Kinetic energy dissipation due to wave breaking

The important role of the turbulent kinetic energy distribution in the dynamics of the upper ocean has led to numerous attempts to measure and parameterize it in terms of air-sea interaction variables. Most results are expressed in wall layer coordinates and, although the scatter of data points is usually about an order of magnitude, agreement with the wall layer value $\varepsilon = \frac{u_*^3}{\kappa z}$ is generally claimed (Jones, 1985; Soloviev et al., 1988). Some near surface measurements in Lake Ontario (Kitaigorodskii et al., 1983) show dissipation rates that are 2 or 3 orders of magnitude above $\frac{u_*^3}{\kappa z}$. New measurements of such high dissipation rates are to be reported at this conference (Drennan et al., 1991). It is shown that these high dissipation rates occur near the surface in breaking wave conditions and are entirely consistent with the expected energy transfer to a wide range of steep waves at and above the spectral peak.

Heat and mass transfer

Of obvious geophysical importance, the parameterisation of mass and heat transfer across the air-sea interface has attracted considerable interest. Despite the intrinsic difficulties of measuring transfer rates within very thin conduction sublayers adjacent to the interface, recent investigators have made substantial progress in identifying wind-wave influences on these processes. In an authoritative recent paper, Jähne *et al.* (1987) review the status of the field and present detailed measurements that improve our knowledge on the underlying role of the wave field. They report that the mass transfer rates are enhanced by the wave field, closely tracking the mean squared wave slope of the whole wave spectrum, rather than just the capillary waves previously thought to be dominant. Interestingly, the photographs [Figures 6 and 7] of the wave field shown in their paper reveal the presence of microscale breaking, even at the low wind speed of 2.7 m/s. Thus while this was not noted explicitly by Jähne *et al.* in their discussion,

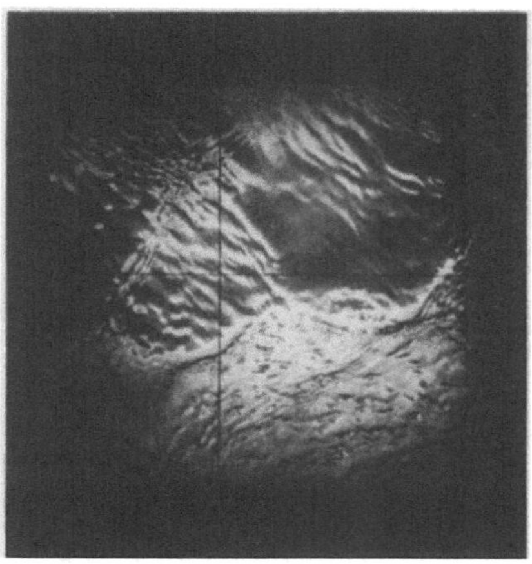

Fig. 6. Photograph of wind waves at 21 m fetch in the large IMST wind/wave facility at 2.7 m/s wind speed. The sector shown is 40 cm times 40 cm.

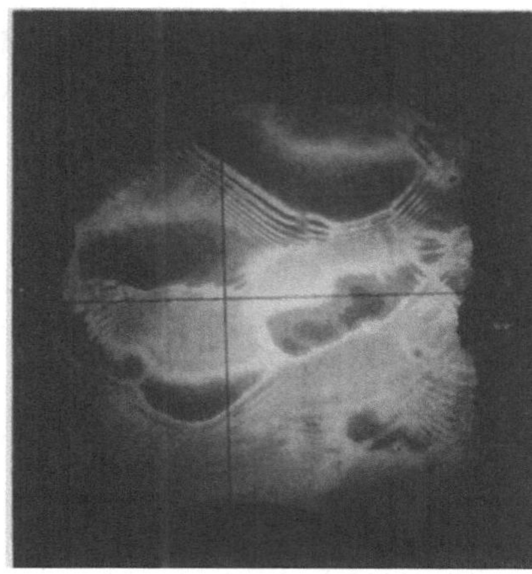

Fig. 7. Photograph of the wind waves at 21 m fetch in the large IMST facility at 8.1 m/s wind speed. See below for detailed explanation.

N.B.: The pictures have been taken using a special slope visualization technique indicating the slope in different colours or a gray scale. The picture shown is a black and white copy of a color slide, so that the dark parts correspond to upwind slopes, the light parts to small slopes, and the medium gray colors to downwind slopes. The wind is blowing from the left to the right. After Jähne et al. (1987).

it suggests the importance of wave breaking on interfacial scalar transfers, especially as the spilling zones disrupt the very thin conduction sublayers. In an experiment aimed at isolating the influence of breaking, Banner and Wilkinson (1991) have found that the sensible heat transfer rate is augmented considerably following the onset of breaking. Thus, in a spectral context, the distribution of microscale breaking is likely to be a significant factor in improving estimates of scalar interfacial fluxes. The importance of microscale breaking in controlling the transfer of heat and mass has added impetus to the search for the wavenumber spectrum of capillary-gravity waves and its wind sensitivity. Kahma and Donelan (1987) investigated the growth rates of initial wavelets and found that a (10 m) wind speed of 0.7 m/s ($u_* \sim 2$ cm/s) is sufficient to begin the growth process and that both shear flow instability (Valenzuela, 1976; Kawai, 1979 and van Gastel et al., 1985) and resonance with advected pressure fluctuations (Phillips, 1957) could be effective.

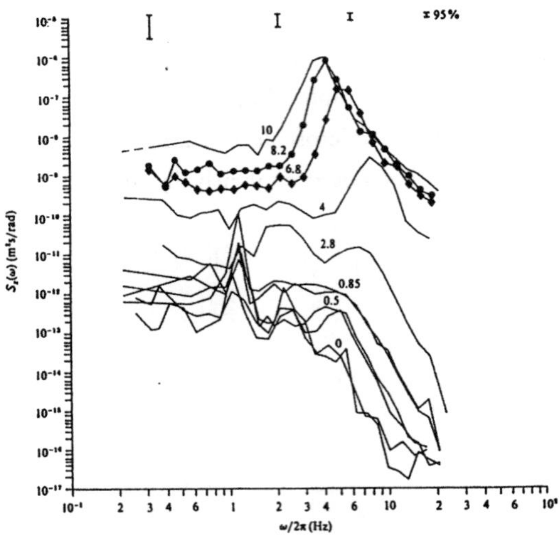

Fig. 8. The evolution of the frequency spectrum with wind speed (indicated on the figure). The spectra at 0, 0.5, 0.85 and 2.8 m/s wind speed are composites derived from surface displacement spectra and transformed slope spectra. From Kahma and Donelan (1988).

The development of the spectrum at various wind speeds and fixed fetch (4.3 m) is shown in Figure 8. The first waves to grow ($\sim$ 5 Hz) exhibit strong wind sensitivity at high winds and quickly approach saturation at wind speeds in excess of 4 m/s. Kahma and Donelan (1988) suggest that the approach to saturation depends on available fetch and, indeed, the slope measurements of Jähne and Riemer (1990) at much longer fetch (90 m) show that the 5 Hz waves are already nearly saturated at 2.7 m/s (Figure 9). Their wavenumber spectra (Figure 10) demonstrate that although the downwind slopes tend to saturation above 12 m/s, continued growth at off-wind angles extend the range of wind sensitivity of the overall mean-square slope. This wind sensitivity of the small waves and their connection with microscale-breaking helps to explain the observed strong wind dependence of mass transfer velocity of liquid phase limited compounds across the air-water interface (Liss, 1973).

Water vapour, on the other hand, is gas phase limited and no particularly dramatic increase of the evaporation coefficient (Dalton number, C_E) is expected when microscale breaking intensifies. Field measurements of C_E show the usual geophysical and sam-

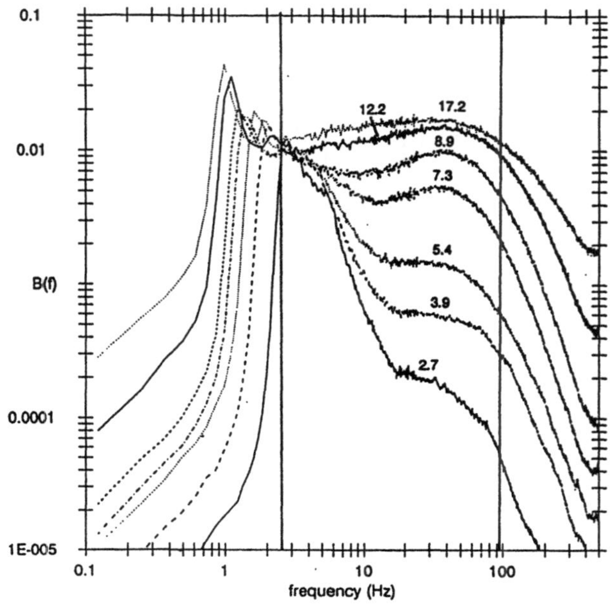

Fig. 9. Wave frequency spectra weighted by f^5 measured at several wind speeds and about 90 m fetch with the laser slope gauge. The vertical lines mark the approximate range contained in the wave number spectra shown in Fig. 10. The curves are labelled with wind speed U_{10} in m/s. From Jähne and Riemer (1990).

pling variability, but are, in general, quite insensitive to wind speed. The breaking of large waves and the production of spray in the boundary layer might be expected to cause a substantial increase in C_E at high winds ($> 12m/s$, say). Indeed, in a comprehensive treatment of the subject, Bortkovskii (1987) suggests that spray evaporation will double C_E between 9 and 18 m/s. There is no evidence for this in the data. The recent experiment to investigate the Humidity Exchange over the Sea (HEXOS, Smith and Anderson, 1988, and DeCosmo et al., 1988) extended C_E values to 18 m/s and found no appreciable increase. The study of heat and mass transfer at the air-sea interface and the effects of water temperature, slicks and wave breaking is being approached with all the modern tools required to examine the properties of short waves, the breaking characteristics of all waves and the turbulence near the interface. There is much to be learned here and the rewards, in terms of improved modeling of the energy balance of the ocean atmosphere system, are substantial.

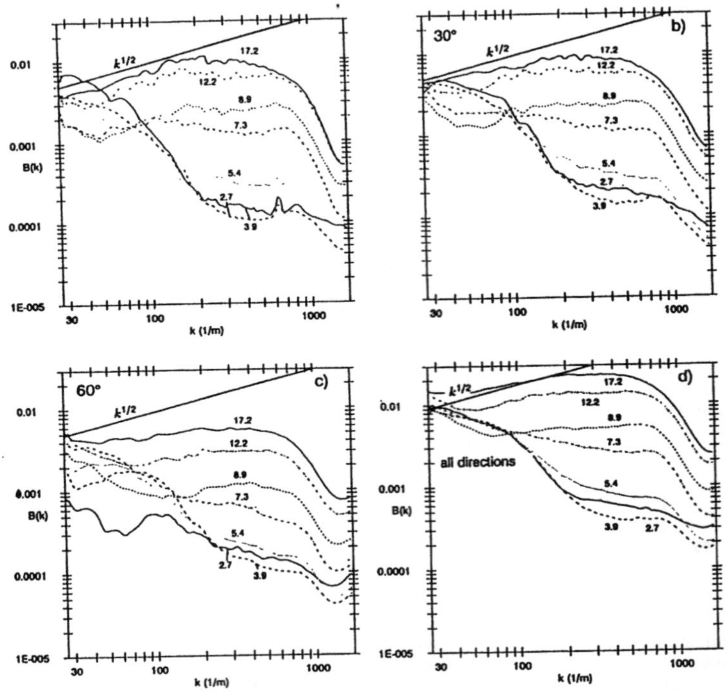

Fig. 10. Wave number spectra weighted by k^4 in different directions: (a) along-wind ($\pm 5°$); (b) $30 \pm 5°$; (c) $60 \pm 5°$; (d) unidirectional spectra integrated over all angles; wind speeds (U_{10} in ms^{-1}) as indicated. From Jähne and Riemer (1990).

Allied processes

Microwave reflectivity

Techniques utilising active microwave backscatter from the sea surface have been investigated intensively for satellite remote sensing of fundamental ocean surface variables, such as ocean currents, sea state and wind stress. Reliable interpretation of the microwave returns requires a detailed understanding of the backscattering mechanisms. Wave tank investigations have established that Bragg backscatter plays a major role in the backscattering process but in the field context, where a broader wave spectrum often modulates the short wave components responsible for the backscatter, additional mechanisms may be operative. Microwave return signals are often characterised by intermittent spikey signatures, known as sea spikes. In particular, associated with steep waves and the onset of breaking, it has been demonstrated that specular and wedge scattering may occur [e.g. Kwoh *et al.* (1988)], while Banner and Fooks (1985) found the breaking regions of short gravity waves to be a significant source of strong microwave reflections, owing to the concentration of very short wave energy there, plausibly arising from the instability of the intense shear layer formed by the spilling water. It is likely that these regions enhance the local Bragg backscatter, but the detailed backscattering mechanism was not resolved conclusively. Strong microwave returns from breaking waves in the field have been documented by Chaudhry and Moore (1984), Keller *et al.* (1986) and Jessup *et al.* (1990). These studies illustrate the intrinsic difficulties in deciding the threshold for the presence of spikes. Phillips (1988) investigated theoretically the relative contributions to the radar backscatter cross section from Bragg and sea spike returns, on the basis of his revised equilibrium range model [Phillips (1985)]. The choice of fixed threshold level as in Phillips (1988) and Jessup *et al.* (1990) detects spikes associated with larger-scale breaking waves, but misses contributions from smaller-scale breaking waves and thereby underestimates non-Bragg contributions. Nevertheless, the fractional sea spike power contribution reported by Jessup *et al.* (1990) was significant, particularly for HH polarisation. However, Plant and Keller (1990) argued that the dominant process is Bragg scattering via Doppler spectral broadening of the return signal rather than amplitude characteristics, and concluded that model calculations based on Bragg scattering were generally very close to the observations. Further research in this area is required to resolve the importance of sea spike contributions to the return signal, requiring an improved knowledge of the wavenumber spectrum for the very short wind wave components for checking the proportionality with the radar return power spectrum.

Bubble clouds

The entrainment of air in the wave breaking process produces a distinct visual signature and also alters the microwave emissivity and reflectivity in characteristic ways, many of which are exploited by downward looking electromagnetic remote sensing methods. The submerged bubbles alter the optical and acoustic properties of the near surface layer and various possibilities for upward looking remote sensing are being actively explored. A wide spectrum of air bubbles is entrained in a typical whitecap and the smallest of these remain to mark the passage of a breaker long after the surface, viewed from above, retains no further memory of it. These "bubble clouds" are significant acoustic scatterers, and may reveal various dispersive properties of the turbulence and trace the organized motions in the surface layers. The collection of papers in the first two NATO symposia on surface sound [Kerman (1988), (1991)] explore various aspects of bubble clouds including their effect on gas transfer and noise generation.

Underwater ambient noise generation

The physics of ambient noise generation by sea surface sources has received much attention in recent years, with whitecapping identified as a primary contributor. There is now a growing literature on both observations and models investigating the detailed mechanisms associated with sound generation by breaking waves. A major goal in this research is the prediction of the influence of wave breaking on the underwater ambient noise spectrum. While it has long been appreciated that noise generation is associated with the air bubbles in whitecaps, the bubble motions actually contributing significantly to noise generation had not been resolved observationally. Banner and Cato (1988) investigated the detailed configurations of bubbles in the spilling zones which were involved in active noise generation, and concluded that the process was dominated by noise bursts from only a subset of the bubbles present. The predominant noise bursts arose when bubbles relaxed after air was entrained at the toe of the breaker, when bubbles coalesced and when they collided impulsively. Interestingly, bursting bubbles at the surface were not a strong noise source, nor were most of the bubbles present, which were convected passively in the spilling zone. More recent related work by Pumphrey and Ffowcs Williams (1990) correlated bubble size with underwater noise bursts, and concluded that the lowest order bubble oscillation mode was dominant. Further discussion of related studies on breaking wave influence on noise generation are given in the recent symposium volumes [Kerman (1988, 1991)], with more recent contributions described in this symposium.

Conclusions

Wave breaking is a highly nonlinear aspect of ocean wave dynamics with a significant impact on the evolution of the wave spectrum as well as on fundamental transfers of momentum, energy and scalar quantities across the air-sea interface. It also enhances radar reflectivity of the sea surface and ambient underwater noise generation. A more complete understanding of the underlying physics of this conspicuous but incompletely understood process is required to improve the reliability of predictions from ocean wave, general circulation and climatological models.

References

1. Banner, M.L. and Fooks, E.H. 1985 On the microwave reflectivity of small-scale breaking water waves. Proc. R. Soc. Lond. A 399, 93-109.

2. Banner, M.L. 1990a The influence of wave breaking on the surface pressure distribution in wind wave interactions. J. Fluid Mech. 211, 463-495.

3. Banner, M.L. 1990b On the influence of the sea state on the wind stress. Internat. TOGA Scient. Conf. Proc. July 1990, WCRP-43 (WMO/TD-No. 379), Hawaii, 139-144.

4. Banner, M.L. 1990c Equilibrium spectra of wind waves. J. Phys. Oceanogr. 20, 966-984.

5. Banner, M.L. and Wilkinson, D.W. 1991 Breaking wave enhancement of the sensible heat flux at an air-water interface (in progress).

6. Bortkovskii, R.S., 1987 Air-Sea Exchange of Heat and Moisture during Storms. Reidel, Dordrecht.

7. DeCosmo, J., Katsaros, K.B. and Lind, R.J. 1988 Surface Layer Measurements During HEXMAX by the University of Washington. Tech. Rpt. (Hexos Contr. #16) Eds. Oost, W.A., Smith, S.D. and Katsaros, K.B., Dept. Atmos. Sci., Uni. of Washington, 29-39..

8. Donelan, M.A., 1979 On the fraction of wind momentum retained by waves. In Marine Forecasting, Predictability and Modelling in Ocean Hydrodynamics. J.C.Nihoul, ed. Elsevier, Amsterdam.

9. Donelan, M.A. 1982 The dependence of the aerodynamic drag coefficient on wave parameters. In the First International Conference on Meteorology and Air-Sea Interaction of the Coastal Zone. 381-387, American. Met. Soc., Boston.

10. Donelan, M.A. 1987 The effect of swell on the growth of wind waves. Johns Hopkins APL Technical Digest 8, 1, 18-23.

11. Donelan, A.A. and Kahma, K.K. 1987 Observations of velocities beneath wind driven waves. Proc. Int'l. Workshop on Wave Hindcasting and Forecasting, Halifax, N.S. Sept. 23-26, 1986. Env. Studies Rev. Fund, Ottawa, 243-252.

12. Donelan, M.A. 1990 Air-Sea Interaction. The Sea: Vol.9, 239-292. J. Wiley & Sons, Inc.

13. Donelan, M.A. and W.J.Pierson, 1987 Radar scattering and equilibrium ranges in wind-generated waves with application to scatterometry. J. Geophys. Res., 92 (C5), 4971-5029.

14. Drennan, W.J., K.K. Kahma, E.A.Terray, M.A. Donelan, and S.A. Kitaigorodskii, 1991. Observations of the enhancement of kinetic energy dissipation beneath breaking wind waves. Proc. IUTAM Breaking Waves Symposium, Sydney, Australia, July, 1991.

15. Duncan, J.H. 1981 An experimental investigation of breaking waves produced by a towed hydrofoil. Proc. R. Soc. Lond. A 377, 331-348.

16. Duncan, J.H. 1983 The breaking and non-breaking wave resistance of a two-dimensional hydrofoil. J. Fluid. Mech. 126, 507-520.

17. Gastel, K. van, P.A.E.M. Janssen and G.J. Komen 1985 On phase velocity and growth rate of wind-induced gravity-capillary waves. J. Fluid Mech., 161, 199-216.

18. Geernaert, G.L., Katsaros, K.B. and Richter, K. 1986 Variation of the drag coefficient and its dependence on sea state. J. Geophys. Res. 91, 7667-7679.

19. Hasselmann, K., T.P. Barnett, E. Bouws, H. Carlson, D.E. Cartwright, K. Enke, J.A. Ewing, H. Gienapp, D.E. Hasselmann, P. Kruseman, A. Meerburg, P. Muller, D.J. Olbers, K. Richter, W. Sell, and H. Walden, 1973 Measurements of wind-wave growth and swell decay during the Joint North Sea Wave Project (JONSWAP). Deut. Hydrogr. Z., Suppl. A, 8 (12).

20. Jähne, B., Munnich, K.O., Bosinger, R., Dutzi, A., Huber, W. and Libner, P. 1987 On the parameters influencing air-water gas exchange. J. Geophys. Res. 92, 1937-1949.

21. Jahne, B., and K.S. Riemer, 1990. Two dimensional wavenumber spectra of small-scale water surface waves. J. Geophys. Res. 95, C7, 11531-11546.

22. Janssen, P.A.E.M. 1989 Wave-induced stress and the drag of air flow over sea waves. J. Phys. Oceanogr. 19, 745-754.

23. Jessup, A.T., Keller, W.C. and Melville, W.K. 1990 Measurements of sea spikes in microwave backscatter at moderate incidence. J. Geophys. Res. 95, 9679-9688.

24. Jones, I.S.F., 1985 Turbulence below wind waves. In the Ocean Surface - Wave Breaking, Turbulent Mixing and Radio Probing. Reidel, Dordrecht, 437-442.

25. Kahma, K.K., and M.A. Donelan 1988. A laboratory study of the minimum wind speed for wind wave generation. J. Fluid Mech., 192, 332-364.

26. Kawai, S. 1979 Generation of initial wavelets by instability of a coupled shear flow and their evolution to wind waves. J. Fluid. Mech., 9, 661-703.

27. Kawai, S. 1981 Visualisation of airflow separation over wind wave crests under moderate wind. Boundary Layer Met. 21, 93-104.

28. Kawai, S. 1982 Structure of the air flow over wind wave crests revealed by flow visualisation techniques. Boundary-Layer Met. 23, 503-521.

29. Kerman, B.R. 1988 (Ed.) Sea Surface Sound — Natural Mechanisms of Surface Generated Noise in the Ocean. Kluwer, Dordrecht, 639pp.

30. Kerman, B.R. 1991 (Ed.) "Natural Physical Sources of Underwater Sound". Kluwer, Dordrecht (in Press).

31. Kitaigorodskii, S.A., 1970. The physics of air-sea interaction. Israel Program for Scientific Translations, Jerusalem. Translation dated 1973.

32. Kitaigorodskii, S.A., 1983. On the theory of the equilibrium range in the spectrum of wind generated gravity waves. J. Phys. Oceanogr., 13, 816-827.

33. Kitaigorodskii, S.A., M.A. Donelan, J.L. Lumley, and E.A. Terray. 1983 Wave-turbulence interactions in the upper ocean. Part II: Statistical characteristics of wave and turbulence components of the random velocity field in the marine surface layer. J. Phys. Oceanogr., 13, 1988-1999.

34. Lake, B.M. and Yuen, H.C. 1978 A new model for nonlinear wind waves. Part 1. Physical model and experimental evidence. J. Fluid Mech. 88, 33-62.

35. Liss, P.S., 1973 Processes of gas exchange across an air-water interface. Deep Sea Research, 20, 221-238.

36. Longuet-Higgins, M.S. 1987 A stochastic model of sea surface roughness. I: wave crests. Proc. R. Soc. Lond. A 410, 19-34.

37. Okuda, K., Kawai, K. and Toba, Y. 1977 Measurements of the skin friction distribution along the surface of wind waves. J. Oceanogr. Soc. Japan, 33, 190-198.

38. Okuda, K. 1982 Internal flow structure of short wind waves I. On the internal vorticity structure. J. Oceanogr. Soc. Japan, 38, 28-42.

39. Phillips, O.M., 1957. On the generation of waves by turbulent wind. J. Fluid. Mech., 2, 417-495.

40. Phillips, O.M. 1985 Spectral and statistical properties of the equilibrium range in wind-generated gravity waves. J. Fluid Mech. 156, 505-531.

41. Phillips, O.M. 1988 Radar returns from the sea surface-Bragg scattering and breaking waves. J. Phys. Oceanogr. 18, 1065-1074.

42. Phillips, O.M. and Banner, M.L. 1974 Wave breaking in the presence of wind drift and swell. J. Fluid Mech. 66, 625-640.

43. Plant, W.J. and Keller, W.C. 1990 Evidence of Bragg scattering in microwave doppler spectra. J. Geophys. Res. 95, 16299-16310.

44. Plant, W.J., 1991 Wave influences on wind profiles over water. J. Phys. Oceanogr., (in press).

45. Pumphrey, H.C. and Ffowcs Williams, J. 1990 Bubbles as sources of ambient noise. IEEE J. Ocean Eng. 15, 268-274.

46. Rapp, R.J. and Melville, W.K. 1990 Laboratory measurements of deep water breaking waves. Phil. Trans. R. Soc. Lond. A 331, 735-780.

47. Smith, S.D. and Anderson, R.J. 1988 Bedford Institute of Oceanography Eddy Flux Measurements During HEXMAX, Tech. Rpt. (Hexos Contr. 16) Eds. Oost, W.A., Smith, S.D. and Katsaros, K.B., Dept. Atmos. Sci., Uni. of Washington, 14-21.

48. Soloviev, A.V., N.V. Vershinsky and V.A. Bezverchnii, 1988. Small-scale turbulence measurements in the thin surface layer of the ocean. Deep Sea Res. 35, 1859-1874.

49. Su, M-Y. and Green, A.W. 1986 Experimental studies of strong nonlinear interactions of deep-water gravity waves. In Wave Dynamics and Radio Probing of the Sea Surface, Eds. O.M. Phillips and K. Hasselmann, Plenum Press, N.Y., 231-253.

50. Toba, Y., Iida, I., Kawamura, H., Ebuchi, N. and Jones, I.S.F. 1990 The wave dependence of the sea surface wind stress. J. Phys. Oceanogr. 20, 705-721.

51. Trulsen, K. and Dysthe, K. 1990 Frequency downshift through self-modulation and breaking. In "Water Wave Kinematics", Eds. A. Torum and O.T. Gudmestad, Kluwer Academic, Dordrecht.

52. Valenzuela, G.R., 1976. The growth of gravity-capillary waves in a coupled shear flow. J. Fluid.Mech., 76, 229-250.

53. Wright, J.W. 1976 The wind drift and wave breaking. J. Phys. Oceanogr. 6, 404-405.

Capillary Rollers and Bores

MICHAEL S. LONGUET-HIGGINS

Institute for Nonlinear Science, University of California, San Diego
La Jolla, California 92093-0402

1. Introduction and summary

A very intriguing phenomenon, which throws much light on the small-scale structure of the sea surface, is the occurrence of "parasitic capillaries" on the forward face of moderately short gravity waves, especially those with wavelengths 5 to 50 cm; see Figure 1a. These capillary waves were first studied experimentally by Cox (1958). Evidently their existence depends on the fact that a gravity wave and a much shorter capillary wave may have the same phase speed. The dynamical theory of the generation of parasitic capillaries has been developed by Longuet-Higgins (1963), Crapper (1970) and Ruvinsky et al. (1981, 1985, 1991). This so far takes into account only the first-order effects of viscous damping.

However, an important feature of short, steep gravity waves, especially those with parasitic capillaries, has been pointed out by Okuda, Kawai and Toba (1977) and also by Ebuchi, Kawamura and Toba (1987); see Figure 1b. This is the occurrence of a very strong vortical region, or roller, at the crest of the gravity wave, even in the absence of wind (c.f. Cox 1958). Similar effects can indeed be detected in small-scale capillary bores and from water jets entering a still-water surface (Koga, 1982). The question arises: what is the source of this strong vorticity? The vorticity does not seem to be present in the trough of the gravity wave - only in the crest.

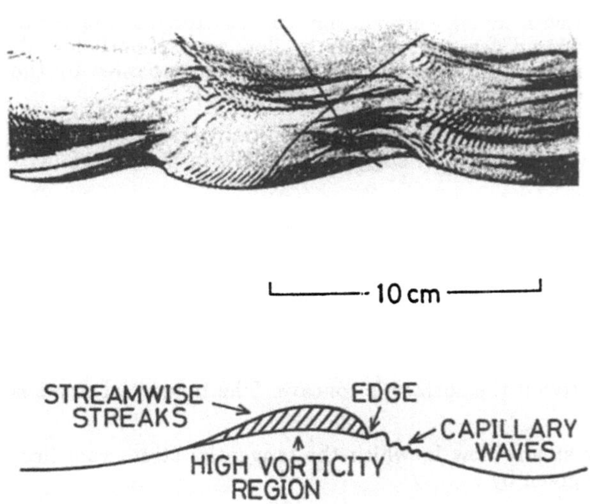

Figure 1. (from Ebuchi, Kawamura and Toba, 1987). A close-up photograph of wind-waves of length 9 cm taken with wind-speed 6 m/s at fetch 6 m. Below is a schematic picture showing the high vorticity region at the gravity wave crests.

M. L. Banner, R. H. J. Grimshaw (Eds.)
Breaking Waves
IUTAM Symposium Sydney/Australia 1991

In the present paper we shall show that a very likely source of the crest vorticity is the parasitic capillaries themselves, especially the steepest capillaries close to the forward edge of the roller. This is a nonlinear effect which can be understood in two steps. First, as proved in Section 2, we note that any curved free surface in a steady flow irrotational or not is necessarily a source of vorticity. The strength of the vorticity is $-2\kappa q$, where κ is the curvature of the streamline at the surface and q is the stream velocity.

It follows that any oscillatory flow must develop a vortical boundary-layer, or "Stokes layer" (see Sections 4 and 5). To first order (Section 4), the vorticity diminishes exponentially inwards. But it is a remarkable fact that at second order, there is a mean "rectified" vorticity just beyond the boundary-layer. This must diffuse into the interior, in a similar way to the vorticity in the flow past a flat plate. For waves of small slope, the strength of this vorticity is double that generated at the surface itself, and is given by

$$\bar{\omega}_\infty = -2(ak)^2\sigma \tag{1.1}$$

independently of the viscosity! Here, ak denotes the capillary wave steepness and σ the radian frequency. The simple result (1) has indeed been verified experimentally for gravity waves (Longuet-Higgins 1960). For capillary waves, however, the frequency σ is much higher than for gravity waves of the same phase speed. Hence the vorticity generated is much greater also. This vorticity from the capillaries accumulates mainly in the gravity wave crests.

The total vorticity diffused into the "stream" which flows backwards under the ripples (in a frame of reference moving with the wave speed) is calculated in Section 7 below. It is found fully capable of producing a vortex beneath the crest of the gravity wave, with a velocity difference of order c (the phase-speed) between top and bottom of the vortex.

This conclusion is confirmed in Section 8 by an integral argument involving the mass transport in the capillary waves. Steep capillary waves have very rounded crests, in which the particle velocity almost equals the phase speed c. They are therefore propogated somewhat like "balloons" of water floating on the undisturbed surface. When these waves are damped out by viscosity, their forwards momentum (mass transport) must be converted into a shearing current, in which the velocity difference between top and bottom is of order c.

For steep capillary waves having very sharply curved troughs, the surface vorticity may be shed into the interior differently, namely by flow separation below the wave troughs; see Section 9. Evidence of this effect, which can be accompanied by the trapping of air bubbles, is to be seen in the observations of Koga (1961).

The development of a capillary roller, from an initially irrotational wave, is discussed in Section 10. Conclusions and further suggestions follow in Section 11.

2. The vorticity generated at a free surface

In this section we prove some general results concerning the generation of vorticity.

Let (s, n) denote coordinates tangential and normal to any streamline in a steady, two dimensional flow, as in Figure 2. Here q denotes the particle speed and κ the curvature of the streamline - positive if the surface is concave. The tangential stress is denoted by τ_{ns}. Then we have

Theorem A. In any steady flow in which the tangential stress vanishes, the vorticity at the surface must be given by

$$\omega_s = -2\kappa q \, . \tag{2.1}$$

Proof. In general, if (u, v) are components of fluid velocity in the directions of any fixed rectangular coordinates (x, y) then

$$\omega = \frac{\partial v}{\partial x} - \frac{\partial u}{\partial y} \quad \text{and} \quad \tau_{ns} = \nu\left(\frac{\partial u}{\partial y} + \frac{\partial v}{\partial x}\right). \tag{2.2}$$

So if $\tau_{ns} = 0$,

$$\omega = 2\frac{\partial v}{\partial x}. \tag{2.3}$$

Take the axes (x, y) tangential and normal to the streamline at the fixed point O, and let θ be the angle between the tangents to the streamline at O and the general point P, as in Figure 2. Then we have

$$v = -q\sin\theta \quad \text{and} \quad \frac{\partial}{\partial x} = \cos\theta\frac{\partial}{\partial s} + \sin\theta\frac{\partial}{\partial n} \tag{2.4}$$

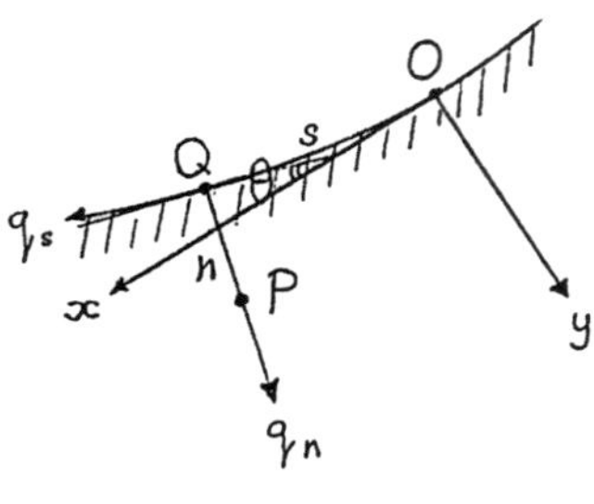

Figure 2.

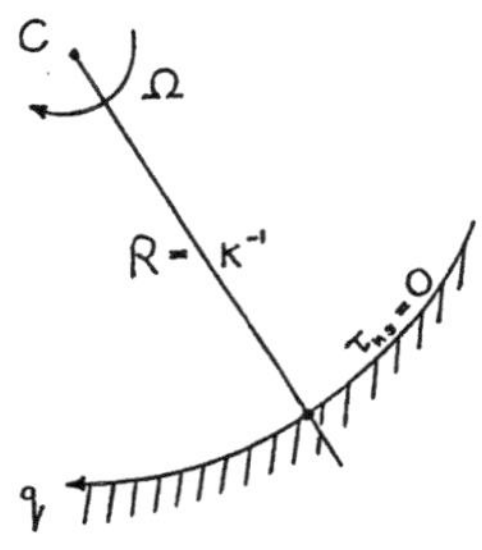

Figure 3.

Performing the differentiation and setting $\theta = 0$ afterwards we get $\frac{\partial v}{\partial x} = -q\frac{\partial \theta}{\partial s} = -\kappa q$. Substituting in equation (2.4) we obtain (2.1).

Physical interpretation. To ensure that $\tau_{ns} = 0$ the fluid must move locally so that there is no relative distortion of the fluid elements. That means that the local flow is a solid rotation, see Figure 3. The particle speed q in this flow is given by $q = -\Omega r$ where Ω is the rate of rotation and r is the distance from the centre of rotation. Since r must be equal $1/\kappa$ where κ is the curvature, we have $\Omega = -\kappa q$. But it is well-known that the local rate of rotation Ω of a fluid element is just $\frac{1}{2}\omega$, where ω is the vorticity. Hence (2.9) is equivalent to (2.1).

Note that for any linear surface wave, given by $y = a\cos ks$ where y is the surface elevation and c the phase-speed, we have

$$\omega = -2\kappa q = -2cy_{xx} = ak\sigma\cos ks \tag{2.5}$$

where $\sigma = ck$, the radian frequency. Thus ω is positive at the wave crests and negative in the wave troughs, as shown in Figure 4.

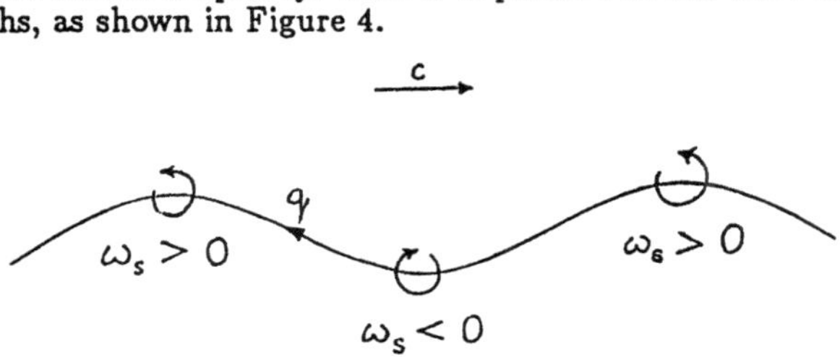

Figure 4. Schematic drawing of the vorticity in a progressive surface wave.

A consequence of Theorem A is that in the absence of surface stresses, waves in real fluids cannot be entirely irrotational. Nevertheless, we can still use the well-known theory of irrotational waves as a first approximation, provided that we add boundary-layers at the free surface (and if necessary at solid boundaries) to accomodate the non-zero values of the vorticity at the boundary (c.f. Longuet-Higgins 1953, 1960). We shall now apply this procedure in the first place to pure capillary waves, that is waves so short that the influence of gravity on the phase-speed can be considered as negligible.

3. Pure capillary waves

To illustrate the nonlinear behaviour of ω_s, we apply the results of Section 2 to Crapper's pure capillary wave solution, given in the form

$$z = \tan w - w \tag{3.1}$$

where $z = x + iy$, $w = \phi + i\psi$ and we have chosen dimensionless units so that the phase speed c and wavenumber k are 1 and 2 respectively (see Figure 5). Then it is easily seen that the free surface condition $q^2 + 2T\kappa = 1$ is satisfied provided that

$$\coth 2\psi_o = 2T \tag{3.2}$$

T being the surface tension in these units. Equation (3.2) serves as the dispersion relation. The wave steepness ak equals $2/\sinh 2\psi_0$.

A salient feature of the profiles in Figure 5 is that as the wave amplitude increases, so the waves become more rounded in the crests and sharply curved in the troughs. In the limiting case $\psi_o = 0.3941$ the ratio of the curvatures is about 50:1.

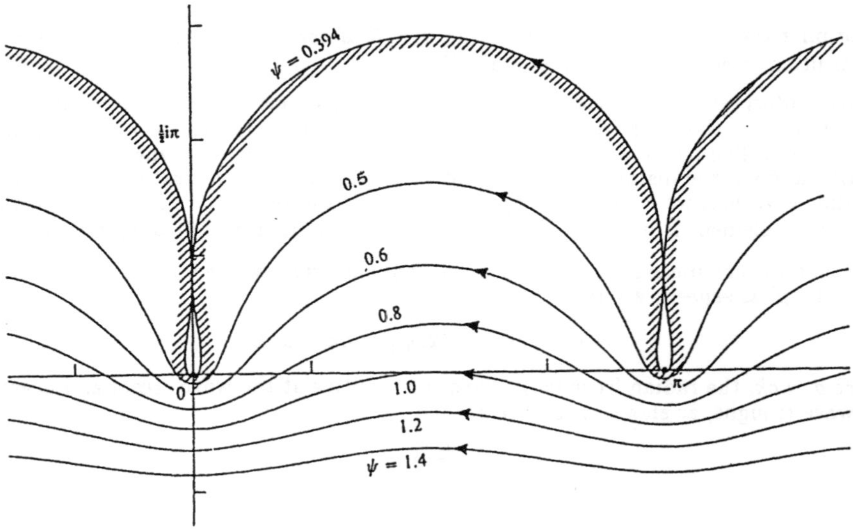

Figure 5. (from Longuet-Higgins 1988). Streamlines and free surfaces in Crapper's solution for an irrotational, pure capillary wave, viewed in a frame of reference moving to the right with the phase-speed c. Units are dimensionless, with $c = 1, k = 2$.

In Figure 6 we show the vorticity ω_s at the free surface as given by equation (2.5), plotted against the distance s along the free surface, $s = 0$ being taken in the wave trough. Figures 6a to 6d show the cases $\psi_o = 1.6$, 1.2, 0.8 and 0.4, corresponding to $ak = 0.163$ to 2.252. Even in case (a), that is for a fairly low wave steepness, it can be seen that ω_s is already quite nonlinear, being almost twice as great in the troughs as in the crests. As ψ_o decreases to 0.4, the vorticity becomes almost entirely concentrated in the troughs, and elsewhere is relatively very small.

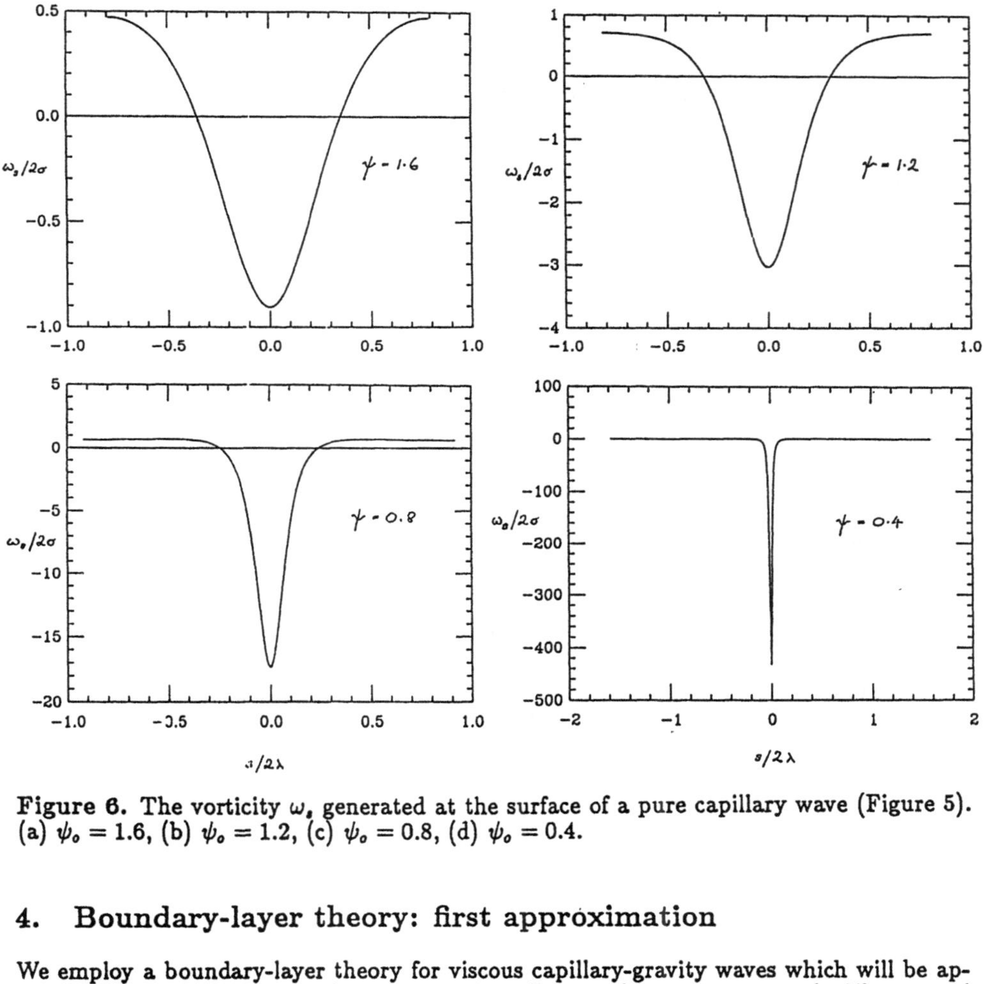

Figure 6. The vorticity ω_s generated at the surface of a pure capillary wave (Figure 5). (a) $\psi_o = 1.6$, (b) $\psi_o = 1.2$, (c) $\psi_o = 0.8$, (d) $\psi_o = 0.4$.

4. Boundary-layer theory: first approximation

We employ a boundary-layer theory for viscous capillary-gravity waves which will be applicable, in the first place, only to waves of small or moderate steepness ak. The general trend of the results as the steepness increases will, however, be indicative of the results to be expected when ak is of order unity.

In any oscillatory flow, vorticity tends to diffuse inwards from the boundary on a length scale of order $\delta = (2\nu/\sigma)^{1/2}$ where ν is the kinematic viscosity and σ the radian frequency. Our approximation will be based on the assumption that $\gamma \equiv k\delta << 1$ where k is the wavenumber for the surface waves.

Table 1. Values of the boundary-layer parameter γ

λ (cm)	k $(\text{cm})^{-1}$	σ (s^{-1})	δ (mm)	γ
100	0.0628	7.85	.575	.0036
10	0.628	25.2	.321	.0202
1.74	3.617	84.2	.176	.0635
1.0	6.283	157.4	.0129	.0808
0.1	62.83	4320.	.0024	.1541

Some numerical values of γ are shown in Table 2, in which we took $g = 981\ \text{cm/s}^2$, $T = 75$ dyne/cm and $\nu = 0.013\ \text{cm}^2/\text{s}$. For very small wavelengths ($k >> 1$) γ varies only as $k^{1/4}$.

The profile of the free surface will be affected to order γ^2, not γ. For, the ratio of the viscous stress τ_{ns} to the normal stress τ_{nn} is of order

$$\frac{\tau_{ns}}{\tau_{nn}} \sim \frac{2\nu\kappa q}{T\kappa} = \frac{2\nu q}{T} \tag{4.1}$$

for capillary waves. But q is of the same order as the phase-speed c, that is $(Tk)^{1/2}$. Hence

$$\frac{\tau_{ns}}{\tau_{nn}} \sim \frac{2\nu c}{c^2/k} = \frac{2\nu k^2}{\sigma} = k^2\,\delta^2 = \gamma^2\ . \tag{4.2}$$

To first order in γ we may therefore use the inviscid streamlines as if they were the real streamlines.

We write

$$(q_s,\ q_n) = (q_{sI},\ q_{nI}) + (q_{sB},\ q_{nB}) \tag{4.3}$$

where $(q_{sI},\ q_{sN})$ are the tangential and normal components of velocity in the irrotational wave (seen in a frame of reference moving horizontally with the wave speed c) and (q_{sB}, q_{nB}) are velocity components of vortical flow induced by the diffusion of vorticity from the boundary. It is easy to see that $q_{sI}/c = 0(1)$, $q_{sB}/c = 0(\gamma)$ and that to lowest order in γ

$$q_{nI} = -n\,\frac{\partial q_{sI}}{\partial s} = 0(\gamma)c \quad \text{and} \quad q_{nB} = -n\,\frac{\partial q_{sB}}{\partial s} = 0(\gamma^2)c\ . \tag{4.4}$$

Also the vorticity ω is given to lowest order by

$$\omega = \frac{\partial q_{sB}}{\partial n} \tag{4.5}$$

Since in two-dimensional flow there is no stretching of the vortex lines, the basic equation for the vorticity is

$$\frac{D\omega}{Dt} = \nu\,\nabla^2\,\omega \tag{4.6}$$

where D/Dt denotes differentiation following the motion and ∇^2 is the two-dimensional

Laplacian. Within the boundary-layer this reduces to

$$\nu \frac{\partial^2 \omega}{\partial n^2} = \left(q_{sI} \frac{\partial}{\partial s} + q_{nI} \frac{\partial}{\partial n} \right) \omega \, . \tag{4.7}$$

Now let $\bar{q}$ denote the mean speed of a particle along the boundary, defined as

$$\bar{q} = \frac{1}{s_{max}} \int_0^{s_{max}} q_{sI} \, ds \tag{4.8}$$

Then we may write

$$\bar{q}_{sI} = \bar{q} + q'_{sI} \tag{4.9}$$

where $\bar{q}'_{sI} = 0$, in an obvious notation. For low waves, $\bar{q}$ equals the phase-speed c and q'_{sI} represents the tangential component of the orbital velocity. Then on substituting in (4.7) our basic vorticity equation becomes

$$\left(\nu \frac{\partial^2}{\partial n^2} - \bar{q} \frac{\partial}{\partial s} \right) \omega = q'_{sI} \frac{\partial \omega}{\partial s} - n \frac{\partial q'_{sI}}{\partial s} \frac{\partial \omega}{\partial n} \tag{4.10}$$

The boundary condition at $\omega = 0$ is that

$$\omega = -2\kappa q_{sI} = -2\kappa(\bar{q} + q'_{sI}) \quad \text{when } n = 0. \tag{4.11}$$

Also $\nabla\omega \to 0$ as $n \to \infty$. To solve these equations we now make a perturbation expansion in powers of an ordering parameter ϵ. Thus we set

$$q'_{sI} = \epsilon \, q_{sI1} + \epsilon^2 q_{sI2} + - - - \text{ and } \kappa = \epsilon \, \kappa_1 \, ; \tag{4.12}$$

we absorb all higher powers of ϵ into κ_1. We then seek a solution for ω in the form $\omega = \epsilon\omega_1 + \epsilon^2\omega_2 + - - -$ by successive approximation.
On substituting in equations (4.10) and (4.11) and equating the coefficients of ϵ we have

$$\left(\nu \frac{\partial^2}{\partial n^2} - \bar{q} \frac{\partial}{\partial s} \right) \omega_1 = 0 \tag{4.13}$$

$$\omega_1 = -2 \, \kappa_1 \, \bar{q} \quad \text{when } n = 0 \tag{4.14}$$

$$\frac{\partial \omega_1}{\partial n} \to 0 \quad \text{when } n \to \infty. \tag{4.15}$$

Suppose now that κ_1 is given as a Fourier series in s:

$$\kappa_1 = \sum_{\ell=1}^{\infty} C_\ell \, e^{i\ell K s}, \quad K = 2\pi / s_{max} \tag{4.16}$$

equations (4.13 to 4.15) have the unique solution

$$\omega_1 = -2\bar{q} \sum_{\ell=1}^{\infty} C_\ell \, e^{i\ell K s - \ell^{1/2} \alpha n} \tag{4.17}$$

where

$$\alpha = \frac{1+i}{\delta}, \quad \delta = (2\nu/\sigma)^{1/2}. \tag{4.18}$$

Each term on the right of (4.17) represents a term decaying inwards exponentially, but with a complex exponent α. The corresponding velocity components are calculated from

$$q_{sB1} = \int_n^\infty \omega_1 \, dn, \qquad q_{nB1} = -\int_0^n \frac{\partial q_{sB}}{\partial s}\Big| \, dn. \tag{4.19}$$

A typical example is shown in Figure 7a. The tangential component q_{sB1} tends to be greatest, and in the direction of propagation, at a point lagging 45° behind the wave trough. There is an eddy, due to the varying normal velocity q_{nB1}, with center approximately 90° behind the trough. For the very steep wave shown in Figure 7b, these features are shifted closer to the trough.

In this first approximation the vorticity ω is confined to the Stokes boundary-layer, and none escapes beyond a distance of order δ. In the next approximation we shall see that the situation is quite different.

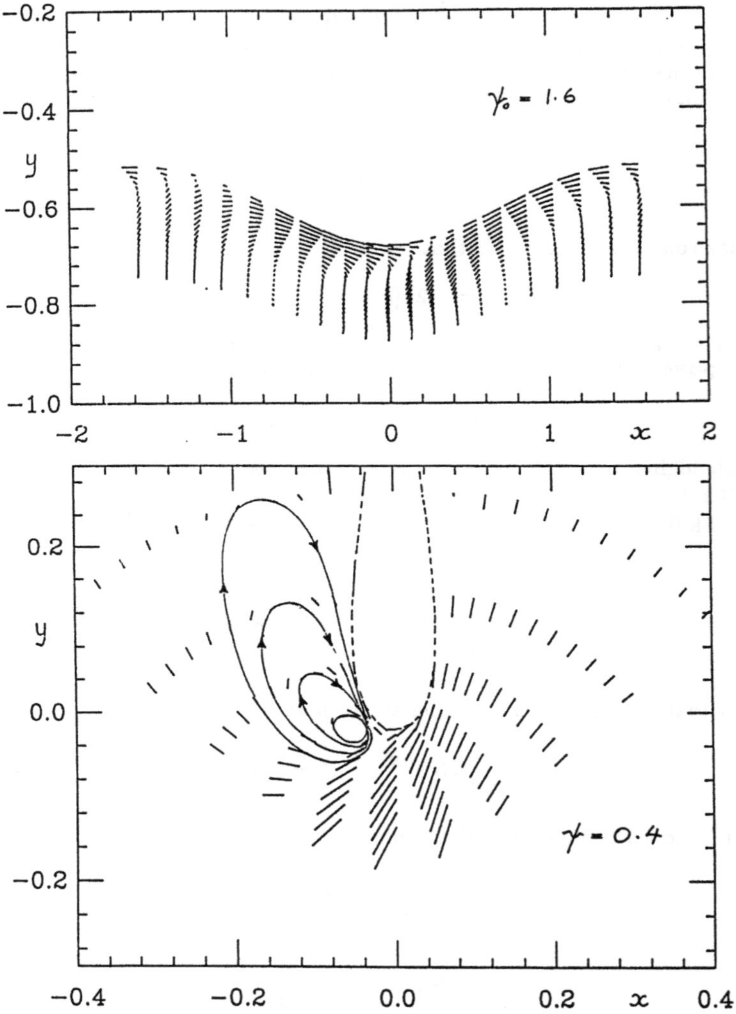

Figure 7. The first-order flow (q_{sB1}, q_{nB1}) in the boundary-layer for the capillary waves in Figure 5 (a) $\psi_o = 1.6$, (b) $\psi_o = 0.4$.

5. The mean vorticity

By Theorem A, the mean vorticity $\bar{\omega}_s$ at the free surface is given by

$$\bar{\omega}_s = -2\,\overline{\kappa q} = -2\,\overline{\kappa_1\, q'_{sI}} \tag{5.1}$$

since $\bar{\kappa} = \overline{\partial\theta/\partial s} = 0$ by the periodicity. Now for any sinusoidal wave such as (2.5) in deep water we have

$$q'_{sI1} = -a\sigma \cos ks \tag{5.2}$$

where $\sigma = kc$ is the radian frequency. Thus the tangential velocity q'_{sI1} correlates with the curvature κ_1, and from (5.1) we find

$$\bar{\omega}_s = -(ak)^2\, \sigma \tag{5.3}$$

when ak is small. In other words, the mean vorticity is negative. We show in Figure 7 the value of $-\bar{\omega}_s$ for the finite-amplitude capillary waves of Figure 3, plotted against $(ak)^2$. It can be seen that this increases like $(ak)^2$ at first, but then less rapidly.

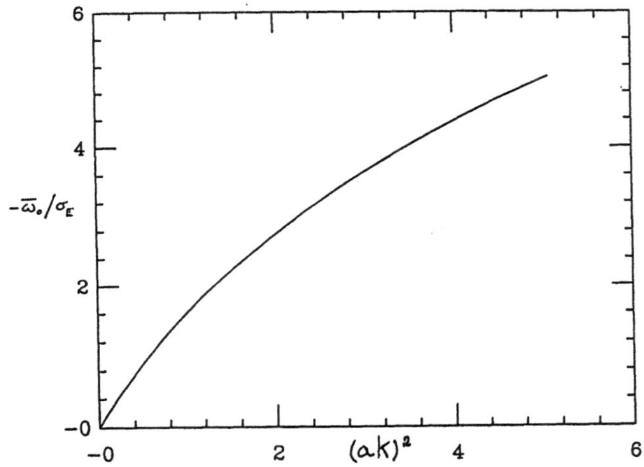

Figure 8. The mean vorticity $\bar{\omega}_o = -2\overline{\kappa q}$ generated at the surface of the pure capillary wave of Figure 5, as a function of $(ak)^2$.

We shall now prove a remarkable result:

Theorem B. The mean vorticity $\bar{\omega}_\infty$ just beyond the Stokes layer is just double the mean vorticity $\bar{\omega}_s$ at the free surface.

Proof. Taking averages with respect to S in equations (4.10) and (4.11) we have, to order ϵ^2,

$$\nu \frac{\partial^2 \bar{\omega}}{\partial n^2} = \overline{q'_{sI}\frac{\partial \omega}{\partial s}} - \overline{\frac{\partial q_{sI}}{\partial s}\cdot n \frac{\partial \omega}{\partial n}} \tag{5.4}$$

$$\bar{\omega} = -2\overline{\kappa q_{sI}} \qquad \text{when } n = 0 \tag{5.5}$$

$$\frac{\partial \bar{\omega}}{\partial n} \to 0 \qquad \text{when } n \to \infty \tag{5.6}$$

(for brevity, the suffices 1 and 2 are omitted). Now in any two periodic quantities A and

B we have $\overline{\partial A/\partial s\, B} = \overline{A\, \partial B/\partial s}$. Hence

$$\nu \frac{\partial^2 \bar{\omega}}{\partial n^2} = \overline{q'_{sI} \frac{\partial \omega}{\partial s}} + \overline{q'_{sI}\, n\, \frac{\partial^2 \omega}{\partial s \partial n}} = \frac{\partial}{\partial n}\overline{\left(q'_{sI}\, n\, \frac{\partial \omega}{\partial s}\right)} = \overline{\frac{\partial q_{sI}}{\partial n}\, q'_{sI} \cdot n\, \frac{\nu}{c}\, \frac{\partial^2 \omega}{\partial n^2}} \tag{5.7}$$

by (4.13). Integrating from $n = \infty$, where both $\partial \bar{\omega}/\partial n$ and $\partial \omega/\partial n$ vanish, we have

$$\frac{\partial \bar{\omega}}{\partial n} = \overline{q'_{sI} \cdot n \frac{\partial^2 \omega}{\partial n^2}} / \bar{q} \tag{5.8}$$

A second integration from $n = 0$ gives

$$\bar{\omega} - \bar{\omega}_s = \overline{q'_{sI} \left(n \frac{\partial \omega}{\partial n} - \omega + \omega_o \right)} / \bar{q} \tag{5.9}$$

where a suffix zero denotes the values at $n = 0$. Letting $n \to \infty$ we find

$$\bar{\omega}_\infty - \bar{\omega}_s = \overline{q'_{sI}\, \omega_o} / \bar{q} = \overline{q'_{sI} (-2\, \kappa \bar{q})} / \bar{q} = -2\kappa \overline{q'_{sI}} = \bar{\omega}_s\, . \tag{5.10}$$

Thus $\bar{\omega}_\infty = 2\bar{\omega}_s$, the result to be proved.

From equation (5.3) it follows that for low waves

$$\bar{\omega}_\infty = -2(ak)^2 \sigma \tag{5.11}$$

independently of the viscosity or of the dispersion relation.

This result has been checked experimentally for low gravity waves of periods 0.65s to 1.2s, see Longuet-Higgins (1960).

6. Diffusion of the mean vorticity

Unlike the periodic components, the mean vorticity $\bar{\omega}_\infty$ will diffuse into the interior of the fluid on a longer time-scale than the wave period, according to the diffusion equation

$$\frac{\partial \bar{\omega}}{\partial t} = \nu \frac{\partial^2 \bar{\omega}}{\partial n^2}\, . \tag{6.1}$$

When n becomes comparable with k^{-1}, it may be replaced by z, the mean depth of a particle below the mean surface level. On this time and length scale, if $t = 0$ is the time at which the motion is started, we find from (6.1) that

$$\bar{\omega}(z,t) = \frac{2\bar{\omega}_\infty}{\pi^{1/2}} \int_Z^\infty e^{-\lambda^2} d\lambda \quad \text{where} \quad Z = z/2(\nu t)^{1/2}\, . \tag{6.2}$$

The horizontal velocity $\bar{u}$ associated with this mean vorticity is given by

$$\bar{u} = \frac{4}{\pi^{1/2}} (\nu t)^{1/2}\, \bar{\omega}_\infty \left[\frac{1}{2} e^{-Z^2} - Z \int_Z^\infty e^{-\lambda^2}\, d\lambda \right] \tag{6.3}$$

In particular at the mean surface level $z = 0$ we have

$$\bar{u}_o = -\frac{2}{\pi^{1/2}} (\nu t)^{1/2}\, \bar{\omega}_\infty = -2\, N^{1/2}\, \delta\, \bar{\omega}_\infty \tag{6.4}$$

where $N = \sigma t/2\pi$ is the number of wave cycles after starting the motion and δ is the boundary-layer thickness, defined by equation (4.1). A graph of $\bar{u}$ as a function of k_z after N cycles is shown in Figure 8, for various values of N.

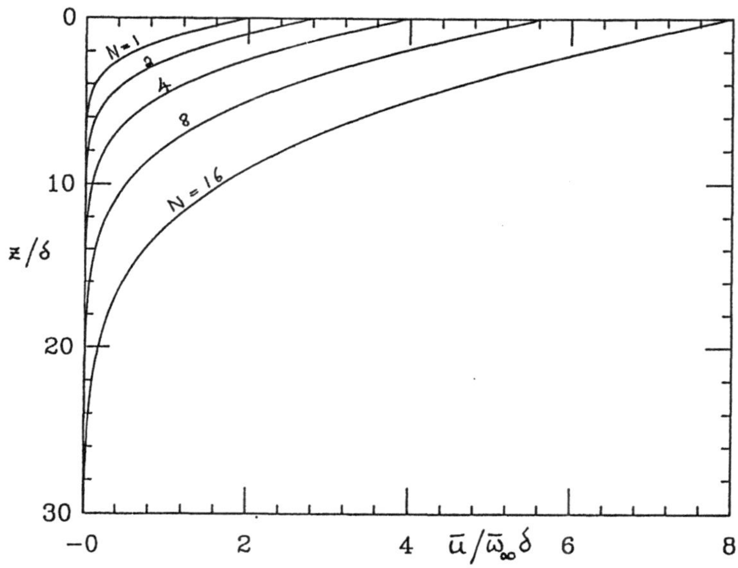

Figure 9. The mean horizontal velocity $\bar{u}$ after N cycles, shown as a function of the mean depth z.

7. Capillary rollers

As seen in Figure 1, a finite train of capillary waves often forms just ahead of the crest of a short gravity wave on which there is a strong "roller", or concentration of vorticity. In a frame of reference moving with the phase speed c, the particles at the surface of the crest travel slightly forwards whereas the particles beneath the roller travel backwards with nearly the phase speed. Thus there is in the roller itself a strong negative vorticity. This strong shearing flow appears also to produce 3-dimensional instabilities in the form of vortices having axes aligned in the direction of wave propagation, similar to a Langmuir circulation. We regard the latter as a secondary, though important, effect, but enquire here only into the source of the primary, two-dimensional vorticity.

We remark that the rollers associated with parasitic capillaries are but one example of a more general class of flows which we may call "capillary bores"; see Figure 10. A capillary bore may be defined as a small jump in surface elevation, where the surface may be locally vertical, ahead of which is a train of short capillary waves, and behind which is found a strong negative vortex. Such flows are seen frequently on a disturbed water surface, with or without the presence of wind, and they occur on a range of scales. Generally the smaller scales appear to be laminar, the larger scales turbulent. The model we propose assumes that the flow is laminar.

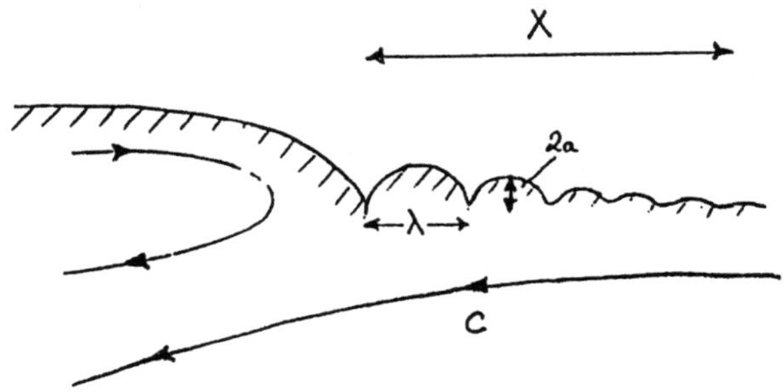

Figure 10. Schematic drawing of a capillary bore, with no separation of the flow.

We shall first estimate the total amount of the vorticity shed by a finite train of capillary waves. Separation of the flow will not be assumed. In Sections 3 to 8 above we neglected the decay of the wave train with horizontal distance x in the direction of wave propagation. For a uniform wave train the time-rate of decay of the amplitude a is given by

$$\frac{1}{a}\frac{\partial a}{\partial t} = -2\nu k^2 . \tag{7.1}$$

(see Lamb 1932, 348). In a capillary roller, however, the source of the capillary wave energy is localized near the gravity wave crest. Since the group-velocity of capillary waves equals $\frac{3}{2}c$, greater than the phase velocity, the wave energy is propagated upstream of the source, that is on the forward face of the gravity wave, and decays with distance x from the source, while remaining steady in time. If we neglect the work done by the current against the radiation stress (see Longuet-Higgins, 1963) we can determine the horizontal rate of decay by dividing the right-hand side of (8.1) by the group velocity $3c/2$. Hence

$$\frac{1}{a}\frac{\partial a}{\partial x} = -\frac{4}{3}\,\nu k^2/c . \tag{7.2}$$

Since the mean vorticity is proportional to a^2, the effective length of the wave train is therefore $X \sim 3/\nu k^2$. A stream flowing beneath the wave train with speed $-c$ passes the wave train in a time t of order $t \sim 3/\nu k^2$.

A wave train of initial steepness $(ak)_o$ at $x = 0$ will, over the decay distance X, have a mean-squared amplitude $(1 - e^{-1})(ak_o)^2$. We can estimate the magnitude of the vorticity $\bar{\omega}$ shed into the stream by supposing that it is comparable to the total vorticity shed by a uniform wave train of this mean-square steepness over the given time t. Substituting in (6.4) we find altogether

$$\bar{u}_o = 0.87(ak)_o^2 c . \tag{7.3}$$

Now from Figure 3 it is clear that $(ak)_o$ can be of order 1 so that $\bar{u}_o$, by (7.3), is of the same order as c. In other words, the vorticity generated beneath the capillary waves appears quite capable of generating the velocity-differences above and below the roller, even in the absence of wind.

In the case of rollers on a short gravity wave, the velocity difference between top and bottom of the roller is further augmented by the vertical gradient of the orbital velocity in the gravity wave. In addition, work is done on the parasitic capillaries by the orbital motion in the gravity wave (Longuet-Higgins 1963; Longuet-Higgins and Stewart, 1964), causing them to become both shorter and steeper, thus increasing the vorticity which is shed.

8. Vorticity and mass transport

The above conclusions are corroborated by considering the creation of vorticity from the point of view of the capillary wave momentum. For, the total horizontal momentum density I in a surface wave on deep water is given

$$I = 2KE/c \tag{8.1}$$

where KE is the kinetic energy density (see for example Longuet-Higgins, 1975). For a pure capillary wave

$$I = \frac{1}{2}(ak)^2\ (T/k)^{1/2} \tag{8.2}$$

at small wave steepnesses. At the maximum wave steepness $ak = 2.293$ this value is increased by a further factor of 1.233; see Hogan (1979).

When the capillary wave is damped out by viscosity, the wave momentum I is converted mainly to a horizontal current (though some may go to produce a depression of the mean surface level). From equation (6.2), the depth d to which the momentum is diffused is of order $(\nu t)^{1/2}$ where $t \sim 3/\nu k^2$. Thus we have

$$d \sim (3/8)^{1/2}\ k^{-1} \tag{8.3}$$

The mean current velocity is of order I/d. Using (8.2) we have

$$u_o \sim \frac{1}{2}(ak)^2\ (8/3)^{1/2}\ c = 0.82(ak)^2c \tag{8.4}$$

in rough agreement with (7.3).

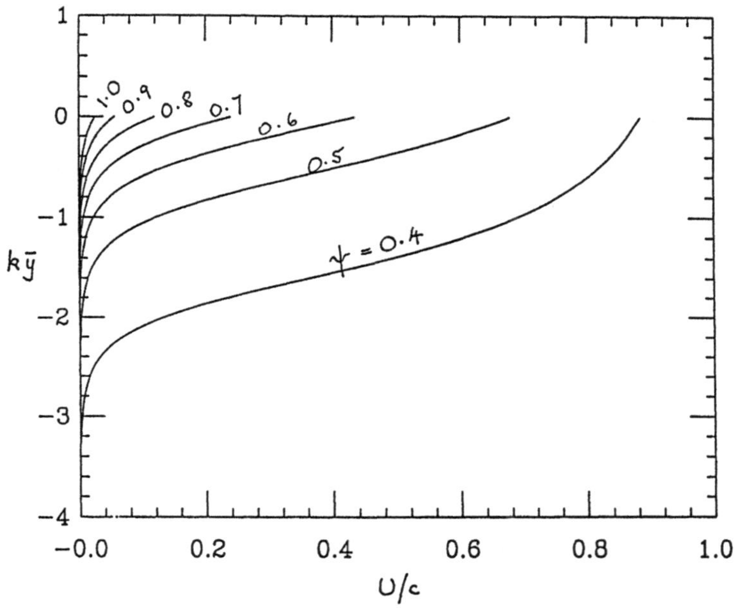

Figure 11. The mass-transport velocity U for pure capillary waves, shown as a function of the mean depth $\bar{y}$ of each streamline in Figure 5.

If the capillary wave were dissipated instantaneously by internal friction, the profile of the horizontal momentum would presumably be given by the mass-transport velocity U in the original wave. In Figure 11 we show U/c plotted as a function of the mean depth $\bar{y}$ of a streamline below the mean surface level $\bar{y}$, for waves of various steepnesses ak. It is notable that for the steepest waves, U/c is close to unity over an appreciable thickness of fluid nearest the free surface. This reflects the fact that the fluid in the rounded crests of the wave all travels forward with nearly the phase-speed c. The momentum of this fluid has only to be released into a configuration where the surface is flattened, in order to create a positive current moving very nearly with speed c.

9. Steep capillary waves

In very steep capillary waves having sharply curved troughs, such as shown in Figures 5d and 7b, it seems inevitable that the streamlines will separate at the wave troughs. The high vorticity $-2\kappa q$ generated at a trough will give rise directly to a free streamline as in Figure 12. The initial velocity-difference Δu across the layer will be of the same order as the vorticity ω integrated round the sharp bend that is

$$\Delta u \sim \int \omega \, ds = \int (-2\kappa\, q) \, ds \, . \tag{9.1}$$

Since

$$\int \kappa \, ds = \Delta\theta = \pi \tag{9.2}$$

we see that velocity-differences of the same order as q or c are to be expected.

The enclosed space, or "bubble", in the trough of the limiting wave in Figure 5d has long suggested that bubbles of air may be trapped by such waves (see Crapper 1957, Schooley 1958). This no doubt explains the injection of air bubbles beneath capillary-gravity waves as seen in the laboratory by Toba (1961), Koga (1982) and others, although we may also expect the density and viscosity of the air to be of significance in this process.

We recall that this limiting form of steep waves, in which adjacent wave crests come into contact and trap air, is not confined to pure capillary waves, but is found over a whole range of the parameter (Tk^2/g), as shown by Schwartz and Vanden-Broeck (1980), Chen and Saffman (1979, 1980) and Hogan (1980, 1981). The range of (Tk^2/g) extends even to zero, to include capillary-gravity waves of solitary type on deep water (Longuet-Higgins 1989). Boundary-layer calculations similar to those that we have carried out for pure capillary waves could equally well be performed for all such waves.

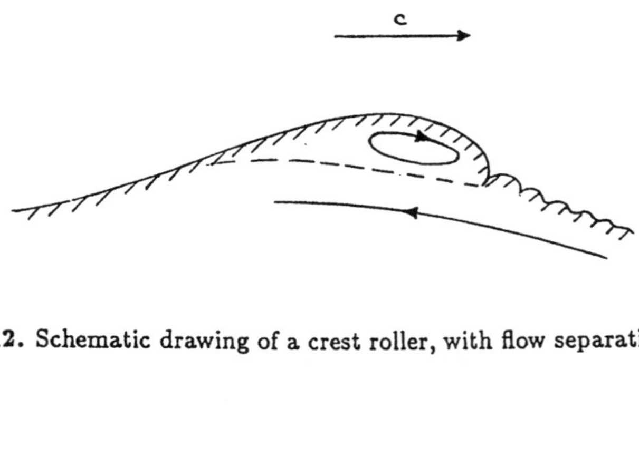

Figure 12. Schematic drawing of a crest roller, with flow separation and free shear layer.

10. The development of rollers and bores

The development of a capillary roller at the crest of a short gravity wave appears to be a consequence of the existence of the parasitic capillaries. This in turn may be a phenomenon of irrotational flow, in the first place.

For example, we know the theoretical form of the wave crest in a steep gravity wave (the "almost-highest wave") as the limiting 120° corner-flow is approached (see Longuet-Higgins and Fox, 1976). At "infinity" the flow approaches asymptotically the 120° corner-flow, and the only scale-length in this solution is the radius of curvature at the crest.

The corresponding irrotational solution including capillarity has not yet been calculated, but we may conjecture that it will feature capillary waves ahead of the gravity wave crest. These must have a speed which enables them to propagate against the backward flow. The source of the capillary wave energy is presumably connected with nonlinear features of the flow near the gravity wave crests, where the length-scale of the flow is comparable to the wavelength of the capillaries. A first theory for such parasitic capillaries was given by Longuet-Higgins (1963). This assumed the wave amplitude to be due to the perturbation in the pressure brought about by capillarity in the neighbourhood of the gravity wave crest, where the curvature is relatively large. The theory was further developed by Vanden-Broeck (1960) and Crapper (1970), with qualitative experimental confirmation by Chang et al. (1978) and Yermakoo et al. (1986); see also Ebuchi et al. (1987). Ruvinsky et al. (1981, 1985, 1991) have developed a theory which takes fuller account of the dynamics of the wave crest, and includes the damping of the capillaries, to first order, but ignores the existence of the roller at the crest.

Gravity waves in a random sea are essentially transitory, so that steep or breaking waves may appear in a short time of the order of one or two wave periods. The wave crests may initially be irrotational, or relatively so. However, very soon after the first appearance of the parasitic capillaries, the shed vorticity will begin to build up a vortex at the gravity wave crest. This vortex will tend to accentuate the curvature immediately behind (i.e. downstream of) the first capillary wave, and so increase its amplitude. The dissipation of energy may also reduce the steepness of the gravity wave. But it seems likely that, over a short time at least, there is a positive feed-back between the crest roller and the parasitic capillaries. In other words, the configuration of roller and capillaries is self-sustaining, the energy for the flow being drawn from the much larger energy of the gravity wave. A schematic picture is shown in Figure 12.

The flow may break down in various ways. One is by dissipation of the energy of the gravity wave (see Longuet-Higgins 1963, Section 10), or the gravity wave may flatten by the familiar process of moving forwards through a wave group. A second possibility is for the capillary waves to trap air bubbles in the sharp wave troughs as observed by Koga (1982). This process is accompanied by a characteristic circulation near the gravity-wave crest, a partial model for which has been suggested by Longuet-Higgins (1990).

A third possibility, which is always present when a laminar viscous flow is pushed to larger Reynolds numbers, is that the flow will pass via instability into a turbulent flow. In such a case the vorticity generated by the capillaries may be replaced entirely by an unstable shear layer between the roller and the underlying flow, and then the parasitic capillaries may altogether disappear. This appears to have happened in the microbreaker shown for example by Banner and Cato (1988); see also Banner and Phillips (1974).

11. Conclusions and suggestions

We have shown that the mean vorticity shed from parasitic capillaries on the forward face of a short gravity wave may contribute substantially towards the vorticity in the roller at the gravity wave crest. Under some circumstances, in the absence of wind, this mechanism may account for all the observed vorticity.

The crest roller and the train of parasitic capillaries form a cooperative system, each sustaining the other. The capillaries may become so steep as to trap air bubbles beneath the surface, as seen by Koga (1982).

The scale of the phenomenon is limited. At larger Reynolds numbers the capillary wave train disappears and is replaced by an unstable shear layer between the roller and the fluid beneath.

At small scales, the role of surface tension and viscosity are all-important. In this paper we have estimated the effect of viscosity in a rough way, as a perturbation to an otherwise irrotational flow. Ultimately, a more radical approach is needed, incorporating viscous boundary conditions from the start.

We have considered the dynamics of the boundary-layer mainly from the point of view of vorticity. It may also be fruitful to consider the balance of tangential momentum in the layer (Longuet-Higgins, 1969a; Csanady 1989), especially over a wavelength of the gravity wave. The generation of mean vorticity $\bar{\omega}_\infty$ by the parasitic capillaries is equivalent in its effects to applying a tangential stress $-\nu\bar{\omega}_\infty$ at the surface of the gravity wave. This tends to pile up an additional mass of fluid lagging 90° behind the applied stress (see Longuet-Higgins 1969a). Since the capillaries are on the forward face of the wave, we expect the additional mass to appear near the crest, as is observed. The virtual tangential stress will also contain a second harmonic which will in general cause the gravity wave to become asymmetric, in the horizontal sense. Ordinary, second-harmonic viscous damping will of course have a similar effect, but the effect of dissipation by the capillaries is much stronger. In fact it may lead to a kind of dissipative coupling between the gravity wave and its second and higher harmonics.

REFERENCES

Banner, M.L. and Cato, D., 1988, "Physical mechanisms of noise generation by breaking waves - a laboratory study," pp. 429-436 *in Sea Surface Sound*, ed. B.R. Kerman, Dordrecht, Kluwer Acad. Publ., 639 pp.

Chang, J.H., Wagner, R.N. and Yuen, H.C., 1978, "Measurement of high frequency capillary waves on steep gravity waves," *J. Fluid Mech.* 86, 401-413.

Chen, B. and Saffman, P.G., 1979, "Steady gravity-capillary waves on deep water - I. Weakly nonlinear waves," *Stud. Appl. Math.* 60, 183-210.

Chen, B. and Saffman, P.G., 1980, "Steady gravity-capillary waves on deep water - II. Numerical results for finite amplitude," *Stud. Appl. Math.* 62, 95-111.

Cox, C.S., 1958, "Measurements of slopes of high-frequency wind waves," *J. Mar. Res.* 16, 199-225.

Crapper, G.B. 1957, "An exact solution for progressive capillary waves of arbitrary amplitude," *J. Fluid Mech.*, 2, 532-540.

Crapper, G.D. 1970, "Non-linear capillary waves generated by steep gravity waves," *J. Fluid Mech.* 40, 149-159.

Hogan, S.J., 1979, "Some effects of surface tension on steep water waves," *J. Fluid Mech.*, **91**, 167-180.

Hogan, S.J., 1980, "Some effects of surface tension on steep water waves," Part 2. *J. Fluid Mech.* **96**, 417-445.

Hogan, S.J., 1981, "Some effects of surface tension on steep water waves." Part 3. *J. Fluid Mech.* **110**, 381-410.

Koga, M., 1982, "Bubble entrainment in breaking wind waves," *Tellus* **34**, 481-489.

Lamb, H., 1932, Hydrodynamics, 6th ed., Cambridge Univ. Press, 738 pp.

Longuet-Higgins, M.S., 1953, "Mass transport in water waves," *Phil. Trans. R. Soc. Lond.* A **245**, 535-581.

Longuet-Higgins, M.S., 1960, "Mass-transport in the boundary-layer at a free oscillating surface," *J. Fluid Mech.* **8**, 293-306.

Longuet-Higgins, M.S., 1963, "The generation of capillary waves by steep gravity waves," *J. Fluid Mech.* **16**, 138-159.

Longuet-Higgins, M.S., 1969a, "Action of a variable stress at the surface of water waves," *Phys. Fluids* **12**, 737-740.

Longuet-Higgins, M.S., 1988, "Limiting forms for capillary-gravity waves," *J. Fluid Mech.* **194**, 351-375.

Longuet-Higgins, M.S., 1989, "Capillary-gravity waves of solitary type on deep water," *J. Fluid Mech.* **200**, 451-470.

Longuet-Higgins, M.S., 1990, "Flow separation near the crests of short gravity waves," *J. Phys. Oceanogr.* **20**, 595-599.

Longuet-Higgins, M.S., 1991, "Theory of weakly damped Stokes waves: a new formulation and its physical interpretation, *J. Fluid Mech.* (to appear).

Okuda, K., Kawai, S., and Toba, Y., 1977, "Measurements of skin friction distribution along the surface of wind waves," *J. Oceanogr. Soc. Japan* **33**, 190-198.

Ruvinski, K.D. and Freidman, G.I., 1981, "The generation of capillary-gravity waves by steep gravity waves," *Izv. Atmos. Ocean Phys.* **19**, n:7.

Ruvinski, K.D. and Freidman, G.I., 1985, "Improvement of first Stokes method for the investigation of finite-amplitude potential gravity-capillary waves. *IX All-Union Symp. on Diffraction and Propagation Waves*, Theses of Reports. Tbilisi, **2**, 22-25.

Ruvinski, K.D., Feldstein, F.I., and Freidman, G.I., 1991, "Numerical simulation of the quasi-stationary stage of ripple excitation by steep gravity-capillary waves," *J. Fluid Mech.*, (to appear).

Schooley, A.H., 1958, "Profiles of wind-created water waves in the capillary-gravity transition region," *J. Mar. Res.* **16**, 100-108.

Toba, Y., 1961, "Drop production by bursting of air bubbles on the sea surface III. Study by use of a wind flume," *Mem. Coll. Sci: Univ. Kyoto* **A29**, 313-343.

Vanden-Broek, J.M., 1960, "Ph.D. Thesis, University of Liege.

Yermakov, S.A., Ruvinski, K.D., Salashin, S.G. and Freydman, G.I., 1986, "Experimental investigation of the generation of capillary-gravity ripples by strongly nonlinear waves on the surface of a deep fluid," *Izv. Atmos. Ocean. Phys.* **22**, 835-842.

Mechanisms of Water-Wave Breaking

D.H. Peregrine,
Department of Mathematics,
University of Bristol, Bristol BS8 1TW, England.

Abstract
The fluid mechanics of wave breaking is discussed, selecting only aspects involved directly in the wave breaking process rather than wave evolution towards breaking or the effects of breaking. As well as plunging and spilling breakers other less usual topics are considered including examples where the free surface "breaks" at a trough rather than a crest.

Introduction

There are many papers on the *causes* of wave breaking, such as instabilities, shoaling water, forcing, or simple evolution of long waves in shallow water. Similarly there is much discussion of the *effects* of wave breaking, such as dissipation, mixing and the generation of currents. Here, however, I discuss the actual process of wave breaking and try to give a little insight into some of the various mechanisms involved. Only the breaking of surface waves is considered though there is reference to other circumstances and some comments are more widely applicable. Also, a rather wide interpretation of breaking is taken, beyond the case of the usual propagating waves, which might plunge or spill, though these are considered. In particular, some attention is given to Rayleigh-Taylor instability and to the often inconspicuous breaking associated with turbulent flow beneath a free surface. The overall aim of this presentation is to stimulate further work to clarify the fluid mechanics of breaking. When the strong influence of wave breaking on the design of marine structures, the evolution of coastlines and on mass and heat transfer between oceans and atmosphere is considered, we can argue that wave-breaking is the most important feature of surface waves.

M. L. Banner, R. H. J. Grimshaw (Eds.)
Breaking Waves
IUTAM Symposium Sydney/Australia 1991

Jets

The phrase "a breaking wave" will usually stimulate a mental picture of a plunging breaker. The most characteristic feature of a plunging breaker is the sheet, or two-dimensional jet, of water that forms near the crest of the wave and plunges forward into the water. The falling jet has the simplest mechanics. It is usually thin and has the same pressure around it so that to a good approximation the thinner parts are simply falling under gravity. Longuet-Higgins (1980) has described a solution for a rotating hyperbola in free-fall which simulates the motion of a jet's tip.

It is more difficult to interpret the initial growth of a jet. Peregrine, Cokelet and McIver (1980) considered the details of the fluid motion for some typical breaking waves. The most remarkable feature is that a region of large fluid accelerations and corresponding large pressure gradients, develops on the front face of a wave. Some "extra" acceleration of fluid particles is clearly a necessary precursor of breaking since the jet must attain a forward velocity greater than that of the bulk of the wave.

Similarly some convergence of the flow is necessary to form a jet. However, convergence of the surface is insufficient. This is readily seen from the exact solution for flow above a horizontal plane $y = 0$, given for $t < 0$ by velocity potential

$$\phi(x,y,t) = (x^2 - y^2)/2t$$

and surface elevation $\eta(z,t) = -H/t$

for constant H. This flow has the peculiarity of singular behaviour as $t \rightarrow 0$ for which the infinite extent of the flow is relevant. On the other hand, a solution which leads to a jet like flow, in the absence of gravity, is the Dirichlet ellipse (Longuet-Higgins, 1972). An initially extended but converging half ellipse in $y > 0$ becomes a long "jet" along the y axis, where again $y = 0$ is taken as a bounding plane.

If we widen our view of jet formation at a free surface then very violent jets are found. For over-forced standing waves in a beaker Longuet-Higgins (1983) shows photographs of a violent jet emerging from the central trough of the wave and Cooker and Peregrine (1990, and 1991 in this volume) compute similarly violent two-dimensional jets arising from a trough in the free surface. These computations indicate that very high pressures and accelerations are involved in the jet's formation. They seem to be similar to the jets formed when cavitation bubbles collapse in a non-uniform environment. The formation of these jets is an "active"

Figure 1. Turbulent jet entering the River Avon, Bristol

process in which the fluid flow appears to be organised towards their creation.

One might on the other hand argue the free surface is entirely passive. The two boundary conditions at the free surface are firstly the kinematic boundary condition which states that the free surface is a material surface, and secondly the dynamic boundary condition which for the inviscid model being considered is that pressure is constant. Both of these appear unrelated to active jet formation. In fact jet-like structures in a material surface are commonplace in unsteady flows. For example, see the dye streaks in figure 10.3 of Van Dyke (1982) which shows a replica of Reynolds' experiment on the development of turbulence in pipe flow. A further related example is the development of filamentation in contour dynamics, the two dimensional flows with regions of constant vorticity. Thin strands of the interface between regions of differing vorticity are convected away, e.g. see Pullin (1981) or Dritschel (1988).

When flows that are not directly related to the free surface are considered then it is easy to see that if they are energetic enough surface breaking occurs. An example of a turbulent jet near the free surface is shown in figure 1 where breaking can be seen over the more energetic part of the jet. Here the free surface behaviour is almost incidentally caused by the turbulent jet flow. As long as gravity fails to restrict the excursions of the free surface this may be interpreted as passive breaking.

We can examine the pressure field in a wave that is just about to break in a conventional manner. Figure 2 shows a wave breaking because of a shoaling bed. Two plots of pressure are shown. The first shows the strong pressure gradient at the front of the wave accelerating water towards the incipient jet. The second is designed to indicate the "activity" of the flow and shows pressure contours after hydrostatic pressure has been subtracted from the flow field. The contours are concentrated in the front of the wave and are sparse at and beyond the crest. Comparison with previously mentioned examples suggests that this typical breaking-wave flow has both active and passive aspects, corresponding to flow beneath and beyond the area from which the jet is about to emerge.

Splashing

After a plunging jet hits the water in front of the breaking wave a sequence of splashes is created. The study of this aspect of wave breaking is only in its infancy. It is at the stage where observation and description are of value in order to stimulate theoretical developments.

(a)

(b)

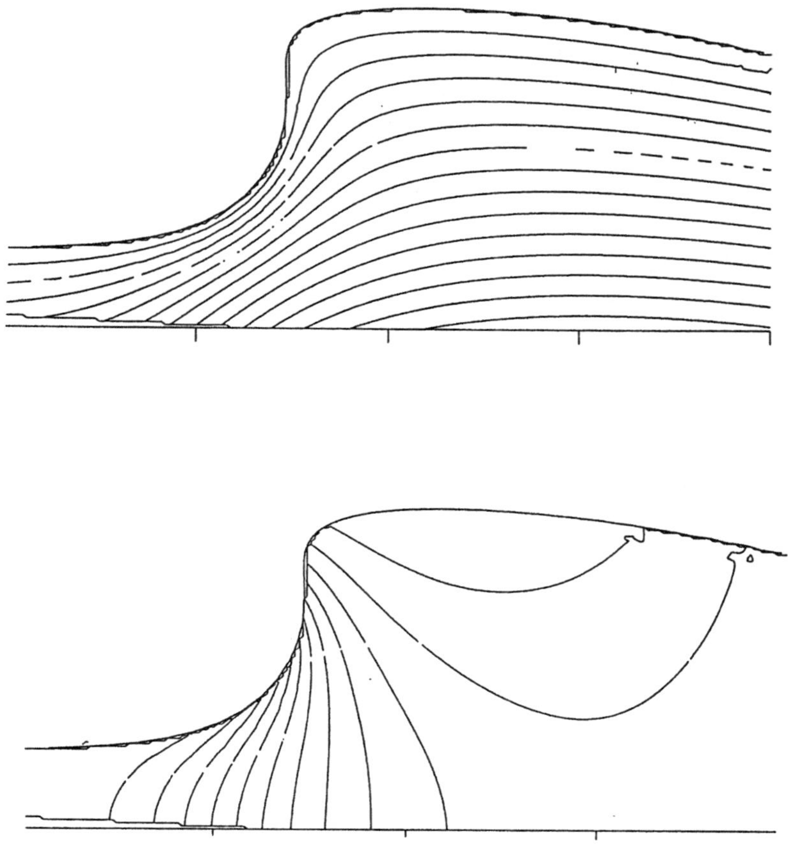

Figure 2 Pressure contours in a wave which is about to break
The wave is propagating toward the left
(a) Total pressure
(b) Pressure minus hydrostatic pressure.

Excellent examples of the complex, but repeatable, splashing that occur in breaking waves are Tallent *et al*'s (1990) illustrations drawn from careful study of sequential video frames.

A first attempt at splash modelling is described in Peregrine (1981). More progress has been made in modelling the splash caused by the impact of solid bodies, see Greenhow (1988) for a recent example.

"Waterfall breaking"

This type of breaking seems not to have been remarked upon before, yet it occurs in many different ways. As an introductory example consider the flow of a stream of water over a bed in which there is a substantial downward step. If the water level downstream of the step is well above the top of the step, the step causes no more than a slight disturbance. On the other hand if the water level downstream of the step is appreciably below the level of the top of the step then a waterfall forms and the descending stream of water plunges beneath the lower free surface. If the downstream water level is unsteady and descends, not too slowly, there is a sudden and rapid transition between smooth flow along the free surface and a plunging flow. The free surface "breaks" downwards in a manner which is altogether different to a conventional breaking wave. Ohtsu and Yasuda (1991) give a comprehensive description of the many different steady flows that occur over a step, though they do not report the significant hysteresis that occurs if a given downstream depth is reached by lowering or raising the water level.

Similar waterfall breaking can occur if a large bluff body is raised towards the free surface, as sketched in figure 3. Perhaps the most common form of waterfall breaking occurs when a region of vorticity approaches a free surface. The region of vorticity can be a turbulent eddy, a vortex pair or a vortex ring such as occurs if an upward pointing hose is turned on beneath the water surface and is turned on.

The motion due to a vortex pair approaching the surface is described in computations by Telste (1989). Figure 10, in that paper shows the final stage of an inviscid calculation where a depression in the free surface has almost formed an angle or cusp at the bottom. A close-up of streamlines of a similar flow we have recently computed, with a single vortex are shown in figure 4. Careful examination of the flow field shows an extreme variation of the normal pressure gradient at the surface at the bottom of the trough. It is -0.35, 0.99 and 11.98 ρg at successive discretization points. Similarly the rate of strain and the angular velocity of the free surface

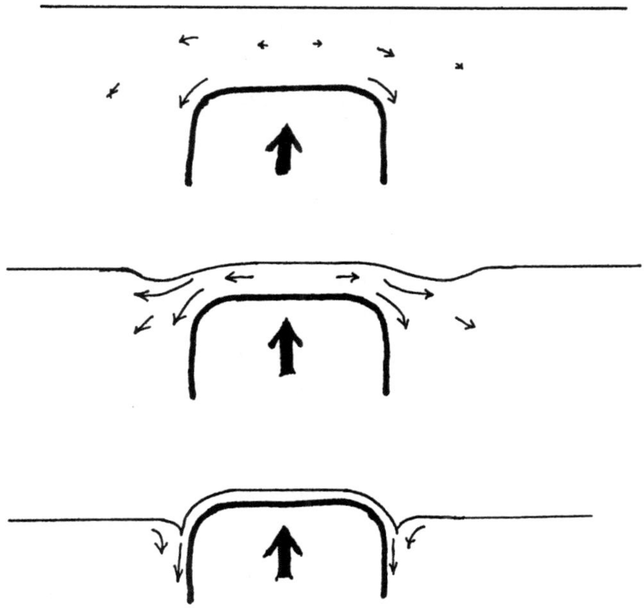

Figure 3. Waterfall breaking: as a bluff body approaches the free surface the flow can pass from (a) through (b) to (c). Between (b) and (c) a "break" in the free surface occurs.

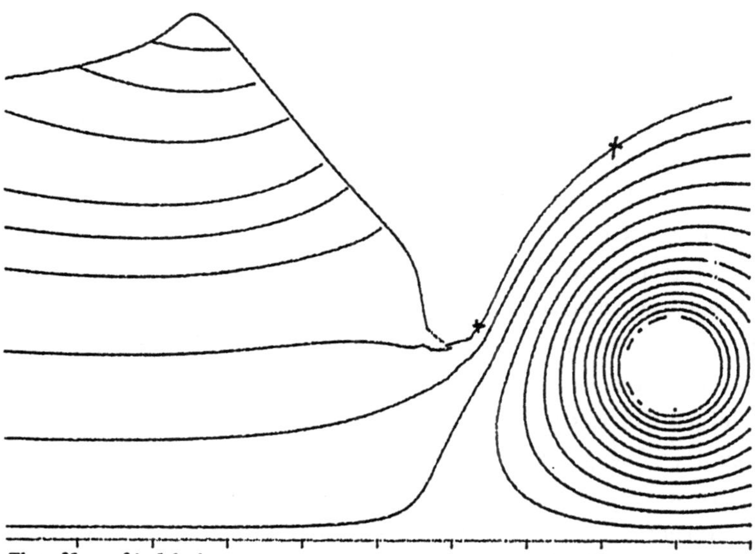

Figure 4. The flow field due to a strong vortex near the free surface. The portion of the free surface between the two crosses, x—x, has sub-atmospheric pressure beneath it. Courtesy of R. Tong.

are also large with strong gradients. It is hard to extrapolate from numerical results to determine what sort of singularity of the free surface might occur. It seems likely that an angle, cusp or pinching off of the trough will occur leading to the development of a shear layer within the water.

The vortex pair computation has also been performed for a viscous liquid. Ohring and Lugt (1991) report computations for a Reynolds number of 100, with and without surface tension. In this case it is clear that vorticity generated at the bounding free surface plays a very significant role and leads to a free surface structure which briefly resembles a spilling breaker. See also Longuet-Higgins (1991, in this volume).

The steady or quasi-steady flows which follow waterfall breaking involve a shear layer descending into the water, usually with a noticeable trough or line on the free surface. For example, see the far field of the turbulent jet in figure 1. In the context of a ship's wake which can be modelled as a vortex pair such lines or troughs have been called "scars". These flow features, on a much smaller scale, may be crucial to a full understanding of heat and mass transfer across a gas-liquid interface whether on the ocean surface or in industrial equipment.

Rayleigh-Taylor instability

Although Rayleigh-Taylor instability is sometimes proposed as a cause of wave breaking, I do not consider that it is relevant to ordinary propagating waves. This viewpoint is based on both observation and theory. On the theoretical side we, at Bristol, have been computing many unsteady water wave solutions for about a dozen years. Despite deliberate attempts to cause waves to develop Rayleigh-Taylor instability is has not been observed in ordinary waves.

There are, however, cases where Rayleigh-Taylor (R-T) instability does occur and can be observed, but the type of flow disturbance that arises has characteristics which differ from ordinary plunging or spilling breakers.

Before discussing flows where R-T instability occurs let us briefly review it. The primary example of R-T instability is where a heavier fluid overlies a less dense fluid in uniform gravitational field. For an infinite plane interface between semi infinite fluids it is readily found that all wavelengths are unstable in the absence of surface tension. The same analysis applies to an accelerated interface and clearly unsteady motions of a free surface do involve accelerations of the surface. Since these accelerations are caused by the pressure gradient which must be perpendicular to the free surface it is most convenient to summarise the likelihood of R-T instability in the following way. If the normal pressure gradient at a free surface is such that pressure decreases on passing from less dense to denser fluid then R-T instability is possible.

If a region of free surface has a normal pressure gradient tending to R-T instability such instabilities do not necessarily develop. For example the flow above a vortex illustrated in figure 4 has a substantial length of "unstable" pressure gradients. However, the rapid evolution of the flow appears to be inhibiting growth of instabilities.

A case where growth of disturbances does occur in such an unstable region is on large amplitude waves on certain flows with shear. The most easily observed example is where large waves are nearly stationary on a strongly sheared flow such as the backwash down a beach. Simple models (Peregrine, 1974) or models with constant vorticity in the flow (Teles da Silva and Peregrine, 1988) can be used to show that the crest of these large waves is an "unstable" region. Indeed turbulent fluctuations of the surface can be seen to grow with a transverse wave number dominant. They also have the rounded shape typical of the evolution of R-T instability, but look different from other breaking waves. See figure 18 of Teles da Silva and Peregrine (1988).

Waves with shear

As has just been remarked waves propagating against a shear flow are liable to R-T instability, however shear in the vertical also affects the shape of the basic wave solution. The above mentioned waves have more rounded crests than irrotational waves and appear less likely to break in the conventional way. On the other hand if the shear has the opposite sense waves have sharper crests and seem more likely to break. This is the case when wind induces a surface drift current and waves propagate in the same direction. Banner and Phillips (1974) and Phillips and Banner (1974) discuss the effect of such a thin shear layer and illustrate how it can greatly increase the likelihood of wave breaking. By considering a wave in a frame of reference moving with the wave profile it is easily seen how such shear reduces the total head on the surface stream line.

It is readily appreciated that ocean waves are not steady and individual wave crests grow and decay in amplitude. The effect of wind drift is that for a growing wave crest a thin layer of surface water may be raised to its "stagnation level" and start falling down the wave face. This is likely to commence on the scale of micro-breakers. The small mixing that occurs can reduce the shear in the flow increasing the "stagnation level". If the wave is large enough further slight mixing occurs at a higher level until eventually a clearly spilling breaker may develop. This type of initially small scale mixing may be the cause of many spilling breakers which appear to arise spontaneously with no prior overturning jet however small.

Very little experimental work appears to have been done on the effects of shear on breaking. The only experiments I am aware of are those of Douglas and Weggel (1988) which although relatively simple did show a very strong influence of wind shear on wave breaking properties.

Spilling breakers

Unlike plunging breakers which are essentially unsteady, spilling breakers can occur on flows which are steady except for turbulent fluctuations. Good examples are the hydraulic jump and waves behind obstacles. Examples of the latter are the waves generated by a submerged airfoil as studied experimentally by Duncan (1981, 1983) and Battjes and Sakai (1981). Propagating waves with spilling breakers are quasi-steady in the sense that the evolution of the breaker is long compared with the time scale for fluid particles to pass through the breakers.

Although mean streamlines, as in figure 5(a), indicate fluid may stay in the roller, it is found in experiments, as described by Peregrine and Svendsen (1978), that a drop of dye on landing in the roller is immediately dispersed throughout the turbulent region sketched in figure 5(b).

Madsen and Svendsen (1983) followed up this observation and used a simple model of the turbulent layer to provide a remarkably effective description of the mean flow in a hydraulic jump. Their model was extended to unsteady shallow-water flow in Svendsen and Madsen (1984) and illustrated by a bore propagating up a beach.

Many other models of spilling breakers, treat the roller simply as a passive body of water supported by the wave. However, given the strong effects of shear mentioned above it seems inappropriate to ignore the bulk of the turbulent flow. On the other hand we are sufficiently far from having a good model that even the simplest models have their uses.

One of the key aspects to both understanding and modelling a spilling breaker is the toe of the spiller. That is the point where the previously

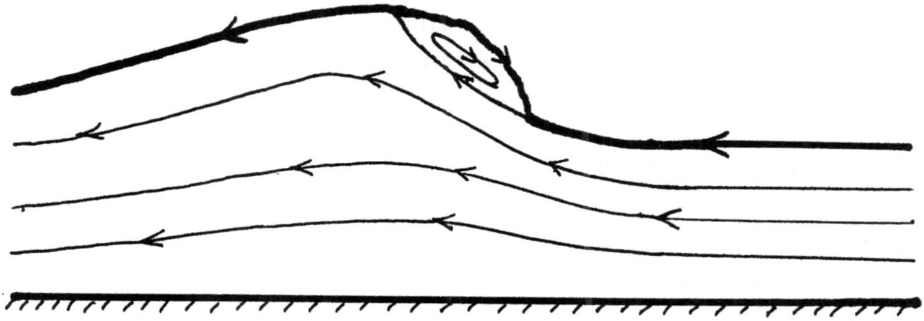

Figure 5(a) The traditional view of a spilling breaker in a reference frame moving with the wave: sketch of streamlines.

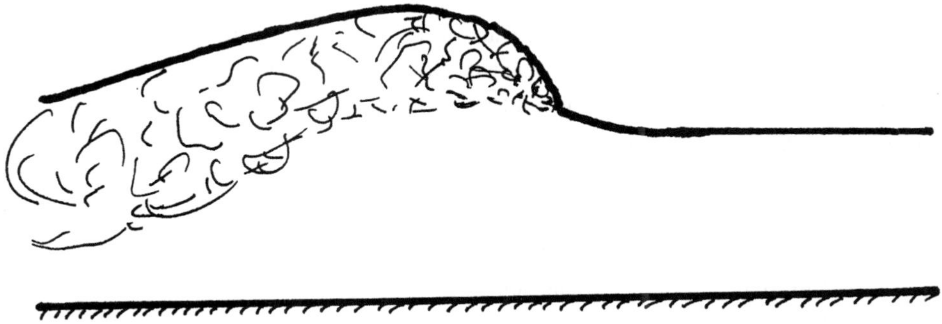

Figure 5(b) Spilling breaker: turbulent region emphasised by Peregrine & Svendsen (1978).

undisturbed water first meets the turbulent roller and entrainment of new fluid into the turbulent region comments. Longuet-Higgins (1973) identified this as a key region and proposed a model in which the turbulence is confined to the roller and the flow is dominated by the Reynolds stresses. On the other hand, Madsen and Svensden's (1984) model has singular behaviour in the shear at the toe. This is not inconsistent with Peregrine and Svendsen's (1978) suggestion that the flow is like the classical turbulent mixing layer between two streams of differing velocity. In this case the upper layer has a velocity equal to that of the toe, or greater since no entrainment through the upper part of the turbulent layer is possible.

There is a previously unexamined aspect of flow in a spiller which indicates that a singularity in the mean-flow at the toe may be appropriate. Down the front of the spiller the mean free surface is *not*, in general, a streamline of the mean flow. Those parts of the fluctuating free surface which are outside the mean surface are likely to have a downward velocity on average. A net flux of mass and momentum through the mean free surface to the "outer surface layer" should occur. This is equivalent to the mean drift in irrotational Stokes waves and corresponding terms can be seen in Hasselmann's (1971) averaging of the equations of motion. At the toe of the spiller the mass and momentum flux associated with the outer surface layer is all returned to the mean flow.

Exactly what type of behaviour and flow streamlines one obtains depends on how the flow is averaged at the toe. The sketch of figure 5(a) shows an angle in the mean surface at the toe, at a point corresponding to the toe's mean position. However the fluctuating position of the toe could be included in the average yielding a smooth transition between roller and undisturbed flow. In the sketch of figure 5(a) there must be a singularity at the toe. Whereas, if the surface is averaged to a smooth transition, then mass and momentum in the outer surface layer is deposited into the transition. My view is that we should consider this surface-layer mass and momentum impinging on the undisturbed flow as the originator of the spilling breaker's turbulence.

In most of the well-studied examples of turbulent flow mean properties have been well defined and coherent structures identified. These coherent structures often appear to have a dominant contribution to the Reynolds stresses and hence to the mechanics of the flow. Some work in this direction has been presented by Nadaoka et al. (1989) where obliquely descending eddies have been observed frequently in the flow well behind the spiller. From many years of visual observation I recognise such eddies, but also consider that they may not be entirely typical, due to effects of scale.

In considering water waves, especially breaking waves we are looking at a two-phase flow. Two-phase flows are notoriously difficult to scale and there is no exception for breakers. The smallest spilling breakers entrain no air. Surface tension acts to keep the gas and liquid phases separated by a single simply connected surface. Once air entrainment occurs there can be a large difference depending at least on factors such as bubble size and fraction of air entrained. As is now well documented (Scott, 1975a,b) there is a substantial difference in typical bubble size, between fresh water and sea water, though the reasons are not understood. In laboratory scale waves (say up to 30 cm high) there is no obvious effect of the lower mean density of the air-water mixtures obtained. However in larger waves the buoyancy of the bubbles does appear to lead to a structure with many regions of upwelling fluid being visible after the passage of a breaking wave. This aspect of breakers is significant since buoyant turbulent fluid will have a different effect on sediment transport from the less buoyant fluid. Similarly in the deep ocean buoyant turbulence is less efficient than non buoyant turbulence in deepening or maintaining a mixed layer.

Conclusion

This review ranges widely including topics which are not usually embraced by the concept of breaking. For the most part it describes the initial stages of the study of the mechanisms of wave breaking and has the aim of stimulating further study.

I wish to acknowledge the support of many colleagues in this area especially younger colleagues at Bristol whose support from the U.K. Science and Engineering Research Council has been and is much appreciated.

References

Banner, M.L. and Phillips, O.M. (1974) On the incipient breaking of small-scale waves. J.Fluid Mech. **65**, 647-656.

Battjes, J.A. and Sakai, T. (1981) Velocity field in a steady breaker J.Fluid Mech. **111**, 421-437.

Cooker, M.J. and Peregrine, D.H. (1991) Violent motion as near-breaking water waves meet a vertical wall. Proc. IUTAM Symposium.

Douglas, S.L. and Weggel, J.R. (1988) A laboratory experiment on the influence of wind on nearshore breaking waves. Proc. 21st Internat. Conf. Coastal Engng. A.S.C.E. **1**, 632-643.

Dritschel, D.G. (1988) The repeated filamentation of two-dimensional vorticity interfaces. J.Fluid Mech. **194**, 511-547.

Duncan, J.H. (1981) An experimental investigation of breaking waves produced by a towed hydrofoil. Proc.Roy.Soc.Lond.A **377**, 331-348.

Duncan, J.H. (1983) The breaking and non-breaking wave resistance of a two-dimensional hydrofoil. J.Fluid Mech. **126**, 507-520.

Greenhow, M. (1988) Water-entry and -exit of a horizontal circular cylinder. Appl. Ocean Res. **10**, 199-206.

Hasselmann, K. (1971) On the mass and momentum transfer between short gravity and larger-scale motions. J.Fluid Mech., **50**, 189-205.

Longuet-Higgins, M.S. (1972) A class of exact, time-dependent, free surface flows. J.Fluid Mech. **55**, 529-543.

Longuet-Higgins, M.S. (1973) A model of flow separation at a free surface. J.Fluid Mech. **57**, 129-148.

Longuet-Higgins, M.S. (1980) On the forming of sharp corners at a free surface. Proc.Roy.Soc.Lond.A **371**, 453-478.

Longuet Higgins, M.S. (1983) Bubbles, breaking waves and hyperbolic jets at a free surface. J.Fluid Mech. **127**, 103-121.

Longuet-Higgins, M.S. (1991) Capillary rollers and bores, in Proc. IUTAM Sympos. "Breaking Waves.

Madsen, P.A. and Svendsen, I.A. (1983) Turbulent bores and hydraulic jumps. J.Fluid Mech. **129**, 1-25.

Nadaoka, K., Hino, M. and Koyano, Y. (1989) Structure of the turbulent flow field under breaking waves in the surf zone. J.Fluid Mech. **204**, 359-387.

Ohring, S. and Lugt, H.J. (1991) Interaction of a viscous vortex pair with the free surface. J.Fluid Mech. **227**, 47-70.

Ohtsa, I. and Yasuda, Y. (1991) Transition from supercritical to subcritical flow at an abrupt drop. J.Hydraulic Res. **29**, 309-328.

Peregrine, D.H. (1974) Surface shear waves. J.Hydraul.Div., Proc. Amer. Soc. Civil Eng. **100**, 1215-1227. [Discussion in **101**, 1032-1034 (1975)].

Peregrine, D.H. (1981) The fascination of fluid mechanics. J.Fluid Mech. **106**, 59-80.

Peregrine, D.H. and Svendsen, I.A. (1978) Spilling breakers, bores and hydraulic jumps. Proc. 16th Coastal Engng.Conf. ASCE Hamburg, **1**, 540-550.

Peregrine, D.H., Cokelet, E.D. and McIver, P. (1980) The fluid mechanics of waves approaching breaking. Proc. 17th Coastal Engng.Conf. ASCE, Sydney, Vol. 1, 512-528.

Phillips, O.M. and Banner, M.L. (1974) Wave breaking in the presence of wind drift and swell. J.Fluid Mech. **66**, 625-640.

Pullin, D.I. (1981) The nonlinear behaviour of a constant vorticity layer at a wall. J.Fluid Mech. **108**, 401-421.

Scott, J.C. (1975a) The preparation of water for surface-clean fluid mechanics. J.Fluid Mech. **69**(2), 339-351.

Scott, J.C. (1975b) The role of salt in whitecap persistence. Deep-Sea Res. **22**, 653-657.

Svendsen, I.A. and Madsen, P.A (1984) A turbulent bore on a beach. J.Fluid Mech. **148**, 73-96.

Tallent, J.R., Yamashita, T. and Tsuchiya, Y. (1990) Transformation characteristics of breaking water waves, in "Water-Wave Kinematics" eds. A. Torum and O.T. Gudmestad. Kluwer Academic, 509-523.

Teles da Silva, A.F. and Peregrine, D.H. (1988) Steep, steady, surface waves on water of finite depth with constant vorticity. J.Fluid Mech. **195**, 281-302.

Telste, J.G. (1989) Potential flow about two conter-rotating vortices approaching a free surface. J.Fluid Mech. **201**, 259-278.

Van Dyke, M. (1982) An album of fluid motion. Parabolic Press, Stanford, Calif., 176pp.

Surf Zone Dynamics

Ib A. Svendsen
University of Delaware
Department of Civil Engineering
Newark, DE 19716, USA

Introduction

Surf zone dynamics is a highly complicated topic in hydrodynamics which deals with the waves and wave generated phenomena in the region between the breaker line on a beach and the shoreline.

When waves break on a gently sloping beach large amounts of energy are released and turned into turbulence. As the waves keep breaking and interacting with the bottom topography, the momentum of the waves also decreases along with the decrease in wave height. The forcing this represents causes the generation of both currents and longer waves.

The proper analysis of the dynamics of the surf zone requires a detailed knowledge of the breaking waves and the turbulence they create, that is not yet available. However, significant progress has been made over the last decade or two, and the following is a brief account of some of the phenomena discussed in the literature and the results available.

General Description

Fig. 1 shows a schematic of the wave motion from the breaker point of a gently sloping beach, as most littoral beaches are, to the shoreline. The waves may initially break in a range of different patterns that have become known to reach from the relatively controlled "spilling" to the violent and relatively sudden "plunging" breaker type. In any type of breaking there will be a rapid and substantial change in the shape of the wave immediately following the initiation of breaking. This region has been termed the outer or transition region, which covers a distance of, say, 8–10 water depths after the breaker point (Svendsen et al., 1978).

Shoreward of the transition region, the wave shape and the general velocity field induced by the wave will change much more slowly. In this region the broken waves have many features in common with bores. This is the so-called inner or bore region which stretches all the way to the shore (or, if the breaking occurred on a longshore bar, till the waves stop breaking by passing into the deeper water shoreward of the bar).

On many natural beaches the foreshore is much steeper than the rest of the beach.

M. L. Banner, R. H. J. Grimshaw (Eds.)
Breaking Waves
IUTAM Symposium Sydney/Australia 1991

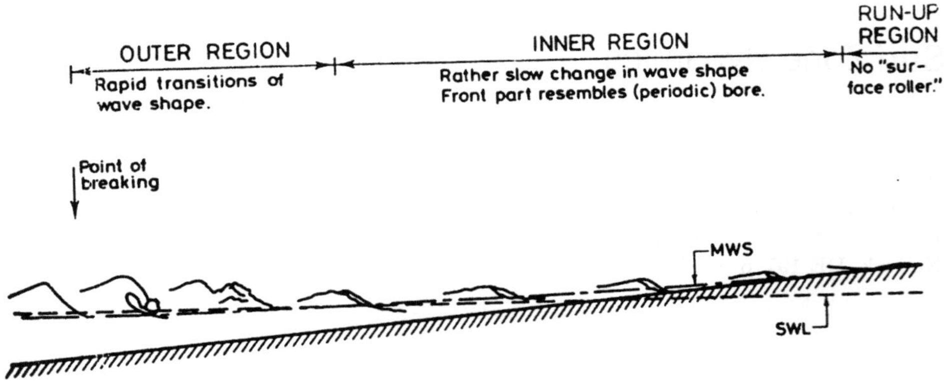

Figure 1: Wave characteristics in the surf zone (from Svendsen et al., 1978).

In the run up on the shore on such beaches (termed the swash zone), the wave motion often shows a different pattern from that of the rest of the surf zone.

The Transition Region

Very little has been published in the literature about the transition region. The results are almost entirely descriptive and based on photographic and optical methods. Basco and Yamashita (1986) gives an interpretation of the flow based on such information particularly for a plunging breaker and shows how the overturning of the wave creates patterns that look chaotic but are nevertheless largely repeated from wave to wave. Similar interpretations are given by Tallent et al. (1989), and Jansen (1986) has mapped the variation of the free surface in this region through high speed video recordings of fluorescent tracers. Finally, Okayasu (1989) gives detailed measurements of the entire velocity field in the transition region from experiments using laser doppler velocimetry. Those results have been obtained, however, by repeating the same experiments many times and each time averaging over several waves and therefore cannot be regarded as a picture of the instantaneous velocity field in a particular wave.

Wave Properties

The information about surf zone wave properties is also almost entirely empirical since no predictive models of the actual wave motion have been developed so far.

Thus Svendsen et al. (1978) showed experimentally that in the bore region the wave surface profiles would develop a relatively steep front with a much more gently sloping rear side. The profile of the rear will develop from a concave towards an almost linear variation, so that the entire wave near the shore is close to a sawtooth shape.

Measurements of velocity fields using laser doppler velocimetry have been reported by Stive (1980), Stive and Wind (1982), Nadaoka (1986) and Okayasu (1989). In all get significant contributions from those regions, can only be determined with limited accuracy. Stive and Wind (1982) gives a detailed account of the problem.

Stive (1984) also analyzed data from his experiments to determine the energy dissipation D in surf zone waves extending a theoretical result developed by Svendsen et al. (1978), and confirmed the dissipation is likely to be up to 50% larger than in a bore of the same height.

In many of the wave models various characteristics of the wave motion are used as parameters. Examples are the rms of the surface profile $B_o = \overline{\eta^2}/H^2$, the wave propagation speed c, the vertical skewness given as relative crest elevation η_c/H, in addition to breaker data. Hansen (1990) analyzed original data from most of the detailed experiments quoted above and developed empirical representations that in most cases fit the data remarkably well.

Turbulence

Peregrine and Svendsen (1978) found experimentally that the turbulence generated by the breaking, while initiated at the toe of the turbulent wave front, spreads downwards and continues to do so long after the breaker has passed. Pointing to the resemblance between spilling breakers and waves in the bore region of a surf zone, and bores and hydraulic jumps, they speculated that the spread mechanism was similar to that in a shear layer. Later, measurements by Battjes and Sakai (1981) indicated closer resemblance with the turbulence characteristics in a wake. The distribution of turbulent intensities below wave MWL was reported by Stive and Wind (1982), Nadaoka (1986), and in more detail by Okayasu (1989).

Data for breaker generated turbulence has also been provided by Hattori and Aono (1985) who found the turbulent energy spectra have large proportions of the energy at frequencies only somewhat higher than the wave frequency indicating the existence of large scale vortices. Nadaoka (1986) identified a regular system of vortices with axes sloping downwards from the free surface and developing at some distance behind the front.

Battjes (1975) and later Svendsen (1987) analyzed turbulent kinetic energies under breaking waves, and the latter found that most of the energy is dissipated in the crest above the MWL.

Surf Zone Wave Models

No convincing or predictive models have been developed for surf zone wave motion. The range of "modelling" efforts goes from assuming that in the surf zone the wave height H a constant fraction γ of the water depth ("saturated breaker") and invoking linear wave results for all other wave properties to complicated turbulence models of which there are a few.

Several of the wave models aim at predicting the correct cross-shore variation of wave height, H, and set-up, b, which corresponds to correctly modelling cross-shore momentum and energy balance in the surf zone (H-b models). These models operate only with wave averaged quantities and cannot predict details such as wave shape or particle velocities. Examples are Svendsen (1984a), Dally et al. (1984), the first of which introduces surf zone characteristics for the waves. The latter includes an empirical threshold in the energy dissipation that let the waves stop breaking when their height to depth ratio becomes too small (such as waves passing into deeper water behind a longshore bar).

In H-b models, irregular waves have only been dealt with on a statistical basis (Battjes and Janssen (1978), Roelvink & Stive (1989), Dally (1990)).

Wave models in the full time domain have primarily been based on the non-linear shallow water (NSW) equations. Normally, these equations predict incorrectly that all waves break, even on a constant depth. However, numerical solutions of the equations using the special dissipative Lax-Wendroff scheme artificially freeze the wave fronts and compensate for this by a numerical dissipation which equals that of a hydraulic jump or bore of the same height as the wave. This has been utilized by Hibberd and Peregrine (1979) and later by Kobayashi and co-authors to study particularly wave motion on steep slopes (such as structures) and in the swash zone. These models can analyze irregular waves as a time series (Kobayashi et al., 1990).

An extension of the NSW-model to include the effect of turbulence and avoiding the above mentioned deficiencies of the ordinary NSW model was developed by Svendsen and Madsen (1984) but only for a single bore incident on a beach.

The details of the highly turbulent area at the front (the so-called "roller") was analyzed by Longuet-Higgins and Turner (1974) who assumed that air entrainment played a vital part in maintaining this roller in position on the sloping front. Later experiments and analysis by Duncan (1981), Svendsen and Madsen (1984), Banner (1987) and Deigaard (1989) all in various ways attribute the support of the roller to turbulent shear stresses. Longuet-Higgins (1973) also analyzed the nature of the flow in the neighborhood of the toe of roller and an alternative interpretation was used in the model by Svendsen and Madsen (1984).

Nearshore Circulation

Nearshore circulation is the term for the currents created by the breaking waves. These currents are essentially generated by the mean forces exerted on the water mass by the waves due to the radiation stress variations.

In fairly general terms there are four mechanisms involved.

i. The radiation stress forcing.

ii. The pressure gradients due to mean water level variations (setup, set down).

iii. The bottom friction due to waves and currents.

iv. Lateral mixing mechanisms usually attributed to the turbulence combined with the horizontal shear of the currents.

In the cases of unsteady flows (such as long waves), the inertia of the water volume constitutes a fifth factor.

The understanding of nearshore circulation dates from the realization of the fact that water waves represent a mean momentum flux, the radiation stress, and derivation of the wave averaged momentum equation (Longuet-Higgins and Stewart, 1960 and subsequent publications). Longuet-Higgins and Stewart (1963) predicted the set-down of nonbreaking waves, and Bowen et al. (1968) measured and computed setup in the surf zone. Another practical result was the analysis of longshore currents (Bowen, 1969; Thornton, 1970; Longuet-Higgins, 1970).

Longshore Currents

The steady longshore currents on a long straight coast with obliquely incident waves represents the balance between longshore radiation stress variation (i), bottom friction (iii) and lateral mixing (iv).

Since the first contributions mentioned above to solving this problem in 1969–1970, many have followed of which only a few can be listed. James (1974) showed convincingly there is no longshore forcing outside the surf zone on a straight coast. Liu and Dalrymple (1978) generalized the simple friction factor approach for the bottom shear stress. Visser (1980, 1982, 1984) conducted detailed experiments with longshore currents on a straight beach.

Finally, a procedure by Putrevu and Svendsen (1991) for extracting cross-shore radiation stress forcing from empirical data can be used for longshore current forcing as well.

In spite of these and many more contributions to the literature on longshore currents, we are still not able to accurately predict the longshore current forcing.Our knowledge of the mechanisms in bottom boundary layer under breaking waves with a current is virtually nil, and bottom shear stresses in longshore current computations are still based on using a friction coefficient to fit the measurements. Finally, the lateral mixing needed to explain the cross-shore distribution of the longshore current is (in particular outside the surf zone) several orders of magnitude larger than what can be justified by other turbulence arguments (Svendsen and Putrevu, 1990). Thus the knowledge about the mechanisms responsible for longshore currents still needs substantial clarification.

It can be added that Thornton and Guza (1986) have pointed to the fact that on natural beaches the randomness of the waves will create a time variation of the break point which will contribute to the spreading of longshore currents in the same way as lateral mixing. This, however, does not eliminate the need for explaining e.g. Visser's experimental results.

Putrevu and Svendsen (1991) suggest that the interaction between cross-shore and longshore currents may substantially contribute to the mixing both inside and outside the surf zone.

Cross-Shore Circulation

Cross-shore currents on long straight coasts with no longshore variation entirely has the character of a circulation in the vertical plane: substantial amounts of water are carried shoreward as mass transport in the breaking waves and this volume is returned as the seaward going undertow currents essentially below trough level of the waves. These currents have been found to be very strong, generally 8–10% of $\sqrt{gh}$ near the bottom. The mechanism was described qualitatively by Dyhr-Nielsen and Sorensen (1970) and analyzed by Svendsen (1984b).

Also, Dally and Dean (1984), Hansen and Svendsen (1984), Stive and Wind (1986), Svendsen et al. (1987), Okayasu et al. (1988) and Deigaard and Fredsoe (1989) have, among others, contributed to the explanation of the phenomenon. It is generally accepted that the undertow is formed as a balance between radiation stresses, the pressure gradient from the sloping mean water surface, and turbulent shear stresses. Important, however, are also the higher turbulent intensities in the main part of the water column relative to the weak boundary layer turbulence and damping of the breaker turbulence near the bottom (Svendsen et al., 1987, Okayasu et al., 1988). Furthermore, the disturbance of wave motion by variation of depth and wave height will modify the shear stress distribution (the effect of wave height variation was addressed by Deigaard and Fredsoe, 1989).

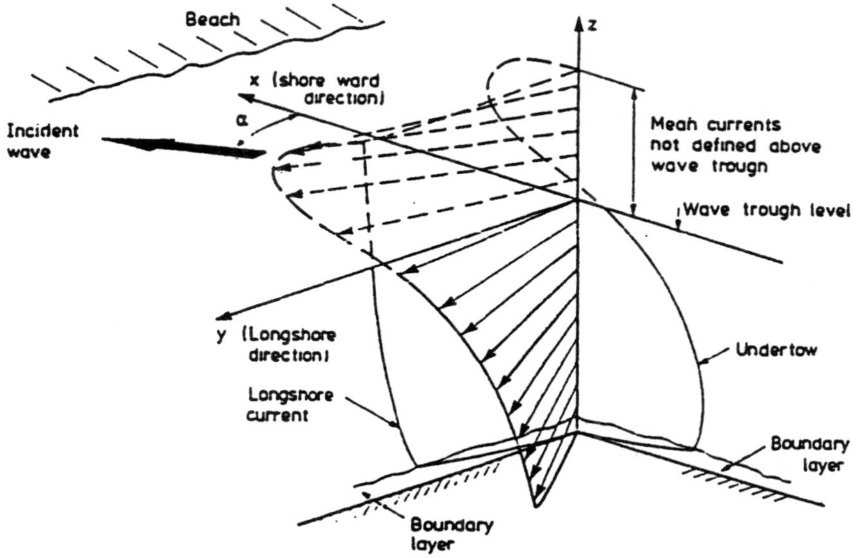

Figure 2: The three dimensional structure of surf zone current profiles (from Svendsen & Lorenz, 1989).

The cross-shore circulation and particularly the seaward oriented undertow is thought to be instrumental in coastal erosion during heavy storms.

3-D Currents

The simultaneous existence of cross-shore and longshore currents together combine to form a vertical distribution of wave generated currents in the surf zone which has a spiral shape as shown in Fig. 2. This was analyzed by Svendsen and Lorenz (1989) and Svendsen and Putrevu (1990).

General Circulation Models

In cases of longshore (as well as cross-shore) variations in bottom topography, the net cross-shore flows are no longer zero and horizontal circulation patterns such as rip currents develop.

This was acknowledged early and a large number of comprehensive circulation models were developed. Based on purely depth averaged equations (and hence neglecting the undertow and vertical cross-shore circulation), these models analyze only net flows. In its most general form, such a model encompasses

(a) A wave propagation model that determines wave patterns due to topography and geometry (refraction, diffraction, interaction with structures) and predicts wave height variation, including breaking.

(b) A current generation model component based on the wave-averaged momentum equations.

However, models capable of dealing with all these phenomena have yet to be developed. In the earlier models the wave component (a) was limited to specifying linear shoaling outside breaking and $H = \gamma h$ (saturated breaker) inside the surf zone with γ constant or given by Miche's formula. Refraction was incorporated using Munk and Arthur's (1952) theory for ray tracing (Noda, 1972,1974) or limiting the models to long straight coasts and using Snell's law (Birkemeier and Dalrymple, 1976; Ebersole and Dalrymple, 1979). Later models often use more advanced models for the pattern of wave propagation but still the simple saturation model for the wave height inside the surf zone. A few examples are Watanabe (1985) (modified mild slope equation) and Winer (1988) (parabolic wave model with energy dissipation).

Long Waves

Long waves (or infra-gravity waves) are waves with significantly longer period than the peak frequency of the incident wave spectrum. Field measurements show that such waves occur very frequently and different mechanisms have been considered for their generation. One is resonant interaction between ordinary waves (Gallagher, 1971; Bowen and Guza, 1978). Another is the effect of wave height variation in the incident wave trains, sometimes called "surf beat" (Munk, 1949; Symonds et al., 1982; Schäffer and Svendsen, 1988).

The long waves occur both as waves bound to the incident wave train and as free waves which develop by either direct energy transfer from the short wave train or are formerly bound waves released from the short wave train by changes in that wave train due to shoaling and breaking. The free waves are often trapped along the coast as edge waves. Numerous references are omitted here for brevity.

Very Low Frequency Motion

Recently, some field experiments have shown signs of very long period oscillations in the horizontal velocity field (Tang and Dalrymple, 1988; Oltman-Shay et al., 1989). These oscillations are of relative short length and propagate along the shore at a speed comparable to that of the longshore current. They have been attributed to instabilities in that current (Bowen and Holman, 1989) but the phenomenon is still under investigation.

Concluding Remarks

In the relatively brief list of surf zone phenomena outlined above, many important topics have been omitted. Thus it has been impossible to cover the effects of different coastal topographies, rip currents and 3D circulation patterns, wave interaction with shear currents, random wave phenomena and time varying currents, the effects of winds and tides, tidal inlets and the analysis of field experiments, to mention a few. Also, within each topic the list of references is highly incomplete and only meant to show examples. For recent, more complete reviews and alternative perspectives reference is made to Battjes (1988) and Battjes et al. (1990).

Acknowledgements

This work was partially sponsored by NOAA Office of Sea Grant, Department of Commerce, under Grant No. AA-D-SG040 (Project No. R/OE 6). Discussion and comments from U. Putrevu are gratefully acknowledged.

References

Banner, M.L. (1987). "Surging characteristics of spilling zones on quasi-steady breaking water waves," in Horikawa & Maeno, *Nonlinear Water Waves*, IUTAM Symp., Springer, Tokyo.

Basco, D.R. and T. Yamashita (1986). "Toward a simple model of the wave breaking transition region in surf zone," *20th ICCE*, Taipei, Chap. 72, 955.

Battjes, J.A. (1975). "Modeling of turbulence in the surf zone," *Proc. Symp. on Modeling Techniques*, San Francisco, 1050–1061.

Battjes, J.A. (1988). "Surf zone dynamics," *Annual Review of Fluid Mechanics*, **20**, 257–293.

Battjes, J.A. and T. Sakai (1981). "Velocity field in a steady breaker," *J. Fluid Mech.*, **111**, 121–137.

Battjes, J.A. and J.P.F.M. Janssen (1978). "Energy loss and set-up due to breaking of random waves," *Proc. 16th ICCE*, Hamburg, Chap. 32, 569–587.

Battjes, J.A., R.J. Sobey and M.J.F. Stive (1990). "Nearshore circulation," in *The Sea: Ocean Engineering Science*, **9**, 468–493.

Birkemeier, W.A. and R.A. Dalrymple (1976). "Numerical models for the prediction of wave set-up and nearshore circulation," *Ocean Eng. Rep. 3*, Dept. of Civil Engineering, University of Delaware.

Bowen, A.J. (1969). "The generation of longshore currents on a plane beach," *J. Marine Res.*, **27**, 206–215.

Bowen, A.J. and R.A. Holman (1989). "Shear instabilities of the mean longshore current. 1. Theory," *J. Geophys. Res.*, C12, 18023–18030.

Bowen, A.J. and R.T. Guza (1978). "Edge waves and surf beat," *J. Geophys. Res.*, **83**, 1913–1920.

Bowen, A.J., D.L. Inman and V.P. Simons (1968). "Wave 'set-down' and set-up," *J. Geophys. Res.*, **73**, 2569–2577.

Dally, W.R. (1990). "Random breaking wave: A closed form solution for planar beaches," *Coast. Eng.*, **14**, 233–263.

Dally, W.R. and R.G. Dean (1984). "Suspended sediment transport and beach profile evolution," ASCE, **110**, WPCOE, 15–33.

Dally, W.R., R.G. Dean and R.A. Dalrymple (1984). "A model for breaker decay on beaches," *Proc. 19th Conf. Coast. Eng.*, Houston, 82–98.

Deigaard, R. (1989). "Mathematical modelling of waves in the surf zone," *ISVA Prog. Rep. 69*, Technical University of Denmark, 47–60.

Deigaard, R. and J. Fredsoe (1989). "Shear stress distribution in dissipative water waves," *Coastal Eng.*, **13**, 4, 357–378.

Duncan, J.H. (1981). "An experimental investigation of breaking waves produced by a towed hydrofoil," *Proc. Roy. Soc. Lond.*, **A377**, 331–348.

Dyhr-Nielsen, M. and T. Sorensen (1970). "Sand transport phenomena on coasts with bars," *Proc. 12th Int. Conf. Coast. Eng.*, Ch. 54, 855–866.

Ebersole, B.A. and R.A. Dalrymple (1979). "A numerical model for nearshore circulation including convective acceleration and lateral mixing," *Ocean Eng. Rep. 21*, Dept. of Civil Engineering, University of Delaware.

Gallagher, B. (1971). "Generation of surf beat by nonlinear wave interactions," *J. Fluid Mech.*, **49**, 1–20.

Hansen, J.B. (1990). "Periodic waves in the surf zone: Analysis of experimental data," *Coastal Eng.*, **14**, 14–41.

Hansen, J.B. and I.A. Svendsen (1984). "A theoretical and experimental study of undertow," *Proc. 19th Int. Conf. Coast. Eng.*, Houston, Ch. 151, 2246–2262.

Hattori, M. and T. Aono (1985). "Experimental study of turbulence structures under spilling breakers," in *The Ocean Surface*, Toba and Mitsuyasu, eds., Reidel Publ. Comp., Dordrecht, 419–424.

Hibberd, S. and D.H. Peregrine (1979). "Surf and runup on a beach: A uniform bore," *J. Fluid Mech.*, **95**, 323- 345.

James, I.D. (1974). "A non-linear theory of longshore currents," *Est. & Coast. Mar. Science*, **2**, 235–249, Part 2.

Jansen, P.C.M. (1986). "Laboratory observations of the kinematics in the aerated region of breaking waves," *Coast. Eng.*, **9**, 5, 453–477.

Kobayashi, N., D.T. Cox and A. Wurjanto (1990). "Irregular wave refraction and run-up on rough impermeable slopes," *WPCOE*, ASCE, **116**, 6, 708–726.

Liu, P.L-F. and R.A. Dalrymple (1978). "Bottom frictional stresses and longshore currents due to waves with large angles of incidence," *J. Marine Res.*, **36**, 2, 357–375.

Longuet-Higgins, M.S. (1970). "Longshore currents generated by obliquely incident sea waves, 1 & 2," *J. Geophys. Res.*, **75**, 6778–6789 & 6790–6801.

Longuet-Higgins, M.S. (1973). "A model of flow separation at a free surface," *J. Fluid Mech.*, **57**, 1, 129–148.

Longuet-Higgins, M.S. and R.W. Stewart (1960). "Changes in the form of short gravity waves on long waves and tidal currents," *J. Fluid Mech.*, **8**, 565–583.

Longuet-Higgins, M.A. and R.W. Stewart (1963). "A note on wave set-up," *J. Mar. Res.*, **21**, 4–10.

Longuet-Higgins, M.S. and J.S. Turner (1974). "An 'entraining plume' model of a spilling breaker," *J. Fluid Mech.*, **63**, 1, 1–20.

Munk, W.H. (1949). "Surf beats," EOS, *Transactions of the American Geophysical Union*, **30**, 849–854.

Munk, W.H. and R.S. Arthur (1952). "Wave intensity along a refracted ray in gravity waves," *Nat. Bur. Stand. Circ. 521*, Washington, D.C.

Nadaoka, K. (1986). "A fundamental study on shoaling and velocity fluid structure of waver waves in the nearshore zone," Ph.D. Diss., Dept. of Civ. Eng., Tokyo Inst. of Tech.

Noda, E.K. (1972). "Rip currents," *Proc. 13 ICCE*, Vancouver, Chap. 35, 653–668.

Noda, E.K. (1974). "Wave induced nearshore circulation," *J. Geophys. Res.*, **79**, 4097–4106.

Okayasu, A. (1989). "Characteristics of turbulence structures and undertow in the surf zone," Ph.D. Diss., Univ. of Tokyo.

Okayasu, A., T. Shibayama and K. Horikawa (1988). "Vertical variation of undertow in the surf zone," *Proc. 21 ICCE*, Ch. 33, 478–491.

Oltman-Shay, J., P.A. Howd and W.A. Birkemeier (1989). "Shear instabilities of the mean longshore current, *J. Geophys. Res.*, **94**, C12, 18031–18042.

Peregrine, D.H. and I.A. Svendsen (1978). "Spilling breaker, bores and hydraulic jumps," *Proc. 16th ICCE*, Hamburg, Ch. 30, 540–550.

Putrevu, U. and I.A. Svendsen (1991). "Wave induced nearshore currents: A study of the forcing, mixing and stability mechanisms," *Center for Applied Coastal Research Report CACR-91-11*, Dept. of Civil Engineering, University of Delaware.

Roelvink, J.A. and M.J.F. Stive (1989). "Bar generating cross shore flow mechanisms on a beach," *J. Geophys. Res.*, **94**, C4, 4785–4800.

Schäffer, H.A. and I.A. Svendsen (1988). "Surf beat generation on a mild slope beach," *Proc. 21st ICCE*, Chap. 79, 1058–1072.

Stive, M.J.F. (1980). "Velocity and pressure field of spilling breakers," *Proc. 17th ICCE*, Chap. 34, pp. 547–566.

Stive, M.J.F. (1984). "Energy dissipation in waves breaking on gentle slopes," *Coastal Eng.*, **8**, 99–127.

Stive, M.J.F. and H.G. Wind (1982). "A study of radiation stress and set-up in the nearshore region," *Coastal Eng.*, **6**, 1, 1–26.

Stive, M.J.F. and H.G. Wind (1986). "Cross-shore mean flow in the surf zone," Coastal Eng., **10**, 325–340.

Svendsen, I.A. (1984a). "Wave heights and set-up in a surf zone," *Coastal Eng.*, **8**, 303–330.

Svendsen, I.A. (1984b). "Mass flux and undertow in a surf zone," *Coastal Eng.*, **8**, 4, 347–365.

Svendsen, I.A. (1987). "Analysis of surf zone turbulence," *J. Geophys. Res.*, **92**, C5, 5115–5124.

Svendsen, I.A. and R.S. Lorenz (1989). "Three dimensional velocity profiles in combined undertow and longshore currents," *Coastal Eng.*, **13**, 1, 55–80.

Svendsen, I.A. and P.A. Madsen (1984). "A turbulent bore on a beach," *J. Fluid Mech.*, **148**, 73–96.

Svendsen, I.A., P.A. Madsen and J. Buhr Hansen (1978). "Wave characteristics in the surf zone," *Proc. 16th ICCE*, Chap. 29, 520–539.

Svendsen, I.A. and U. Putrevu (1990). "Nearshore circulation with 3D profiles," *22 ICCE*, Chap. 18, 241– 254.

Svendsen, I.A., H.A. Schäffer and J.B. Hansen (1987). "The interaction between the undertow and the boundary layer flow on a beach," *J. Geophys. Res.*, **92**, C11, 11845–11856.

Symonds, G., D.A. Huntley and A.J. Bowen (1982). "Two dimensional surf beat: Long wave generation by a time-varying breakpoint," *J. Geophys. Res.*, **87**, C1, 492–498.

Tallent, J.R., T. Yamashita and Y. Tsuchiya (1989). "Field and laboratory investigation of large scale eddy formation by breaking water waves," in *The Water Wave Kinematics*, Torum, A. & O.T. Gudmestad, eds. NATO ASI Series E: Applied Science, Vol. 178, 509–523.

Tang, E. and R.A. Dalrymple (1988). "Rip currents and wave groups," in *Nearshore Sediment Transport*, R.J. Seymour, ed., Plenum Publishing Corp., 205–230.

Thornton, E.G. and R.T. Guza (1986). "Surf zone longshore currents and random waves: Field data and models," *J. Phys. Oceanography*, **16**, 7, 1165–1177.

Thornton, E.G. (1970). "Variation of longshore current across the surf zone," *Proc. 12th ICCE*, Washington, D.C., Chap. 18, 291–308.

Visser, P.J. (1980). "Longshore current flows in a wave basin," *Proc. 17th ICCE*, Sydney, Chap. 29, 462–479.

Visser, P.J. (1982). "The proper longshore current in a wave basin," *Report No. 82-1*, Dept. Civil Eng., Delft Univ. Technology.,

Visser, P.J. (1984). "A mathematical model of uniform longshore currents and the comparison with laboratory data," *Fluid Mech. Lab. Rep. 84-2*, Delft Univ. Tech.

Watanabe, A. (1985). "Three dimensional predictive models of beach evolution around a structure," *Proc. Water Wave Research*, Hanover, 123–141.

Winer, H.S. (1988). "Numerical modelling of wave-induced currents using a parabolic wave equation," Ph.D. thesis, Dept. of Coast. and Oceanograph. Eng., University of Florida.

Inverse and Direct Cascade in the Wind-Driven Surface Wave Turbulence and Wave-Breaking

V.E. Zakharov

Landau Institute for Theoretical Physics
Moscow, USSR. and
Dept. of Mathematics, University of Arizona
Tucson, AZ 85721, USA

Abstract

The problem of formation and evolution of wind-generated wave spectra on a sea surface can be effectively solved due to the fact that the ratio of air density to water density provides a small parameter, with the consequence that characteristic wave amplitudes in typical situations are small. This justifies "a weak turbulent" approach to the problem of a statistical description, using an expansion in powers of nonlinearity.

The simplest variant of such a description is from the kinetic equation for waves, derived first by Hasselman (1962). This equation describes two fundamental processes — direct and inverse cascades of energy. The stationary equation has two remarkable exact solutions giving two energy spectra.

One spectrum describes the direct cascade — energy transport to a region of large wavenumbers (Zakharov and Filonenko, 1966). It is realized for frequencies $\omega \gg \omega_0$ where $\omega_0 = g/V$ and U is a characteristic wind velocity. The other spectrum corresponds to an inverse cascade of energy in which wavelengths increase ("wave aging") (Zakharov and Zaslavsky, 1981). It is realized at $\omega < \omega_0$. The inverse cascade leads to the production of "aged waves" that are common in an open ocean (Glazman, 1989–91). Both spectra were observed experimentally by many authors. Long waves ($\omega < \omega_0$) have a narrow angle distribution that can be explained (Zakharov and Shrira, 1990) by interaction between waves and a wind–generated shear flow, always existing near the sea surface.

Another fundamental phenomenon in surface wave theory is wave breaking. It is a strongly nonlinear effect that is suppressed by surface tension if the wind velocity is small enough. At larger wind speeds weak turbulence coexists with wave breaking, while at still larger wind speeds wave–breaking prevails. A simple model of wave breaking is offered, using a self–similar solution of the Euler equation.

1. Several questions instead of introduction

The study of wind-generated waves on the sea surface is not only a problem of oceanography. It is a problem of physics as well. And just a simple visual observation of this process can give to a thoughtful physicist enough information to formulate first questions. And these questions are far from trivial.

M. L. Banner, R. H. J. Grimshaw (Eds.)
Breaking Waves
IUTAM Symposium Sydney/Australia 1991

First an elementary observation, confirmed by many measurements is the following — a characteristic steepness of wind-driven waves is small. This steepness can be expressed in terms of mean square angle $\langle\theta^2\rangle$ of wavy surface to a horizontal plane. Measurements give for typical cases: $\langle\theta^2\rangle \simeq 10^{-2} - 10^{-3}$. It is important to stress that this value does not depend essentially on the wind velocity V. Both characteristic wave amplitude h and characteristic wavelength λ depend on V dramatically, but their ratio $h/\lambda \sim \theta$ remains more or less constant and small. Actually $\langle\theta^2\rangle$ is a measure of nonlinearity in the wave system.

Small level of nonlinearity implies that there is a certain small parameter in the wave-wind system. It is obvious that knowing this parameter is crucial for developing of any consistent theory. What small parameter provides a small value of wave steepness?

The second question is also stimulated by a very simple observation. It is well known that a characteristic wave length λ increases in time under action of a constant–in–time wind. This phenomenon has in oceanography the special name "wave aging". Waves in the ocean have an "age" that is nothing but the ratio of their phase velocity $V_\phi = \sqrt{2\pi g/\lambda}$ to a wind velocity V.

Hence age is: $a = \sqrt{2\pi g/\lambda V^2}$. In the beginning of the wave process age is small ($a = 0.1$), then it increases to a value of order one: $a \simeq 0.9 - 1.2$. Near the coastline, in inland seas or in lakes this value is limited. But in the open ocean an observer can easily see very "adult" waves with $a \simeq 2 - 5$ or really "old" ones at $a \simeq 10$ (see R.Glazman and S.Pilorz, 1990; R.Glazman, 1991). The process of wave aging is an essentially nonlinear process. It is especially clear for waves moving faster than wind ($a > 1$). These waves cannot take energy from the wind and exist only due to feeding from more short and slow waves, developed in the wave ensemble. The phenomenon of wave aging is so fundamental that any serious theory must explain it *on a qualitive level*, without appealing to complicated computer experiments. What is the physical origin of wave aging? It is interesting, that in the process of wave aging a characteristic wave amplitude displays the tendency to decrease (aged waves are "more linear" than the young ones).

The third question, maybe, is the most elementary (and the "naive") one. It is known that an experienced seafarer can estimate the strength of wind (the wind velocity) just by looking at the sea surface. He will look for the density of "white caps" on this surface. Actually, there is some characteristic velocity of wind $V_c \simeq 6$ meters per second, so that white caps are rare or absent if $V < V_c$. The white-cap density grows fast with the growing of the parameter $b \simeq (V - V_c)/V_c$.

If $b \simeq 4$ ($V \sim 30$ m/sec) almost all sea surface is covered by white foam. How to explain this phenomena, from first principles of physics? What is the formula expressing V_c through any property of water and air (and maybe through gravity acceleration g)?

This series of questions could be continued. It is very interesting, for instance, why a well-developed wave spectrum is so narrow in angle? This is a really difficult question, and we will not discuss it in my lecture. But we will try to give answers on the first three questions.

2. First explanations. Role of surface tension

Let us try to answer the formulated questions. The first one is the simplest. The small parameter in the theory of wind-wave interaction is apparently the ratio of air density ρ_{air} and water density ρ_{water}. This ratio $\epsilon = \rho_{air}/\rho_{water} \sim 10^{-3}$ is really small enough to provide small steepness and weak nonlinearity of wind-driven waves. If our atmosphere would consist of pure hydrogen, wave steepness at the same wind velocity would be several times less then in present conditions. To go further we should do the following imaginary experiment.

Suppose that both water and air are ideal fluids having no viscosity and no surface tension. There is no preferred characteristic velocity in the system in this case. It means that qualitative pictures of wind-driven waves do not depend on a wind velocity. Looking at a snapshot of such a sea, nobody can estimate the value of the wind velocity if the distance from where the snapshot is done is unknown. To understand how a picture of wind-driven waves on the surface of ideal fluid looks like, one can recall in memory a real sea surface at very high wind velocity — in a hurricane, for instance. In this case motion with characteristic scale $\lambda \simeq V^2/g$ is smooth and, as we said before, is a superposition of weakly nonlinear gravity waves. The level of nonlinearity grows with decreasing of scales. At some small scales λ_{min} nonlinearity becomes very strong. For scales $\lambda < \lambda_{min}$, a surface is so rough and fractal that one cannot say anything about "waves", even about a "shape of surface" at these small lengths. The surface is blurred and can be well defined only after averaging over scales $\lambda > \lambda_{min}$. At $\lambda < \lambda_{min}$, there is actually a two-phase mixture of water in the air above and bubbles of air in the water below. A very interesting problem of statistical description of such two phase fluid in a gravity field lies out of the framework of this lecture. It is important only to stress that this "fully developed wave breaking" is a very irreversible process realizing permanent dissipation of mechanical energy to heat. The supply of mechanical energy is coming from more long and regular waves. We will call this flux of energy to small scales as *"direct cascade"* of energy. On the contrary, "wave aging", that is actually the energy transfer to large space scales, can be called *"inverse cascade"* of energy. Both direct and inverse cascades must be explained.

How to estimate λ_{min}? It is very plausible to put:

$$\lambda_{\min} \simeq \epsilon\lambda \simeq 2\pi\frac{\epsilon V^2}{g} \tag{2.1}$$

For wind velocity $V \sim 10m/sec$ we have $\lambda_{min} \sim 6cm$, for $V \sim 100m/sec$, $\lambda \simeq 6m$. Both of these values are quite reasonable.

The picture described above contradicts our everyday experience.

Blowing in a cup of tea, we don't see wave-breaking and two-phase fractal behavior of the tea surface. It means that the ideal fluid model does not work, and "smoothing factors" — viscosity or surface tension should be taken into consideration. Using viscosity and surface tension one can construct some other characteristic wave lengths:

$$\lambda_\nu \simeq 2\pi\left(\frac{\nu^2}{g}\right)^{1/3}$$

$$\lambda_\sigma = 2\pi \left(\frac{\sigma}{g}\right)^{1/2}$$

Here σ is a coefficient of surface tension, and ν is a viscosity coefficient.

For real water $\lambda_\sigma \sim 3cm$ and $\lambda_\nu \sim 0.3cm$. So, surface tension is a more important factor. Fractalization of the sea surface take place only if $\lambda_{min} > \lambda_\sigma$. It gives for the critical velocity:

$$\frac{\epsilon V_c^2}{g} \simeq \left(\frac{\sigma}{g}\right)^{1/2}$$

or

$$V_c \simeq \epsilon^{1/2}(\sigma g)^{1/4} \simeq \left(\frac{\rho_{\text{water}}}{\rho_{\text{air}}}\right)^{1/2} (\sigma g)^{1/4} \tag{2.2}$$

The velocity $V_{min} = (\sigma g)^{1/4}$ is well-known in hydrodynamics. It is the minimal phase velocity of gravity-capillary waves. Approximately $V_{min} \simeq 20cm/sec$. For the critical velocity we get now:

$$V_c \simeq 6\text{m/sec}$$

That is in an excellent accordance with the data of measurements. (see, for instance, V.Bondur and E.Sharkov, 1982; M.Banner, I.Jones and J.Trinder,1989).

3. Origin of wave breaking

Previous considerations lead us to the following conjecture: wave breaking and fractalization of a free surface is a phenomenon typical of the motion of an ideal fluid even in the presence of a gravitational field. This process can be suppressed and even prevented by strong enough surface tension. Let us formulate this conjecture, using more precise mathematical language. Potential flow of an ideal incompressible fluid is described by the Laplace equation:

$$\Delta\phi = 0 \qquad v = \nabla\phi \qquad \left.\frac{\partial\phi}{\partial z}\right|_{z\to-\infty} \to 0 \tag{3.1}$$

and by boundary conditions on the surface of the fluid:

$$\frac{\partial\eta}{\partial t} + \nabla_\perp\phi\nabla_\perp\eta = \left.\frac{\partial\phi}{\partial z}\right|_{z=\eta} \tag{3.2}$$

$$\frac{\partial\phi}{\partial t} + \frac{1}{2}(\nabla\phi)^2\Big|_{z=\eta} + g\eta = 0 \tag{3.3}$$

Here $\eta = \eta(\vec{r}, t)$ is a shape of the surface, $\vec{r} = (x, y)$ and $\nabla_\perp = (\partial/\partial x, \partial/\partial y)$. By fixation of initial data we have:

$$\begin{aligned} \eta|_{t=0} &= \eta_0(\vec{r}) \\ \Psi|_{t=0} &= \Psi_0(\vec{r}) \\ \Psi &= \phi|_{z=\eta} \end{aligned} \tag{3.4}$$

Modern mathematics proves that the Cauchy problem (3.1)–(3.4) for smooth enough and small enough initial data (3.4) is well posed (if $g > 0$) and can be resolved up to some finite time t_{max}, depending on choice of η_0 and Ψ_0. But nobody could prove that such solution exists for a given class of initial data forever, in the half-infinite interval of time $0 < t < \infty$. Moreover, such a statement contradicts our common sense. It is much more relevant to think that evolution of typical initial data (excluding zero-measure sets of special data, like Stoke's progressive periodic waves), will finish by a "splash" or wave-breaking.

This hypothesis also is not proved, of course. But we can formulate some arguments to its support.

First, we may take into account that gravity is a stabilizing factor, and wave-breaking could be achieved easily if $g = 0$. Second, our visual experience shows that wave-breaking is in its main features a two–dimensional process. Hence, it can be studied in the framework of the following system of equations:

$$\phi_{xx} + \phi_{zz} = 0 \tag{3.5}$$

$$\eta_t + \eta_x\phi_x = \phi_z|_{z=\eta} \tag{3.6}$$

$$\phi_t + \frac{1}{2}(\phi_x^2 + \phi_z^2) = 0\Big|_{z=\eta} \tag{3.7}$$

Here $\partial\phi/\partial z \to 0$ at $z \to -\infty$.

This system allows a whole family of self-similar substitutions:

$$\eta = (t_0 - t)^\alpha\,\eta_0\left(\frac{x}{(t_0-t)^\alpha}\right) \tag{3.8}$$

$$\phi = (t_0 - t)^{2\alpha-1}\,\phi_0\left(\frac{x}{(t_0-t)^\alpha}, \frac{z}{(t_0-t)^\alpha}\right) \tag{3.9}$$

Functions $\phi_0(\xi,\zeta)$ and $\eta_0(\xi)$ with:

$$\xi = \frac{x}{(t_0-t)^\alpha}, \zeta = \frac{z}{(t_0-t)^\alpha}$$

satisfy the equations:

$$\alpha(\eta_0 - \xi\eta_{0\xi}) + \eta_{0\xi}\phi_{0\xi} = \phi_{0\zeta}|_{\zeta=\eta_0} \tag{3.10}$$

$$(2\alpha-1)\phi_0 - \alpha(\xi\phi_{0\xi} + \zeta\phi_{0\zeta}) + \frac{1}{2}(\phi_{0\xi}^2 + \phi_{0\zeta}^2)\Big|_{\zeta=\eta_0} = 0 \tag{3.11}$$

$$\frac{\partial^2\phi_0}{\partial\xi^2} + \frac{\partial^2\phi_0}{\partial\zeta^2} = 0 \tag{3.12}$$

Suppose that system (3.10)—(3.12) has a nontrivial, nonsingular solution for some value of α. This solution describes the formation of a singularity on a fluid surface at $t \to t_0$. The nature of this singularity can be found directly from the substitution (3.8). The surface tends at $t \to t_0$ to some limiting shape $\eta_0(x)$. It is possible only if $\eta_0(\xi) \to -c|\xi|$ at $\xi \to \infty$, $\eta_0(x) = -c|x|$. The singularity is a jump of the first derivative. The solution under consideration finishes with formation at $t = t_0$ of a "wedge" with the angle $\theta = 2arctan 1/c$. The value of this angle as well as the value of α are unknown *a priori*. They must be found from the system (3.10)—(3.12) as its nonlinear eigenvalues. It is likely that θ is 90° or 120° (mathematical theory is relatively simpler in these cases).

The discussed solution is a model of wave-breaking on a surface of an ideal fluid. It cannot be extended for $t > t_0$ in the framework of ideal fluid hydrodynamics. Actually at $t = t_0$ it generates a very thin and unstable spray. Behavior of this spray is defined essentially by the viscosity and the surface tension. The self-similar solutions of system (3.5)—(3.7) have not been studied enough. Energy conservation imposes the limitation $\alpha \geq 1/2$. In the interval $1/2 \leq \alpha < 2$ one can neglect the influence of gravity in the vicinity of the wave-breaking point. It is interesting that in the narrow interval $1/2 \leq \alpha < 2/3$ even surface tension does not prevent the wave breaking, that is actually a "collapse of curvature" – formation in a finite time of a point of an infinite second x-derivative. We (together with Dr.V.Cogan and Dr.E.Kuznetzov, to be published in Physics Letters) have carried out numerical calculations of this process. Using the method of boundary integrals we modelled a behavior of an ideal incompressible fluid of infinite density, imposing periodic–in-x boundary conditions. For very broadly chosen initial perturbations of the surface we observed formation of an infinite curvature point in a finite time. A typical behavior of the surface and its second derivative is plotted in Figure 1.

4. Hamiltonian formalism and canonical transformation

Due to small steepness, a wind-driven sea is a stochastically behaving system of weakly interacting waves with strong dispersion. That is a good example of so-called "weak turbulence". For long scales it is an almost Gaussian process, because phases of individual waves with good accuracy may be considered well mixed and statistically independent. A proper tool for description of weak turbulence is the kinetic equation governing evolution of the mean square wave amplitude. For surface waves this equation was derived first by K. Hasselmann (1962). It was indeed the first case of applying the kinetic equation to a weak turbulent problem.

The most economical way of deriving the kinetic equation is using of the Hamiltonian formalism. As it was mentioned, the variables $\eta(\vec{r}, t)$ and $\psi(\vec{r}, t)$ completely describe the potential flow of an incompressible fluid. It is possible to show that these variables are canonically conjugate and their Fourier transforms satisfy the equations (V.Zakharov, 1968):

$$\frac{\partial \psi_k}{\partial t} = -\frac{\delta H}{\delta \eta_k^*} \qquad \frac{\partial \eta_k}{\partial t} = \frac{\delta H}{\delta \psi_k^*} \tag{4.1}$$

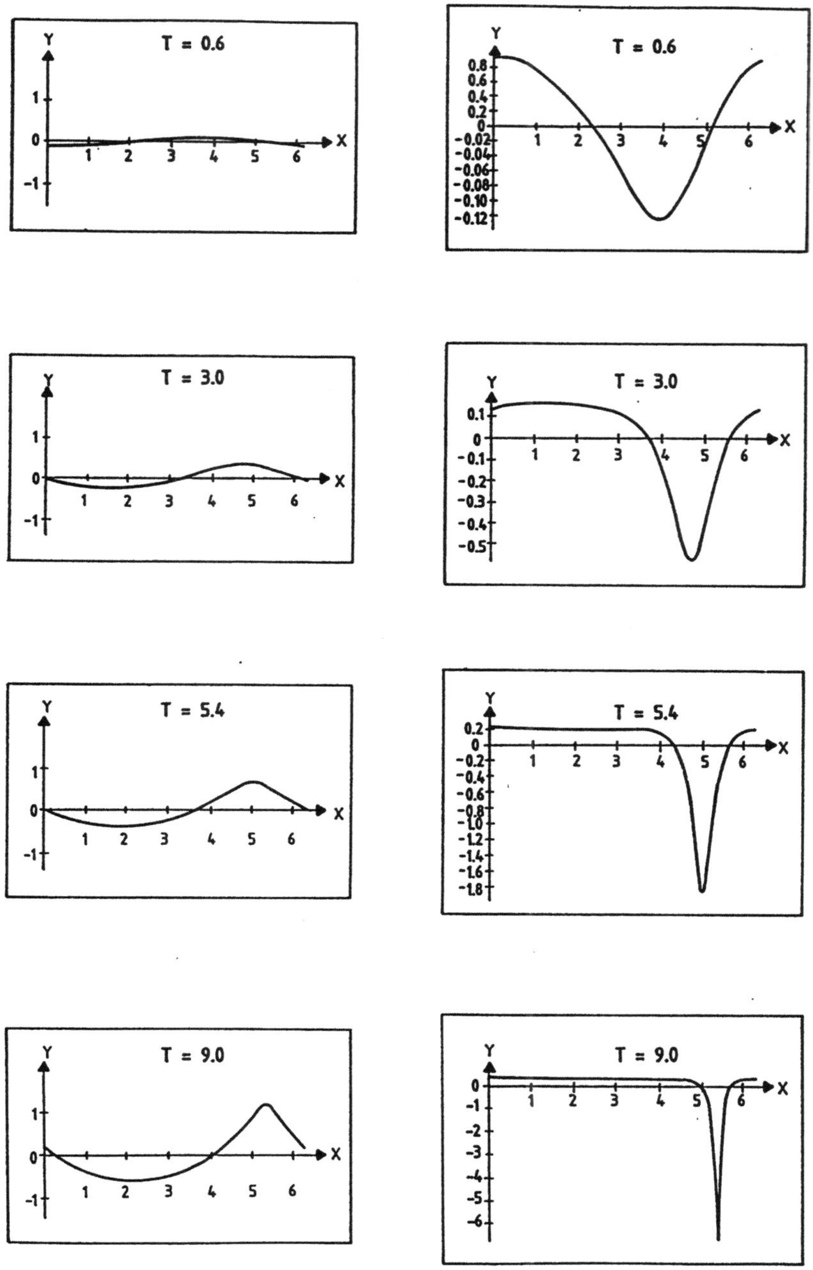

Figure 1.

Numerical simulation of the formation of a singularity on the surface of an ideal fluid. The left side shows the surface shape, the right side shows plots of curvature for successive moments of time. Depth is infinite, boundary conditions are periodic.

Here $H = T + U$ is the total energy of the fluid with the following kinetic and potential energy terms:

$$T = \frac{1}{2}\int d\vec{r}\int_{-\infty}^{\eta} v^2 dz$$

$$U = \frac{1}{2}\int (g\eta^2 + \sigma\nabla\eta^2)d\vec{r} \tag{4.2}$$

The Hamiltonian H can be expanded in the infinite series in powers of a characteristic wave steepness $k\eta_k \sim \theta$ (V.Zakharov, 1968; D.Crawford, H.Yuen and P.Saffman, 1980):

$$H = H_0 + H_1 + H_2 + \cdots \tag{4.3}$$

It is convenient to introduce a normal complex variable a_k:

$$\eta_k = \sqrt{\frac{\omega_k}{2A_k}}(a_k + a^*_{-k}) \qquad \psi_k = -i\sqrt{\frac{A_k}{2\omega_k}}(a_k - a^*_{-k}) \tag{4.4}$$

$$A_k = g + \sigma k^2 \qquad \omega_k = \sqrt{gk + \sigma k^2} \tag{4.5}$$

Here ω_k is the dispersion law.
This variable satisfies the equation:

$$\frac{\partial a_k}{\partial t} + i\frac{\delta H}{\delta a^*_k} = 0, \tag{4.6}$$

where:

$$\begin{aligned} H_0 &= \int \omega_k a_k a^*_k dk \\ H_1 &= \int V_{kk_1k_2}\{a^*_k a_{k_1} a_{k_2} + a_k a^*_{k_1} a^*_{k_2}\}\delta_{k-k_1-k_2} dk dk_1 dk_2 + \\ &+ \frac{1}{3}\int U_{kk_1k_2}\{a_k a_{k_1} a_{k_2} + a^*_k a^*_{k_1} a^*_{k_2}\}\delta_{k+k_1+k_2} dk dk_1 dk_2 \end{aligned}$$

$$\begin{aligned} U_{kk_1k_2} &= \frac{1}{8\pi\sqrt{2}}\left(\left(\frac{\omega_k A_{k_1} A_{k_2}}{\omega_{k_1}\omega_{k_2} A_k}\right)^{\frac{1}{2}} L_{k_1k_2} + \right. \\ &+ \left.\left(\frac{\omega_{k_1} A_k A_{k_2}}{A_{k_1}\omega_k\omega_{k_2}}\right)^{\frac{1}{2}} L_{kk_2} + \left(\frac{\omega_{k_2} A_k A_{k_1}}{A_{k_2}\omega_k\omega_{k_1}}\right)^{\frac{1}{2}} L_{kk_1}\right) \end{aligned} \tag{4.7}$$

$$\begin{aligned} V_{kk_1k_2} &= \frac{1}{8\pi\sqrt{2}}\left(\left(\frac{\omega_k A_{k_1} A_{k_2}}{\omega_{k_1}\omega_{k_2} A_k}\right)^{\frac{1}{2}} L_{k_1k_2} - \right. \\ &- \left.\left(\frac{\omega_{k_2} A_k A_{k_1}}{A_{k_2}\omega_k\omega_{k_1}}\right)^{\frac{1}{2}} L_{-kk_1} - \left(\frac{\omega_{k_1} A_k A_{k_2}}{A_{k_1}\omega_k\omega_{k_2}}\right)^{\frac{1}{2}} L_{-kk_2}\right) \end{aligned} \tag{4.8}$$

Here:

$$L_{kk_1} = (kk_1) + |k||k_1|$$

In the capilliary dominating limit $k > \sqrt{g/\sigma}$:

$$V_{kk_1k_2} = \frac{\sigma^{\frac{1}{4}}}{8\pi\sqrt{2}}\left(\left(\frac{k_1k_2}{k}\right)^{\frac{1}{4}} L_{k_1k_2} - \left(\frac{kk_1}{k_2}\right)^{\frac{1}{4}} L_{-kk_1} - \left(\frac{kk_2}{k_1}\right)^{\frac{1}{4}} L_{-kk_2}\right) \tag{4.9}$$

In the gravity dominating region $k \ll \sqrt{g/\sigma}$:

$$\begin{aligned} V_{kk_1k_2} &= \frac{g^{\frac{1}{4}}}{8\pi\sqrt{2}}\left(\left(\frac{k}{k_1k_2}\right)^{\frac{1}{4}} L_{k_1k_2} - \left(\frac{k_2}{kk_1}\right)^{\frac{1}{4}} L_{-kk_1} - \left(\frac{k_1}{kk_2}\right)^{\frac{1}{4}} L_{-kk_2}\right) \\ U_{kk_1k_2} &= \frac{g^{\frac{1}{4}}}{8\pi\sqrt{2}}\left(\left(\frac{k}{k_1k_2}\right)^{\frac{1}{4}} L_{k_1k_2} + \left(\frac{k_2}{kk_1}\right)^{\frac{1}{4}} L_{kk_1} + \left(\frac{k_1}{kk_2}\right)^{\frac{1}{4}} L_{kk_2}\right) \end{aligned} \tag{4.10}$$

Among the different components of H_2 only one is essential:

$$H_2 = \frac{1}{2}\int W_{k_1k_2,k_3k_4} a^*_{k_1} a^*_{k_2} a_{k_3} a_{k_4} \delta_{k_1+k_2-k_3-k_4} dk_1 dk_2 dk_3 dk_4 \tag{4.11}$$

$$\begin{aligned} W_{k_1k_2,k_3k_4} &= \frac{1}{64\pi^2}\{M_{-k_1,-k_2,k_3,k_4} + M_{k_3,k_4,-k_1,-k_2} - M_{-k_1,k_3,-k_2,k_4} - \\ &- M_{-k_1,k_4,-k_2,k_3} - M_{-k_2,k_3,-k_1,k_4} - M_{-k_2,k_4,-k_1,k_3}\} \end{aligned} \tag{4.12}$$

$$M_{k_1k_2k_3} = k_1k_2\left(\frac{k_3k_4}{k_1k_2}\right)^{\frac{1}{4}}\{-2k_1 - 2k_2 + |k_1+k_3| + |k_1+k_4| + |k_2+k_3| + |k_2+k_4|\}$$

The "natural" variables are good in the capillary dominated region $k > (g/\sigma)^{1/2}$. But in the gravity dominated region $k \ll (g/\sigma)^{1/2}$ they are not so good. The following canonical (up to $O(a^5)$) transformation:

$$\begin{aligned} a_k = b_k &+ \int \Gamma^{(1)}_{kk_1k_2} b_{k_1} b_{k_2} \delta_{k-k_1-k_2} dk_1 dk_2 - 2\int \Gamma^{(1)}_{k_2kk_1} b^*_{k_1} b_{k_2} \delta_{k+k_1-k_2} dk_1 dk_2 + \\ &+ \int \Gamma^{(2)}_{kk_1k_2} b^*_{k_1} b^*_{k_2} \delta_{k+k_1+k_2} dk_1 dk_2 + \int B_{kk_1k_2k_3} b^*_{k_1} b_{k_2} b_{k_3} \delta_{k+k_1-k_2-k_3} dk dk_1 dk dk_2 \end{aligned} \tag{4.13}$$

$$\begin{aligned} B_{kk_1,k_2k_3} &= \Gamma^{(1)}_{k_1,k_2,k_1-k_2}\Gamma^{(1)}_{k_3,k,k_3-k} + \Gamma^{(1)}_{k_1,k_3,k_1-k_3}\Gamma^{(1)}_{k_2,k,k_2-k} - \\ &- \Gamma^{(1)}_{k,k_2,k-k_2}\Gamma^{1}_{k_3,k_1,k_3-k_1} - \Gamma^{1}_{k_1,k_2,k_1-k_2}\Gamma^{(1)}_{k_2,k_1,k_2-k_1} - \\ &- \Gamma^{(1)}_{k+k_1,k,k_1}\Gamma^{(1)}_{k_2+k_3,k_2,k_3} + \Gamma^{(2)}_{-k-k_1k,k_1}\Gamma^{(2)}_{-k_2-k_3,k_2,k_3} \end{aligned} \tag{4.14}$$

$$\Gamma^{(1)}_{kk_1k_2} = -\frac{V_{kk_1k_2}}{\omega_k - \omega_{k_1} - \omega_{k_2}}$$
$$\Gamma^{(2)}_{kk_1k_2} = -\frac{U_{kk_1k_2}}{\omega_k + \omega_{k_1} + \omega_{k_2}} \tag{4.15}$$

transforms the Hamiltonian H to a form not containing cubic contributions:

$$H = \int \omega_k b_k b^*_{k_1} + \frac{1}{2}\int T_{k,k_1,k_2,k_3} b^*_k b^*_{k_1} b_{k_2} b_{k_3} \delta_{k+k_1-k_2-k_3} dk_1 dk_2 dk_3 dk_4 \tag{4.16}$$

Here $T_{kk_1k_2k_3}$ satisfies the symmetry conditions:

$$T_{k,k_1,k_2,k_3} = T_{k_1,k,k_2,k_3} = T_{k,k_1,k_3,k_2} = T_{k_2,k_3,k,k_1} \tag{4.17}$$

and has a form:

$$\begin{aligned}
T_{k,k_1,k_2,k_3} &= W_{k,k_1,k_2,k_3} - \\
&- V_{k,k_2,k-k_2} V_{k_3,k_1,k_3-k_1} \left[\frac{1}{\omega_{k_2} + \omega_{k-k_2} - \omega_k} + \frac{1}{\omega_{k_1} + \omega_{k_3-k_1} - \omega_{k_3}}\right] - \\
&- V_{k_1,k_2,k_1-k_2} V_{k_3,k,k_3-k} \left[\frac{1}{\omega_{k_2} + \omega_{k_1-k_2} - \omega_{k_1}} + \frac{1}{\omega_k + \omega_{k_3-k} - \omega_{k_3}}\right] -
\end{aligned}$$

$$\begin{aligned}
&- V_{k,k_3,k-k_3} V_{k_2,k_1,k_2-k_1} \left[\frac{1}{\omega_{k_3} + \omega_{k-k_3} - \omega_k} + \frac{1}{\omega_k + \omega_{k_2-k} - \omega_{k_2}}\right] - \\
&- V_{k_1,k_3,k_1-k_3} V_{k_2,k,k_2-k} \left[\frac{1}{\omega_{k_3} + \omega_{k_1-k_3} - \omega_{k_1}} + \frac{1}{\omega_k + \omega_{k_2-k} - \omega_{k_2}}\right] -
\end{aligned}$$

$$\begin{aligned}
&- V_{k+k_1,k,k_1} V_{k_2+k_3,k_2,k_3} \left[\frac{1}{\omega_{k+k_1} - \omega_k - \omega_{k_1}} + \frac{1}{\omega_{k_2+k_3} - \omega_{k_2} - \omega_{k_3}}\right] - \\
&- U_{-k-k_1,k,k_1} U_{-k_2-k_3,k_2,k_3} \left[\frac{1}{\omega_{k+k_1} + \omega_k + \omega_{k_1}} + \frac{1}{\omega_{k_2+k_3} + \omega_{k_2} + \omega_{k_3}}\right]
\end{aligned}$$

In spite of its complexity, the expression for T_{k,k_1,k_2,k_3} conceals some inner beauty. For instance, an important function:

$$T_{kk_1} = T_{\vec{k}\vec{k}_1,\vec{k}\vec{k}_1} \qquad \vec{k}\|\vec{k}_1$$

has a very simple form:

$$T_{kk_1} = \frac{1}{8\pi^2}\begin{cases} kk_1^2 & k > k_1 \\ k^2k_1 & k < k_1 \end{cases} \tag{4.18}$$

The new variable b_k satisfies the equation:

$$\begin{aligned} \frac{\partial b_k}{\partial t} &+ i\frac{\delta H}{\delta b_k^*} = 0, \\ \text{or} \quad \frac{\partial b_k}{\partial t} &= -i\omega_k b_k - i\int T_{k,k_1,k_2,k_3} b_{k_1}^* b_{k_2} b_{k_3} \delta_{k+k_1-k_2-k_3} dk_1 dk_2 dk_3 \end{aligned} \tag{4.19}$$

This equation conserves the effective Hamiltonian (4.16) due to satisfaction of the conditions (4.17).

Early, in 1968, another equation was derived (V.Zakharov, 1968). Let us neglect in the expansion (4.13) terms higher than second-order:

$$\begin{aligned} a_k &= \tilde{b}_k + \int \Gamma^{(1)}_{k,k_1,k_2} \tilde{b}_{k_1} \tilde{b}_{k_2} \delta_{k-k_1-k_2} - \\ &- 2\int \Gamma^{(1)}_{k_2,k,k_1} \tilde{b}^*_{k_1} \tilde{b}_{k_2} \delta_{k+k_1-k_2} dk_1 dk_2 + \int \Gamma^{(2)}_{k,k_1,k_2} \tilde{b}^*_{k_1} \tilde{b}^*_{k_2} \delta_{k+k_1+k_2} dk_1 dk_2 \end{aligned}$$

One can show that the variable $\tilde{b}_k$ satisfies the equation:

$$\frac{\partial \tilde{b}_k}{\partial t} = -i\tilde{\omega}_k \tilde{b}_k - i\int \tilde{T}_{k,k_1,k_2,k_3} \tilde{b}^*_{k_1} \tilde{b}_{k_2} \tilde{b}_{k_3} \delta_{k+k_1-k_2-k_3} dk_1 dk_2 dk_3 \tag{4.20}$$

Here

$$\begin{aligned} \tilde{T}_{k,k_1,k_2,k_3} &= W_{k,k_1,k_2,k_3} - 2\frac{V_{k+k_1,k,k_1} V_{k_2+k_3,k_2,k_3}}{\omega_{k_2+k_3} - \omega_{k_2} - \omega_{k_3}} - \\ &- 2\frac{U_{-k-k_1,k,k_1} U_{-k_2-k_3,k_2,k_3}}{\omega_{-k_2-k_3} + \omega_{k_2} + \omega_{k_3}} - 2\frac{V_{k,k_2,k-k_2} V_{k_3,k_3-k_1,k_1}}{\omega_{k_3-k_1} + \omega_{k_1} - \omega_{k_3}} - \\ &- 2\frac{V_{k,k_3,k-k_3} V_{k_2,k_2-k_1,k_1}}{\omega_{k_2-k_1} + \omega_{k_1} - \omega_{k_2}} - 2\frac{V_{k_2,k,k_2-k} V_{k_1,k_1-k_3,k_3}}{\omega_{k_1-k_3} + \omega_{k_3} - \omega_{k_1}} - \\ &- 2\frac{V_{k_3,k,k_3-k} V_{k_1,k_1-k_2,k_2}}{\omega_{k_1-k_2} + \omega_{k_2} - \omega_{k_1}} \end{aligned} \tag{4.21}$$

The equation (4.20) is not Hamiltonian and does not conserve energy, because $\tilde{T}_{k,k_1,k_2,k_3}$ does not satisfies the symmetry conditions (4.17).

It does not mean that the equation (4.20) makes no sense. One can check that on the resonant manifold:

$$\begin{aligned} \omega_k + \omega_{k_1} &= \omega_{k_2} + \omega_{k_3} \\ k + k_1 &= k_2 + k_3 \end{aligned} \tag{4.22}$$

both T_{k,k_1,k_2,k_3} and $\tilde{T}_{k,k_1,k_2,k_3}$ coincide.

In fact T_{k,k_1,k_2,k_3} is nothing but the result of symmetization of $\tilde{T}$:

$$T_{k,k_1,k_2,k_3} = \frac{1}{4}(\tilde{T}_{k,k_1,k_2,k_3} + \tilde{T}_{k_1,k,k_2,k_3} + \tilde{T}_{k_2,k_3,k,k_1} + \tilde{T}_{k_3,k_2,k,k_1}) \tag{4.23}$$

So, the role of the third-order terms in the expansion (4.13) is auxiliary — it just provides symmetrization of T_{k,k_1,k_2,k_3}.

The equations (4.20) are used for direct numerical simulation of weakly nonlinear surface waves. The Hamiltonian equation (4.19) is more convenient for this purpose, because it conserves energy and allows us to realize more effective control over the calculations.

The fundamental property of the gravity wave theory is conserving of an additional motion integral – "wave action":

$$N = \int |b_k|^2 dk \tag{4.24}$$

This integral is conserved by both equations (4.19) and (4.20). But it is not conserved by any other more accurate equation. Continuing the canonical transformation up to fourth-order terms, one can derive a Hamiltonian to the form:

$$H = \int \omega_k b_k b_k^* dk + \frac{1}{2} \int T_{k,k_1,k_2,k_3} b_k^* b_{k_1}^* b_{k_2} b_{k_3} \delta_{k+k_1-k_2-k_3} dk dk_1 dk_2 dk_3 +$$
$$+ \int S_{k,k_1,k_2,k_3,k_4} (b_k^* b_{k_1}^* b_{k_2} b_{k_3} b_{k_4} + b_k b_{k_1} b_{k_2}^* b_{k_3}^* b_{k_4}^*) \delta_{k+k_1-k_2-k_3-k_4} dk dk_1 dk_2 dk_3 dk_4 \tag{4.25}$$

Corresponding equation for b_k is:

$$\begin{aligned} \frac{\partial b_k}{\partial t} &+ i\omega_k b_k + i \int T_{k,k_1,k_2,k_3} b_{k_1}^* b_{k_2} b_{k_3} \delta_{k+k_1-k_2-k_3} dk_1 dk_2 dk_3 + \\ &+ 2i \int S_{k,k_1,k_2,k_3,k_4} b_{k_1}^* b_{k_2} b_{k_3} b_{k_4} \delta_{k+k_1-k_2-k_3-k_4} dk_1 dk_2 dk_3 dk_4 + \\ &+ 3i \int S_{k_1,k_2,k,k_3,k_4} b_{k_1} b_{k_2} bn_{k_3}^* b_{k_4}^* \delta_{k_1+k_2-k_3-k_4-k} dk_1 dk_2 dk_3 dk_4 = 0 \end{aligned} \tag{4.26}$$

One can get from (4.26):

$$\begin{aligned} \frac{\partial}{\partial t} N &= -2\mathrm{Im} \int S_{k_1,k_2,k_3,k_4,k_5} b_{k_1}^* b_{k_2}^* b_{k_3} b_{k_4} b_{k_5} \times \\ &\times \delta_{k_1+k_2-k_3-k_4-k_5} dk_1 dk_2 dk_3 dk_4 dk_5 \end{aligned} \tag{4.27}$$

The formula (4.14) for B_{k,k_1,k_2,k_3} was found by V. Krasitzky (V.Krasitzky, 1991). He also calculated a very cumbesome expression for S_{k,k_1,k_2,k_3,k_4} (private communication).

5. Kinetic equation and Kolmogorov spectra

For statistical description of a stochastic wave field one can use a pair correlation function:

$$\langle a_k a_{k'}^* \rangle = n_k \delta(k - k') \tag{5.1}$$

Introducing:

$$\langle \eta_k \eta^*_{k'} \rangle = I_k \delta(k-k') \tag{5.2}$$

one can get from (4.4) the following:

$$I_k = \frac{\omega_k}{2A_k}(n_k + n_{-k}) \qquad I_{-k} = I_k$$

Thus, in a capillary dominating region we have:

$$I_k = \frac{1}{2(\sigma k)^{\frac{1}{2}}}(n_k + n_{-k}) \tag{5.3}$$

and in a gravity dominating region:

$$I_k = \frac{1}{2}\frac{k^{1/2}}{g}\,(n_k + n_{-k}) \tag{5.4}$$

In the capillary regime, the variable n_k is a natural tool for description of wave turbulence. But in a gravity dominating region it is more suitable to use:

$$\langle b_k b^*_{k'} \rangle = \tilde{n}_k \delta_{k-k'} \tag{5.5}$$

Here n_k and $\tilde{n}_k$ are connected by the relation that could be easily obtained from (4.14):

$$\begin{aligned}
n_k &= \tilde{n}_k + 2\int \frac{|V_{k,k_1,k_2}|^2}{(\omega_k - \omega_{k_1} - \omega_{k_2})^2}\delta_{k-k_1-k_2}(\tilde{n}_{k_1}\tilde{n}_{k_2} - \tilde{n}_k\tilde{n}_{k_1} - \tilde{n}_k\tilde{n}_{k_2})dk_1 dk_2 + \\
&+ 4\int \frac{|V_{k_2,k,k_1}|^2}{(\omega_k + \omega_{k_1} - \omega_{k_2})^2}\delta_{k+k_1-k_2}(\tilde{n}_{k_1}\tilde{n}_{k_2} + \tilde{n}_k\tilde{n}_{k_2} + \tilde{n}_k\tilde{n}_{k_1})dk_1 dk_2 + \\
&+ \int \frac{|U_{k,k_1,k_2}|^2}{(\omega_k + \omega_{k_1} + \omega_{k_2})^2}\delta_{k+k_1+k_2}(\tilde{n}_{k_1}\tilde{n}_{k_2} + \tilde{n}_k\tilde{n}_{k_2} + \tilde{n}_k\tilde{n}_{k_1})dk_1 dk_2 + \\
&+ \cdots
\end{aligned} \tag{5.6}$$

The denominators in (5.6) in the gravity region can't be zero. For the condition of weak steepness $\theta \ll 1$, an infinite set of equations for correlation functions $\langle a_{k_1} \cdots a_{k_n}, a^*_{k_{n+1}} \cdots a^*_{k_{n+n}} \rangle$ could be cut off. To the leading order, the functions $n_k, \tilde{n}_k$ obey the kinetic equations:

$$\frac{\partial n_k}{\partial t} = st(n,n) + f_s(k) - f_a(k) \tag{5.7}$$

$$\frac{\partial \tilde{n}}{\partial t} = st(\tilde{n},\tilde{n},\tilde{n}) + f_s(k) - f_a(k) \tag{5.8}$$

Here

$$\begin{aligned}
st(n,n) &= 2\pi\int |V_{k,k_1,k_2}|^2(n_{k_1}n_{k_2} - n_k n_{k_1} - n_k n_{k_2}) \times \\
&\times \delta_{k-k_1-k_2}\delta_{\omega_k-\omega_{k_1}-\omega_{k_2}} dk_1 dk_2 + \\
&+ 4\pi\int |V_{k_2,k,k_1}|^2(n_{k_1}n_{k_2} + n_k n_{k_2} - n_k n_{k_1}) \times \\
&\times \delta_{k+k_1-k_2}\delta_{\omega_k+\omega_{k_1}-\omega_{k_2}} dk_1 dk_2
\end{aligned} \tag{5.9}$$

and:

$$st(\tilde{n},\tilde{n},\tilde{n}) = 4\pi \int |T_{k,k_1,k_2,k_3}|^2 \left(\tilde{n}_{k_1}\tilde{n}_{k_2}\tilde{n}_{k_3} + \tilde{n}_k\tilde{n}_{k_2}\tilde{n}_{k_3} - \tilde{n}_k\tilde{n}_{k_1}\tilde{n}_{k_2} - \right.$$
$$\left. -\tilde{n}_k\tilde{n}_{k_1}\tilde{n}_{k_3}\right) \delta_{k+k_1-k_2-k_3}\,\delta_{\omega_k+\omega_{k_1}+\omega_{k_2}-\omega_{k_3}}\,dk_1dk_2dk_3 \tag{5.10}$$

It is important to stress that the kinetic equation is written for $\tilde{n}$, not for n. The difference between $\tilde{n}$ and n, expressed by formula (5.6), can be important in comparision of experimental data with theoretical predictions.

In (5.7) and (5.8), the function $f_s(k)$ is a source of waves due to surface–wind interaction, and $f_c(k)$ is an absorption of waves due to viscosity and wave-breaking. None of these functions are known to a sufficient degree. Since the article of Miles (J.Miles, 1957), we believe that wave generation is the result of instability of a wind flow, hence:

$$f_s(k) = \beta(\vec{k})n_k \tag{5.11}$$

Here $\beta(k)$ is a growth rate of instability. From dimensional considerations one can find:

$$\beta(\vec{k}) = \beta(\omega,\theta) = \epsilon F\left(\frac{\omega}{\omega_0},\theta\right)\cdot\omega \tag{5.12}$$

Here $\omega_0 = g/V$, $\xi = \omega/\omega_0$ and V is characteristic wind velocity.

It is impossible to calculate analytically the function $F(\xi,\theta)$ due to strong turbulence in a shear wind flow. Measurements collected and ordered by O.Phillips (O.Phillips, 1985), show that with satisfactory accuracy:

$$F\left(\frac{\omega}{\omega_0},\theta\right) \simeq F\left(\frac{\omega}{\omega_0}\right) cos\theta \tag{5.13}$$

Here θ is an angle with respect to the wind.

The function $F(\xi)$ has a kind of jump at $\xi = 1$ and can be approximated by the following formula:

$$F(\xi) = \begin{cases} 0 & \xi < 1 \\ \alpha\xi^{2-s} & \xi > 1 \end{cases} \tag{5.14}$$

Here $0.10 < \alpha < 0.15$ and $s \ll 1$. This behavior can be followed up to $\xi \simeq 20$, but, according to O.Phillips, data above $\xi = 3$ are not quite reliable.

The situation with the absorption term is even worse. The linear damping is the process competing with wind-driven instability. The existence of such damping just means that $F(\xi)$ comes through the maximum value and changes its sign due to influence of viscosity at definite and large enough ξ:

$$\xi \simeq \alpha\epsilon R_e \qquad R_e \simeq \frac{V^3}{g\nu}$$

Certainly, wave breaking introduces some nonlinear damping into a wave system. But any analytical calculations of this damping is very risky and unreliable. The level and spectral dependence of this damping is very sensitive to the frequency of wave breaking and density of white caps. Meanwhile, observed averaged wave spectra are robust enough with respect to these effects.

One of the goals of my lecture is to show that an entirely self-sustained theory can be developed *without any detailed knowledge about the wave damping.* Just the existence of strong enough damping as $\omega \to \infty$ is enough. The classical theory of well-developed turbulence in incompressible fluid is a paragon for such type of weak–turbulent theories.

The picture of well–developed ($R_e \to \infty$) three–dimensional turbulence is as follows. External factors (perturbing forces or instabilities of a mean flow) generate turbulent energy in the region $k < k_p$. In the "region of transparency" $k \gg k_p$ the picture of turbulence has a universal form. A cascade of energy to the large k viscous region is forming here. This cascade provides step-by-step energy transfer and is defined by a single integral constant - the density of energy dissipation ϵ. The spectrum of turbulence is the Kolmogorov spectrum:

$$\epsilon_k = c\epsilon^{-\frac{2}{3}}k^{-\frac{5}{3}} \tag{5.15}$$

This spectrum can be found from dimensional considerations but still is not proved analytically. The energy cascade to large k region is called traditionally the *direct cascade.* A picture of high-Reynolds number two-dimensional turbulence is more complicated. In this case basic equations have an additional integral of motion - the vorticity:

$$\Omega = \frac{1}{2}\int (rotV)^2 d\vec{r}$$

Any energy-producing factors produce vorticity as well. It is impossible to satisfy the conservation of two integrals of motion by only one cascade of energy — the inverse cascade of energy, transporting energy to the low k region, must be formed inevitably.

Actually in the two-dimensional case the Kolmogorov spectrum (5.15) describes the inverse cascade of energy to large scales, as far as the direct cascade is described by the formula:

$$\epsilon_k = c_1 \frac{\mu^{\frac{2}{3}}}{k^3} \tag{5.16}$$

Here μ is the total flux of vorticity to the viscous region.

The situation in the surface-wave turbulence is in a high degree similar to the situation in the incompressible turbulence.

The kinetic equations (5.7) and (5.8) look similar to Boltzmann's equation for the distribution function in the molecular gas kinetics. This comparison is, in fact, more deceptive and misleading than productive. Wave kinetic equations have rather different properties than particle kinetic equations.

Let us consider equations:

$$st(n,n) = 0 \tag{5.17}$$

and:

$$st(n,n,n) = 0 \tag{5.18}$$

Like Boltzmann's equation they have thermodynamic solutions:

$$h = \frac{P}{\omega_k} \qquad n = \frac{T}{\omega_k + \mu} \tag{5.19}$$

but unlike the Boltzmann case, both basic integrals $N = \int n_k dk$ and $E = \int \omega_k n_k dk$ diverge and become infinite for both of these distributions. As a result these Rayleigh-Jeans spectra only in very special cases could be applied for solutions of real physical problems. Happily, the equations (5.17) and (5.18) have another type of solution.

Let us consider the equation (5.17) in the general case assuming only that:

1. the medium is isotropic with respect to rotations,
2. $\omega = \omega(k)$ is the power function $\omega = k^\alpha$,
3. V_{k,k_1,k_2} is a homogenous function of k, k_1, k_2 and is of order k^β.

Averaging the equation (5.17) over angles in all planes (in a general case — in spaces of dimension d) k, k_1, k_2 and introducing variables $\omega = k^\alpha$ one can get the equation:

$$\int \{R_{\omega,\omega_1,\omega_2} - R_{\omega_1,\omega,\omega_2} - R_{\omega_2,\omega_1,\omega}\} d\omega_1 d\omega_2 = 0 \tag{5.20}$$

Here:

$$R_{\omega,\omega_1,\omega_2} = \frac{S_{\omega,\omega_1,\omega_2}\delta_{\omega-\omega_1-\omega_2}}{n_\omega n_{\omega_1} n_{\omega_2}} \left(\frac{1}{n_\omega} - \frac{1}{n_{\omega_1}} - \frac{1}{n_{\omega_2}}\right) \tag{5.21}$$

$$S_{\omega,\omega_1,\omega_2} = (kk_1k_2)^{\alpha-1} \frac{dk}{d\omega}\frac{dk_1}{d\omega_1}\frac{dk_2}{d\omega_2} \langle |V_{k,k_1,k_2}|^2 \delta_{k-k_1-k_2} \rangle \tag{5.22}$$

$S_{\omega,\omega_1,\omega_2}$ is again a homogeneous function and is of order ω^γ, $\gamma = \frac{2(\beta+d)}{\alpha} - 3$. Let us find the solution of (5.20) in the form of a power function $n(\omega) = \omega^{-x}$. The obvious solution, that makes $R_{\omega,\omega_1,\omega_2} \equiv 0$, is $x = 1$. It is the thermodynamic Rayleigh-Jeans solution. To obtain another solution one can perform in the second term of (5.20) the following "conformal" transformation:

$$\omega_1 = \frac{\omega^2}{\omega_1'};\ \omega_2 = \frac{\omega\omega_2'}{\omega_1'};\ \omega = \frac{\omega\omega_1'}{\omega_1'} \tag{5.23}$$

and in the third term — the transformation:

$$\omega_1 = \frac{\omega\omega_1'}{\omega_2'};\ \omega_2 = \frac{\omega^2}{\omega_2'};\ \omega = \frac{\omega\omega_2'}{\omega_2'} \tag{5.24}$$

Taking into consideration that Jacobians of transformations (5.23), (5.24) are $|\frac{\omega}{\omega_1}|^3$ and $|\frac{\omega}{\omega_2}|^3$ respectively, one can reduce (5.20) to the form:

$$\int R_{\omega,\omega_1,\omega_2} \left(1 - \left(\frac{\omega}{\omega_1}\right)^y - \left(\frac{\omega}{\omega_2}\right)^y\right) d\omega_1 d\omega_2 = 0 \tag{5.25}$$

$$J = \frac{2(\beta + d)}{\alpha} - 1 - 2x \tag{5.26}$$

The integrand in (5.25) becomes zero at $y = -1$. Hence another solution is $x = (\beta + d)/\alpha$ and distributions $n_\omega = c/\omega^{-(\beta+d)/\alpha}$; $n_k = c/k^{(-\beta+d)}$ satisfy the equations (5.20), (5.17) at any value of the constant c. Looking at dimension of c we find ultimately that the discussed solution has a form:

$$n_k = a_1 \frac{P^{\frac{1}{2}}}{k^{(\beta+d)}} \tag{5.27}$$

where P is a flux of energy and α is some absolute constant. More detailed study shows that P is directed to large k if $\beta + d > \alpha$ (usually satisfied).

The described trick (*method of the conformal transformation*) is a convenient tool for finding Kolmogorov-type solutions in different problems of weak and strong turbulence. It was invented by the author (V.Zakharov, 1966) and applied then to various physical situations (see the monograph [12]). In the application to capillary wave turbulence ($\beta = 9/2$, $d = 2$) one gets:

$$n_k = a_1\, \Gamma^{-1/4}\, P^{1/2}\, k^{-17/4} \tag{5.28}$$

Here a_1 is an absolute constant.

For the spectral density of elevations it gives:

$$I_k = \frac{a_1\, P^{1/2}}{\sigma^{3/4}\, k^{19/4}} \tag{5.29}$$

The conformal transformation method, being applied to the equation (5.18) gives two Kolmogorov solutions:

$$n_k^{(1)} = \frac{b_1 P^{\frac{1}{3}}}{k^4}; I_k^{(1)} = \frac{b_1 g^{\frac{1}{2}} P^{\frac{1}{3}}}{k^{\frac{7}{2}}}; E_\omega = \frac{4\pi b_1 g^2 P^{\frac{1}{3}}}{\omega^4} \tag{5.30}$$

$$n_k^{(2)} = \frac{b_2 Q^{\frac{1}{3}} g^{\frac{1}{6}}}{k^{\frac{23}{6}}}; I_k^{(2)} = \frac{b_2 g^{\frac{2}{3}} Q^{\frac{1}{3}}}{k^{\frac{10}{3}}}; E_\omega = \frac{4\pi b_1 g^2 Q^{\frac{1}{3}}}{\omega^{\frac{11}{3}}} \tag{5.31}$$

Here b_1, b_2 are absolute constants, P is the flux of energy to large k and Q is a flux of wave action to small k region. E_ω is frequency spectrum of wave energy.

Existence of two Kolmogorov solutions in this case is the result of existence of the additional integral of motion - wave action. In some sense the capillary turbulence is comparable with three-dimensional incompressible turbulence, as far as the gravity wave turbulence could be comparable with two-dimensional strong turbulence. It is important to stress that Kolmogorov spectra in the incompressible fluid turbulence is still no more than a very plausible hypothesis, satisfactorally confirmed by experiments (in the three dimensional case). In comparison, in the weak turbulent theory, existence of Kolmogorov-type solutions of kinetic equations is an examinable mathematical fact. Its validity limits could be strictly established (see the monograph cited above).

The spectrum $E(\omega) \simeq P^{1/3}\omega^{-4}$ was found first by the author and his graduate student Natasha Filonenko by applying the new-discovered method of conformal transformation (V.Zakharov and N.Filonenko, 1966). In 1973 the same spectrum was found experimentally by Y.Toba (Y.Toba, 1973), who was not aware about previous theoretical achievments in this field. Since the beginning of the eighties this spectrum has been systematically discussed in the scientific press.(See, for instance, Kitaigorodskii, 1983).

The spectrum $E(\omega) \simeq Q^{\frac{1}{3}}\omega^{-\frac{11}{3}}$ in the 1966 publication was missed. It was discovered by the author soon after and included in his thesis but not published in the scientific press until 1982. During 1982-1983 in a series of publications with M.M.Zaslavsky (V.Zakharov and M.Zaslavsky, 1982–1983), this spectrum was exploited for explanation of kinetics of the energy-containing part of surface-wave spectrum.

The spectrum $E(k) \simeq P^{1/2}/k^{17/4}$ for capillary waves was found by the author and N.N.Filonenko in 1967 V.Zakharov and N.Filonenko, 1967).

6. Inverse cascade and wave aging

Now we can answer the second question formulated in the introduction. What is the reason for wave aging? The answer is: wave aging is the inverse cascade of energy stipulated by existence of the additional integral of motion-wave action N.

We would be able to describe the spectrum and kinetics of the inverse cascade in detail, if it were isotropic. In this hypothetical case the spectrum in the low-frequency domain certainly has to be described by the formula (5.31), where:

$$Q = \int n(k)\beta(k)dk \tag{6.1}$$

In this case of course $\beta(k)$ is also symmetric:

$$\beta(b) = \beta(k) = \epsilon\omega F(\frac{\omega}{\omega_0}) \tag{6.2}$$

Substituting (6.2) into (6.1) we can find for Q:

$$Q = \epsilon^{\frac{3}{2}} g^{-1} V^4 q^{\frac{3}{2}} \tag{6.3}$$

$$q = 4\pi b_2 \int_0^\infty F(\xi)\frac{d\xi}{\xi^{\frac{11}{3}}} \tag{6.4}$$

The integral for q converges for $\xi > 1$ if $F(\xi) \sim \xi^2$, so the main "wave action production" take place in frequencies not exeeding in order of measure the critical frequency ω_0.

From (5.31) and (6.3) one can get:

$$E_\omega = 4\pi b_2 q^{\frac{1}{2}} g^{\frac{5}{3}} \epsilon^{\frac{1}{2}} \omega^{-\frac{11}{3}} \tag{6.5}$$

The integral $\int E_\omega d\omega$ diverges at $\omega \to 0$, and the spectrum (6.5) must be cut at some limiting frequency ω_0. The total energy is now:

$$\int E_\omega d\omega \simeq \frac{3\pi}{2} b_2 q^{\frac{1}{2}} \epsilon^{\frac{1}{2}} g^{\frac{5}{3}} V^{\frac{4}{3}} \omega_0^{-\frac{8}{3}} \tag{6.6}$$

For the mean square angle we have ($k_0 = \omega_0^2/g$):

$$\begin{aligned}\langle\theta^2\rangle &= 1/g\,k_0^2 \int E_\omega d\omega \simeq \\ &\simeq (3\pi/2)\,b_2 q^{1/2}\epsilon^{1/2}\left(\frac{V\omega_0}{g}\right)^{4/3} = (3\pi/2)\,b_2 q^{1/2}\,\epsilon^{1/2}\,a^{-4/3}\end{aligned} \tag{6.7}$$

Here $a = g/V\omega$ is "wave age".

The formula (6.7) is remarkable. It shows the smallness of the steepness ($(q\epsilon)^{\frac{1}{2}} \simeq 10^{-2}$) and how it decreases in the inverse cascade, when the wave age is sufficiently large.

If the total action flux Q from a short-wave region is constant in time, the total wave action grows proportionally in time. The minimal frequency ω_0 behaves like:

$$\omega_0 \simeq t^{-\frac{3}{11}} \tag{6.8}$$

and total energy increases like:

$$\int E d\omega \simeq t^{-\frac{3}{11}} \tag{6.9}$$

In a real sea spectrum, the long wave region is very narrow in angle. Reasons for this phenomenon are under discussion now. According to our theory (see V.Zakharov and V.Shrira, 1990; 1992), this compression in angle is the result of nonlinear gravity waves and a current flow, inevitably corresponding to a mature sea. In spite of obvious nonapplicability of an isotropic theory to the real angular-dependent situation, the spectrum $\omega^{-11/3}$ (6.5) and even the dependence of leading frequency on time (6.8) are in satisfactory accordance with observed data (see V.Zakharov and M.Zaslavsky, 1982–1983).

7. Direct cascade and wave-breaking

Direct cascade starts at $\omega = \omega_m = g/V$ and develops to the large frequency region. The spectrum is anisotropic in this region as well, but the degree of anisotropy decreases with increasing frequency. So, here the isotropic approximation works better.

Details of the direct cascade are defined by behavior of the function $F(\xi)$. [See (5.12)]. Let the following integral be convergent:

$$r = \int_0^\infty \frac{F(\xi)}{\xi^3}\,d\xi < \infty \tag{7.1}$$

In this case at $\omega \gg \omega_m$ we will have:

$$E_\omega = \frac{(4\pi b)^{3/2}\,\epsilon^{1/2}\,V g^2}{\omega^4} \tag{7.2}$$

This is a direct Kolmogorov spectrum (5.30) and:

$$P = (4\pi b_1 r)^{3/2}\,\epsilon^{3/2}\,V^3 \tag{7.3}$$

Convergence of (7.1) means that a region of energy exitation is $\omega \geq \omega_\mu$. Convergence implies that $F(\xi)$ grows slower than ξ^2. If $F(\xi) = c\xi^2$ at $\xi > 1$, the spectrum has the following form:

$$E_\omega = (4\pi b_1)^{3/2} \left(\frac{2}{3} log\frac{\omega}{\omega_m}\right)^{1/2} \frac{\epsilon^{1/2} V g^2}{\omega^4} \tag{7.4}$$

In this case energy pumping is distributed uniformly for all frequencies. If $F(\xi)$ grows faster then ξ^2 and the integral (7.1) is divergent, pumping of energy is concentrated at ultimately large ω at the margin of damping. In this case the spectrum goes down more slowly then ω^{-4}.

It is the direct cascade of energy that causes small-scale wave breaking and fractalization of the surface. Let us consider the most well-known case of finite r. The spectral density of elevation of surface :

$$I_k \simeq \frac{\epsilon^{1/2} V}{g^{1/2} k^{7/2}} \tag{7.5}$$

Here a numerical factor is omitted.

This is the spectral density of a fractal surface of dimension 7/4 (R. Glazman, 1986). The formula (7.5) can not be extended for too large k. To find a limit of its validity one can consider the "renormalized" dispersion law:

$$\tilde{\omega}_k = \omega_k + \int T_{kk_1} n_{k_1} \, dk_1 \tag{7.6}$$

A weak turbulence theory is applicable as far as:

$$\int T_{kk'} n'_k \, dk' \ll \omega_k \tag{7.7}$$

For the direct cascade spectrum we have:

$$n_k \simeq \frac{g I_k}{\omega_k} \simeq \frac{\epsilon^{1/2} V}{k^4} \tag{7.8}$$

The condition (7.7) together with $T_{kk'}$ taken from (4.18), gives:

$$\epsilon^{1/2} V k log\frac{k_0}{\epsilon} < (gk)^{1/2} \tag{7.9}$$

or, neglecting a logarithmic factor:

$$k < \frac{k_0}{\epsilon} \qquad k_0 = \frac{V^2}{g} \tag{7.10}$$

in according to the conjecture (2.1).

The condition (7.10) can be interpreted by a different way. The idea, that a typical singularity of a free surface is formation of "wedges" has a long tradition. The famous Phillips' spectrum:

$$I_k \simeq \frac{\alpha}{k^4} \tag{7.11}$$

is obtained exactly using this assumption. Many authors tried to apply Phillips' spectrum for the description of sea-waves in the energy-containing region. Attempts to fit the observations give for α an anomalously small value ($\alpha < 10^{-2}$). It is very natural because a real spectrum contains the small parameter $\epsilon^{1/2} \sim 3 \cdot 10^{-2}$. The small value of α obtained from measurements proves the actual validity of a weak nonlinear perturbative approach to the surface turbulence problem. A small value of α justifies the weak turbulent theory, and undermines the idea of using the ultimately "strongly nonlinear" Phillips' spectrum in this situation. This spectrum is nevertheless very remarkable, but it has quite another physical application.

If the absolute constant α is of order of one or slightly less, it means that each wave has a sharp wedge–type crest. It is a picture of well-developed wave breaking. The conjecture about wedge–type singularity of a free surface, formulated in Chapter 3 of this lecture, has the following consequence for surface turbulent spectra: they cannot exceed the Phillips' spectrum and the inequality:

$$I_k < \frac{\alpha}{k^4} \qquad \alpha \sim 1 \tag{7.12}$$

must hold. Matching a real spectrum with the Phillips' one means fractalization. Comparing (7.5) with (7.12) one gets for the lower margin of fractal behavior $k_m \sim (1/\epsilon)k_0$, which coincides with the upper limit from weak turbulent considerations. So, weak turbulence spectra go to fractal spectra in a natural way.

Let us discuss now the role of surface tension. In the capillarity-dominating region $I_k \sim k^{-19/4}$, and a direct cascade is realised by a smooth free surface. It raises hopes to construct a self-sustained theory that is weakly turbulent throughout, up to a viscous region. To do that, one has to match the direct cascade Kolmogorov spectra (5.29) and (5.30) at the wave number $k \simeq \sqrt{g/\sigma}$. Considering the ratio:

$$\mu \simeq I_k^{cap}/I_k^{grav} \simeq p^{1/6}\,\epsilon^{1/2}\,\sigma^{-3/4}\,k^{-5/4} \simeq \epsilon^{1/4}\,V^{1/2}\,(\rho\sigma)^{-1/8} \tag{7.13}$$

A matching is possible if $\mu < 1$. In this case a fractal behavior (5.30) changes to more smooth at $k \sim \sqrt{g/\sigma}$, producing a "jump" in the spectrum. This jump goes for $k \to \infty$ to a capillary "tail" (5.29), spreading up to viscous scales. But matching is certainly impossible if $\mu > 1$. Putting $\mu \simeq 1$, we get from (7.13) the limiting value of wind velocity, providing smooth behavior of a sea surface:

$$V_c \simeq \epsilon^{-1/2}(\sigma g)^{1/4} \tag{7.14}$$

in according to (2.2).

Calculating a margin of fractal behavior $k_m \sim V^2/\epsilon g$ for the velocity (7.14) one gets now:

$$k_m \sim \left(\frac{g}{\sigma}\right)^{1/2} \tag{7.15}$$

If $\mu > 1$, matching of gravity and capilliary Kolmogorov spectra at $k \sim (g/\sigma)^{1/2}$ is impossible. But in this case $k_m < (g/\sigma)^{1/2}$, and wave–breaking fractalization take place at such large scales that surface tension cannot surpass them.

Concluding the lecture, I would like to admit that I entirely understand the imperfectness of this isotropic approach to the essentially anisotropic problem of wind–driven sea waves. Very important problems, like momentum transfer from wind to water, remain beyond our consideration. Our goal was to show, how a good piece of physics could be understood in framework of a simple isotropic model.

7.1 Bibliography

1. R.E.Glazman and S.H.Pilorz, *Effects of sea maturity on satellite altimeter measurements*, J. Geophys. Res., **95** (1990), 2857–2870.

2. R.E.Glazman, *Statistical problems of wind-generated gravity waves, arising in microwave remote sensing of surface winds*, Trans. Geosci. Rem. Sensing, **29** (1991), 135–142.

3. V.G.Bondur and E.A.Sharkov, *Statistical properties of whitecaps on a rough sea*, Oceanology, **22** (1982), 274–279.

4. M.L.Banner, I.F.Jones and J.C.Trinder, *Wavenumber spectra of short gravity waves*, J. Fluid Mech., **198** (1989), 321–344.

5. K.Hasselmann, *On the nonlinear energy transfer in gravity-wave spectrum*, J. Fluid Mech., **12** (1962), 481–500.

6. V.E.Zakharov, *Stability of periodic waves of finite amplitude on a surface of a deep fluid*, J. Appl.Mech.Tech.Phys.,(1968), No. 2, p.190.

7. D.R.Crawford, H.G.Yuen and P.G.Saffman, J. Wave Motion, (1980), 1–16.

8. V.P.Krasitzky, *On the canonical transformation of the theory of weakly nonlinear waves with non-decay dispersion law*, Sov.Phys. JETP, (1991)

9. J.W.Miles, *On the generation of surface waves by shear flow*, J. Fluid Mech., **3** (1957), 185–204.

10. O.M.Phillips, *Spectral and statistical properties of the equilibrium range in wind-generated gravity waves*, J. Fluid Mech., (1985), 505–531.

11. V.E.Zakharov, *Weak-turbulence in a plasma without magnetic fields*, Zh. Eksp. Teor.Fiz., **51** (1966), p.688 (Sov.Phys. JETP, **24**(1967), 455–459.)

12. V.E.Zakharov, V.S.Lvov and G.Falkovich, *Kolmogorov spectra of wave turbulence*, Springer–Verlag, (1992, in press).

13. V.E.Zakharov and N.N.Filonenko, *The energy spectrum for stochastic oscillation of a fluid's surface*, Doklady Akad.Nauk, **170** (1966), 1292–1295.

14. Y.Toba, *Local balance in the air–sea boundary processes. III. On the spectrum of wind waves*, J. Oceanogr. Soc.Japan, **29** (1973), 209–220.

15. S.A.Kitaigorodskii, *On the theory of the equilibrium range in the spectrum of wave-generated gravity waves*, J. Phys.Oceanogr., **13** (1983), 816–827.

16. V.E.Zakharov and M.M.Zaslavskii, *The kinetic equation and Kolmogorov spectra in the weak–turbulence theory of wind waves*, Izv. Atm.Ocean.Phys., **18** (1982), 747–753.

17. V.E.Zakharov and M.M.Zaslavskii, *Ranges for generation and dissipation in the kinetic equation for a low–turbulence theory of wind waves*, Izv. Atm.Ocean.Phys., **18** (1982), 821–827.

18. V.E.Zakharov and M.M.Zaslavskii, *On the problem of wind–driven waving prognosis*, Doklady Akad. Nauk. SSSR, **256** (1982), 567.

19. V.E.Zakharov and M.M.Zaslavskii, *Shape of spectrum of energy carrying components of a water surface in a weak–turbulence theory of wind waves*, Izv. Atm.Ocean.Phys., **19** (1983), 207–212.

20. V.E.Zakharov and M.M.Zaslavskii, *Dependence of wave parameters of the wind velocity, duration and its action and fetch in the weak–turbulence theory of wind waves*, Izv. Atm.Ocean Phys., **19** (1983), 300–306.

21. V.E.Zakharov and N.N.Filonenko, *Weak turbulence of capillary waves*, J. Appl. Mech. Tech. Phys., **5** (1967). 62–67.

22. V.E.Zakharov and V.I.Shrira, *On the formation of the directional spectrum of wind waves*, Zh. Eksp.Teor.Fiz., **98** (1990), 1941–1958.

23. V.E.Zakharov and V.I.Shrira, *Water wave nonlinear interaction owing to drift current: Formation of the angular spectrum of the waves*, J.Fluid Mech., (1992, in press).

Experiments and Observations

Observations of the Enhancement of Kinetic Energy Dissipation Beneath Breaking Wind Waves

W.M. Drennan[1], K.K. Kahma[2], E.A. Terray[3], M.A. Donelan[1] and S.A. Kitaigorodskii[4]

[1]National Water Research Institute
Canada Centre for Inland Waters
Burlington, Ontario (L7R 4A6) Canada

[2]Finnish Institute for Marine Research
P.O. Box 33
SF-00930 Helsinki, Finland

[3]Dept. of Ocean Engineering
Woods Hole Oceanographic Institution
Woods Hole, Massachusetts 02543 U.S.A.

[4]Dept. of Earth and Planetary Science
Johns Hopkins University
Baltimore, Maryland 21218 U.S.A.

Abstract

Most attempts to characterize the kinetic energy dissipation in the upper 20 metres of the water column revert to simple wall layer scaling - proportional to the cube of the friction velocity u_* and inversely proportional to depth, $\epsilon \propto u_*^3 z^{-1}$. With a concomitant logarithmic velocity profile, this is consistent with a total kinetic energy flux from the wind of ρu_*^3. However, in fully rough flow and strongly forced waves the energy input may be one to two orders of magnitude greater. Where does this energy go? Why is it not reflected in most of the upper layer measurements? This paper attempts to answer these questions and to demonstrate that there are two regimes of kinetic energy dissipation in the near surface layers under breaking waves. Near the surface, the dissipation rate is very high and scales with the wave characteristics. At greater depths the dissipation rate drops quickly and reverts to wall layer scaling. In the intermediate region the dissipation decays more rapidly than z^{-1}. This may be viewed as a transition region between the deeper shear layer and the near surface region, with intense patches of breaking-imposed turbulence. In the absence of density stratification, dissipation in near surface region is analogous to that due to grid-generated turbulence and decays as z^{-4}.

Introduction

The rate of turbulent kinetic energy dissipation (ϵ) in the upper oceanic layers and, in particular, the distribution of ϵ near the surface is of great significance in matters relating to the mixing of near surface waters, mass transfer across the interface, dispersal of buoyant pollutants, testing of similarity hypotheses related to turbulent structure and the modelling of thermocline development, inter alia. It is not surprising that a great deal of effort has been expended in recent years to acquire some observational estimates of its near surface vertical distribution. Estimates of dissipation near the top of the water column have been made from three distinctly different types of platforms, viz: (i) fixed towers (c.f. [1]) ; (ii) horizontally or nearly horizontally moving vehicles (ships, submarines, towed bodies) [2]; (iii) vertically profiling devices driven by a buoyancy difference [3],[4]. Of these three methods, we note that the first two are best suited for

M. L. Banner, R. H. J. Grimshaw (Eds.)
Breaking Waves
IUTAM Symposium Sydney/Australia 1991

the study of dissipation rates in the near surface region. Vertical profilers travel quickly through the wave field, and according to [5], whitecap coverage statistics for a typical wind speed for these experiments (U = 5 m/s) suggest that in the total time to date spent by all profilers in the wave zone (about 10,000 seconds) the chances that a single breaker swept over the randomly chosen locations of the profilers is no better than 5%.

Only in very rare cases has the dissipation been estimated from the smallest scales where conversion of mechanical energy to heat occurs. Much more usually the spectral density at intermediate scales is employed, via the Kolmogorov similarity hypothesis, to estimate the rate at which energy flows from the (large scale) source to the (small scale) sink. Generally speaking the interpretation of surface layer dissipation estimates fall into two classes; viz: (i) general agreement with the structure of a classical wall layer as expressed by similarity scaling (e.g. [2], [3]); (ii) much higher dissipation values than expected from a purely shear driven wall layer, and usually attributed to wave breaking ([1], [4]). If these estimates of enhanced dissipation are correct, then wave breaking and the interaction between waves and turbulence dominate the kinetic energy balance in the surface layers. At sufficient depth the direct wave effects would certainly vanish and the production and dissipation of turbulence would acquire the attributes of a shear driven flow, although some of the momentum supplied to drive the deep shear layer may have been released from the wave field during breaking and wave- turbulence interaction. Evidently, various properties of the wave field must be important in establishing general scaling characteristics of the wind driven surface layer. As one moves away from the surface the influence of the waves quickly diminishes and at some depth it should be no surprise that wall layer scaling appears to describe the observations adequately. How may these two regions be consistently described? What are the appropriate scaling variables or combination of variables for the wave dominated layer?

It was partly to answer these questions that an extensive field observational program was put in place in Lake Ontario in three successive autumns (1985-1987). Velocity measurements were made from a fixed tower with a string of three specially designed drag spheres [6]. The tower was designed for waves and air-water interaction research and thus there is a minimum of interference to flow near the interface. The drag spheres, designed to operate correctly to wave lengths as small as 2 cm, were located at depths of approximately 1, 2 and 4 meters. Calibrations were made in the 100m CCIW towing tank before and after the field season. During the entire experiment a wide range of conditions were observed including: wind speeds from 3 m/s to 17 m/s; significant wave heights from a few centimetres to 2.5 m; inverse wave ages (degree of forcing) from 0.1 (swell) to 4.5 (very young, short-fetched waves); mean current speed from 2 cm/sec to

20 cm/sec. Data from the drag spheres (and all other apparatus - see [7]) were collected at 20 Hz. The resulting time series were windowed and analysed applying standard techniques.

Results

The existence of $-\frac{5}{3}$ regions in the frequency spectra of velocity components in the wave zone has been demonstrated by [1]. From the much larger data base of this experiment we show a clear example in Fig. 1. Although not all velocity spectra show well-established $-\frac{5}{3}$ regions, most do and this is typical of those that do. Both horizontal (u) and vertical (w) velocity spectra have $-\frac{5}{3}$ regions above the wave peak, whereas only the horizontal velocity spectra have $-\frac{5}{3}$ regions below the wave-peak. The likelihood of the condition of isotropy being satisfied is clearly higher in the region above the peak.

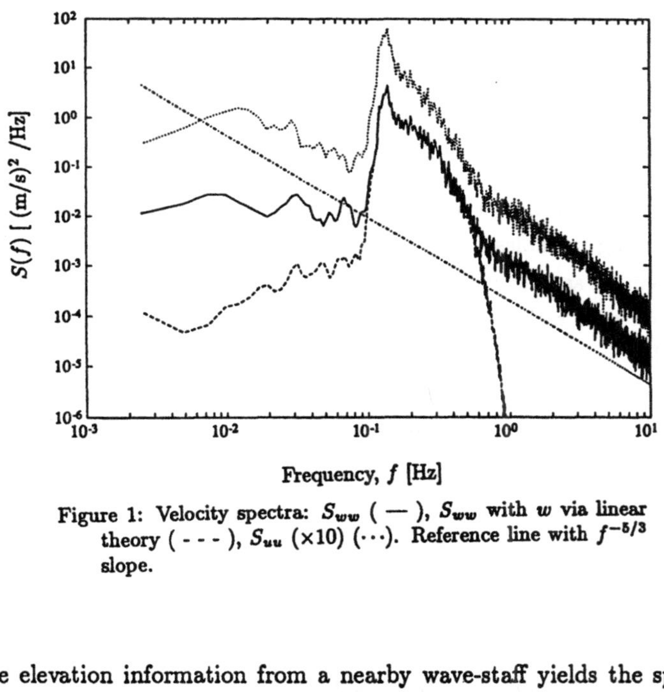

Figure 1: Velocity spectra: S_{ww} (—), S_{ww} with w via linear theory (- - -), S_{uu} (×10) (···). Reference line with $f^{-5/3}$ slope.

The surface elevation information from a nearby wave-staff yields the spectra of expected velocity fluctuations based on linear theory (see Fig. 1). It is clear that the $-\frac{5}{3}$ regions extend well beyond the region of the expected orbital velocities. We may therefore, without recourse to filtering out the wave-induced motion, treat the spectra outside of the zone of direct wave influence as a reflection of the turbulence properties. From Kolmogorov's similarity hypothesis we may obtain ϵ by measuring the spectral density $F(k)$ (k is the wavenumber) in the inertial subrange where $F(k) \propto \epsilon^{2/3}k^{-5/3}$. Although the similarity argument is properly applied to the wavenumber spectrum, the connection to the more easily measured frequency spectrum is made via Taylor's "frozen turbulence" hypothesis with advecting velocity u_d. Then $\epsilon = C_1 S^{\frac{3}{2}} u_d^{-1} \omega^{\frac{5}{2}}$ where $S(\omega)$ is the frequency spectrum of a velocity component evaluated in the inertial subrange and value of the constant C_1 is taken to be 2.4. In calculating dissipation, we employ the inertial subrange at frequencies both above and below the spectral peak, taking u_d to be the rms wave orbital velocity and mean current respectively.

Estimates of ϵ were made from spectra of both horizontal and vertical velocity. Nine pairs of estimates of ϵ from the high frequency region were compared: with the range of ϵ covering $2\frac{1}{2}$ decades, good agreement was seen between the estimates from the two components. In Figure 2, we show dissipation estimates derived from our field data as well as those collected by others. The data are plotted in the wall layer coordinates of [3], and it is evident that our data show dissipation rates several orders of magnitude higher than that would be anticipated by wall layer theory. This will come as no surprise to anyone who has been at sea in storm conditions. The order of magnitude agreement with wall layer theory claimed by [3] and others may simply reflect their inability to measure dissipation near the surface under conditions of pronounced wave breaking.

How then may one estimate the dissipation rate in these conditions? What are the appropriate scaling variables? Neither the significant wave height nor the peak wave length is well suited as a length scale since wave dissipation is distributed over the rear face of the spectrum rather than concentrated near the peak; this is especially true near full development. A better choice of scaling wavenumber would be that associated with the peak of the *slope* spectrum, since the steepest waves are the dissipating waves. Unfortunately, the slope spectrum is very broad, making the choice of a peak rather inprecise. We use instead the mean wavenumber of the slope spectrum

$$\overline{k} = \left[\frac{\int k^2 S_{\eta\eta}(\omega)d\omega}{\int S_{\eta\eta}(\omega)d\omega}\right]^{1/2}. \qquad (1)$$

For full development $\overline{k} \gg k_p$ and as the waves become more and more strongly forced the energy near the peak is enhanced [7] and $\overline{k} \to k_p$. (N.B. In evaluating (1), we follow [8] and append an ω^{-5} tail to $S_{\eta\eta}$ for $\omega > 3.5\omega_p$.)

If the wind input to the surface is being delivered primarily to waves of various lengths i.e., wave drag, as is the case when the flow is aerodynamically rough, then the question to be considered is what is the magnitude of this energy input and how does it scale? As pointed out in [2], the energy input is the wave drag multiplied by an appropriate average

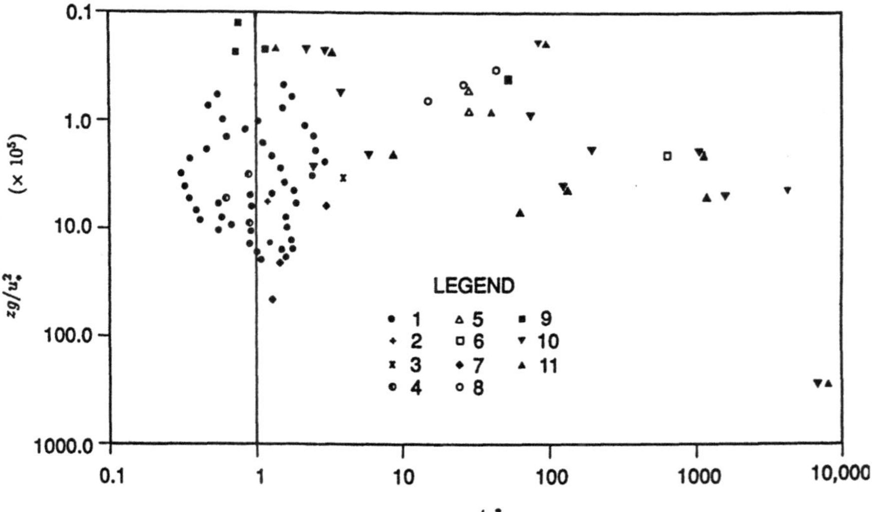

Figure 2: Dissipation in 'wall layer' coordinates, with the 'law of the wall' as a vertical line (—). Symbols : ○ △ □ from [1], × [2], • [3], ■ ▼ ▲ this experiment. + ◆ and ⊙ as given in [3].

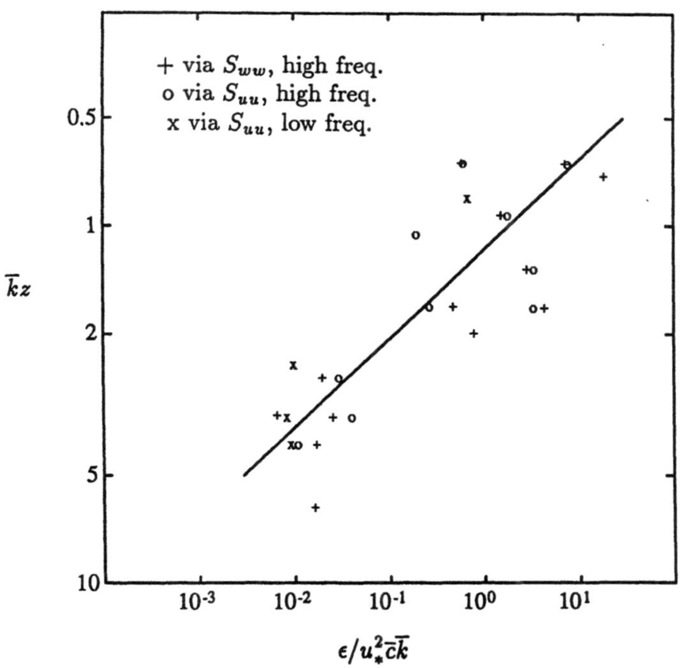

Figure 3: Dissipation in dimensionless coordinates

wave phase speed. The wave drag is delivered preferentially to the steep waves, so that the phase speed $\bar{c}$ associated with $\bar{k}$ would reflect this. We then look for a functional form relating the variables ϵ, $E_{in}(= u_*^2\bar{c})$, $\bar{k}$ and z. In particular, $\epsilon/(E_{in}\bar{k}) = F(\bar{k}z)$. Fig.3 illustrates how the data fall in these coordinates. Although there is substantial scatter, a clear trend of decreasing ϵ with depth is apparent. Since both axes contain variables with some uncertainty, the slope of the linear regression line (on logarithmic axes) was found to lie in the range of -3.0 to -4.6. We propose the following relationship between the non-dimensional variables based on an analogy with grid turbulence:

$$\frac{\epsilon}{u_*^2\bar{c}\bar{k}} = 1.84(\bar{k}z)^{-4} \quad \text{or} \quad \epsilon = 1.84u_*^2\bar{c}\bar{k}^{-3}z^{-4}\,. \qquad (2a,b)$$

Recent results of Gargett [4] include dissipation estimates in the upper 30 metres which also show a depth dependence of close to z^{-4}. We note, however, that no wind speed dependence was observed, although wind speeds of up to 12 m/s were noted. The z^{-4} is also in agreement with observations of a shear-free slab of thickness ~1 metre in the velocity structure beneath a whitecapping sea [9].

The total dissipation may be obtained from the depth integral of ϵ. It should be equal to $u_*^2\bar{c}$, the energy input due to wind. We have postulated that the principal source of the kinetic energy is wave breaking so that energy is injected to a depth of order of the wave height, below which it decays much in the manner of shear-free or grid generated turbulence (z^{-4}) (see [10]). Therefore we evaluate the integral of (2b) from ∞ to βH_s (H_s is the significant wave height) and add a constant value from βH_s to the surface. That is, we assume that the energy is injected in the zone between the surface and depth βH_s and that the dissipation rate is constant in this zone. A similar idea was proposed by Kitaigorodskii in connection with a model for gas transfer [11]. Equating this to the total energy input we find $\beta = 1.2\,(\frac{U}{c_p})^{0.19}$. This suggests that the depth of direct wave mixing extends 16% beyond the significant height at full development and 60% beyond for the most strongly forced case in our data ($\frac{U}{c_p} = 4.2$).

It is of interest to determine the 'transition depth', z_t, at which ϵ equals the conventional wall layer estimate, $\frac{u_*^3}{\kappa z}$. Using an empirical relationship between $\bar{k}$ and k_p and the dependence of H_s on U/c_p given by [7], we have

$$\frac{z_t}{\beta H_s} = 4.5(\frac{U}{c_p})^{-0.24}. \qquad (3)$$

Thus the relationship between z_t and βH_s is nearly constant (3.2 to 4.7 for short fetch to full development). This suggests that the intermediate layer of rapid decrease in ϵ is 3 times the thickness of the source layer (βH_s). This can be of great use in the modelling of mixed layer dynamics.

On the basis of our observations very near the surface, we have proposed a structure in which the uppermost layer is directly mixed by wavebreaking, is relatively shear-free and has a constant (with depth) dissipation rate. Below this there is an intermediate layer into which kinetic energy is diffused from above and in which the dissipation rate decays very rapidly (approximately z^{-4}) with depth. This layer is analogous to that due to grid generated turbulence and eventually merges into a classical wall layer at the transition depth z_t below which the dissipation rate decays more slowly (z^{-1}).

References

[1] Kitaigorodskii, S.A., M.A.Donelan, J.L.Lumley and E.A.Terray; Wave-turbulence interactions in the upper ocean. Part II. J. Phys. Oceanogr. 13 (1983) 1988-1999.

[2] Stewart, R.W. and H.L.Grant; Determination of the rate of dissipation of turbulent energy near the sea surface in the presence of waves. J. Geophys. Res. 67 (1962) 3177-3180.

[3] Soloviev, A.V., N.V.Vershinsky and V.A.Bezverchnii; Small-scale turbulence measurements in the thin surface layer of the ocean. Deep Sea Res. 35 (1988) 1859-1874.

[4] Gargett, A.E.; Ocean turbulence. Ann. Rev. of Fluid Mech. 21 (1989) 419-451.

[5] Monahan, E.C.; Oceanic whitecaps. J. Phys. Oceanogr. 1 (1971) 139-144.

[6] Donelan, M.A. and J.Motycka; Miniature drag sphere velocity probe. Rev. Sci. Instrum. 49 (1978) 298-304.

[7] Donelan, M.A., J.Hamilton and W.H.Hui; Directional spectra of wind-generated waves. Phil. Trans. Roy. Soc. London A315 (1985) 509-562.

[8] Banner, M.L.; Equilibrium spectra of wind waves. J. Phys. Oceanogr. 20 (1990) 966-984.

[9] Terray, E.A., A.J.Williams III, and B.H.Brumley; Observation of shear free turbulence beneath whitecaps in the marine surface layer (in preparation).

[10] Long, R.R.; Theory of turbulence in a homogeneous fluid induced by an oscillating grid. Phys. Fluids 21 (1978) 1887-1888.

[11] Kitaigorodskii, S.A.; On the fluid dynamical theory of turbulent gas transfer across an air-sea interface in the presence of breaking wind waves. J. Phys. Oceanogr. 14 (1984) 960-972.

Microwave Backscattering from Laboratory Wind-Wave Surfaces and its Relation to Wave Breaking with Bubble Entrainment

N. EBUCHI, H. KAWAMURA and Y. TOBA

Department of Geophysics, Faculty of Science, Tohoku University
Sendai 980, Japan.

Abstract

Microwave backscattering from wind-wave surfaces in a wind wave tank is investigated by using an X-band (9.6 GHz) scatterometer. It is found that intensity of the backscattered signature is in phase with the surface displacement. Variation of the Doppler velocity also corresponds to the phase of the individual waves. High Doppler velocity observed at the crests of the individual waves coincides with the propagating speed of the crests. It is concluded that the fine structures of wind-wave surfaces, which are trapped near the crests and are propagating with the crests, mainly contribute to microwave backscattering. It is also shown that the wave breaking with bubble entrainment does not increase the microwave backscattering under the conditions of the present experiment. (This paper is an extended abstract. This study will be published in full elsewhere.)

Experiment

The experiments were conducted in a wind-wave tank which is 60 cm wide, 20 m long and 120 cm high, containing fresh water 70 cm deep. The measurement was made at a fetch of 10.5 m. A schematic picture of the experimental apparatus is shown in Fig.1. The frequency of the microwave is 9.6 GHz, and the wave length is 3.12 cm. In this experiment, we selected the VV polarization. The incident angle is fixed at 45° upwind. The beam of microwave is focused on the water surface 2 m distant by using an ellipsoidal antenna in order to clarify the relationship between the backscattering and the fine structures of the wind-wave surfaces. The -3 dB beamwidth of the antenna is 13 cm at the water surface for the VV polarization.

The amplitude and the Doppler frequency of backscattered signature are measured and recorded. In order to measure the surface displacement and the wave speed, a pair of capacitance gauges is installed at a distance of 1 cm in the fetch direction. We also observed the fine structures of the wave surface by using a video camera from the bottom of the tank.

In this experiment, we selected 7 reference wind speeds from 3.6 m/s to 14.1 m/s. In this paper, results for the case of 14.1 m/s is shown. Parameters of wind and wave fields under this wind speed is summarized in Table 1. Qualitative results were almost same for the other cases in which the wind speeds were larger than 10 m/s. In these cases, the dominant waves were longer than the microwave beamwidth.

Phase Relationship between the Surface Displacement and Backscattered Signature

Figure 2 is an example of the time series at the wind speed 14.1 m/s. Panel(a) and (b) show the surface displacement and the

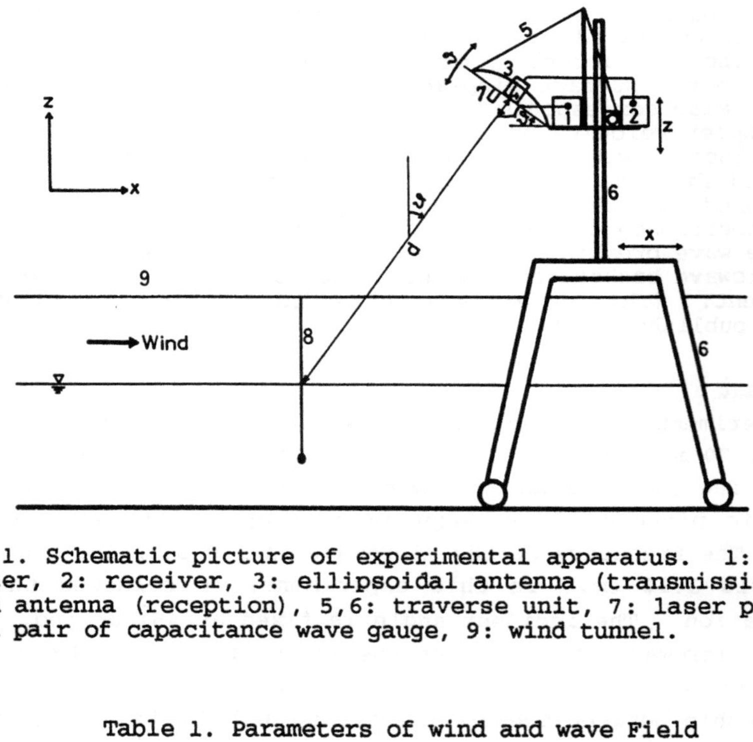

Fig.1. Schematic picture of experimental apparatus. 1: transmitter, 2: receiver, 3: ellipsoidal antenna (transmission), 4: horn antenna (reception), 5,6: traverse unit, 7: laser pointer, 8: a pair of capacitance wave gauge, 9: wind tunnel.

Table 1. Parameters of wind and wave Field

Reference wind speed,	14.1 m/s
Friction velocity,	1.92 m/s
Roughness length,	5.9×10^{-3} m
Significant wave height,	10.4 cm
Significant wave period,	0.71 s
Spectral peak frequency,	1.51 Hz

backscattering intensity relative to the average intensity in a linear coordinate, respectively. The backscattering intensity is large at the crests of individual waves. Panel(c) shows the Doppler spectra calculated from a record for a period of 0.2048 s which is shown by a horizontal bar at the upper right corner of panel(d). The ordinate is the horizontal velocity converted from the Doppler frequency. The curve in panel(d) shows a time series of the Doppler velocity produced from the peaks of the Doppler spectra in panel(c). The Doppler velocity is also in phase with the surface displacement. The circles in panel(d) shows the speed of wave crests measured by the pair of wave gauges. The high velocity observed near the crests in the Doppler spectra coincides with the speed of wave crests.

In order to examine the phase relationship between the surface displacement and the backscattered signature seen in the time series, we calculated phase averages of these data. The time series are divided into 20 blocks along phase of the individual waves. The one-third highest waves are selected and the phase averages are calculated for each of the 20 blocks.

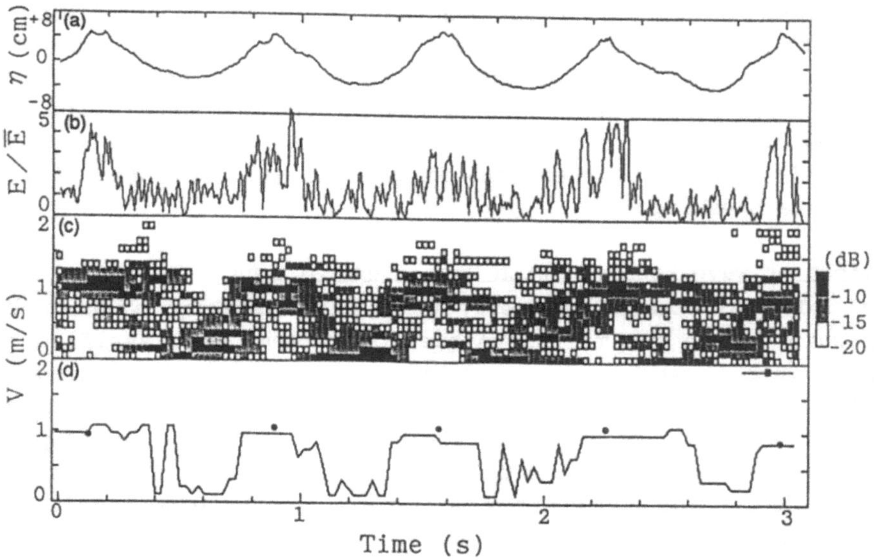

Fig.2. An example of time series of backscattered signature. (a) surface displacement, (b) backscattering intensity, (c) Doppler spectra, (d) Doppler velocity (line) and crest speed (circles).

Figure 3 shows the result. In the upper panel the average of wave profile (η), the average ($\tilde{E}$) and standard deviation (E') of the backscattering intensity are shown. It is confirmed that the backscattering intensity is in phase with the wave profile. The contribution of the crest region to the total power of the backscattering is about 70 % for this case. The lower panel shows the phase average of the Doppler spectra. The Doppler velocity is also in phase with the wave profile. At the crest the Doppler velocity coincides with the phase velocity of waves (1.03 m/s).

In Fig.4, the Doppler velocity at the crests of individual waves is plotted against the crest speed measured by the pair of wave gauges for the all waves observed under winds larger than 10 m/s. The size of the circles show the number of data located at the same point, since the resolutions of the both speeds are not so good. The Doppler velocity agrees well with the crest speed.

From these results, it can be concluded that fine structures of the wind-wave surface, which are trapped near the crests of the dominant waves and which are propagating with the crests, mainly contribute to the microwave backscattering. According to the experimental studies of our research group [1], it is expected that the fine structures near the crests [2] are generated associating with the mechanisms of turbulent boundary layer above and below the water surface such as separation and reattachment of air flow [3], high shear region beneath the crest [4], bursting phenomena in the air [5] and water [6],[7] and wave breaking.

Effect of wave breaking with bubble entrainment

In order to examine the effect of wave breaking with bubble entrainment on the microwave backscattering, we classified the waves into three classes: non-breaking, partially breaking and fully breaking waves, according to the video pictures. In each class, averages of the backscattering intensity and Doppler velocity at the crest are calculated. The result is summarized in Table 2. The backscattering intensities have no significant difference between the classes. The Doppler velocity for fully breaking waves is slightly faster than the others.

From this result, we may conclude that the wave breaking with bubble entrainment has no effect on the intensity of

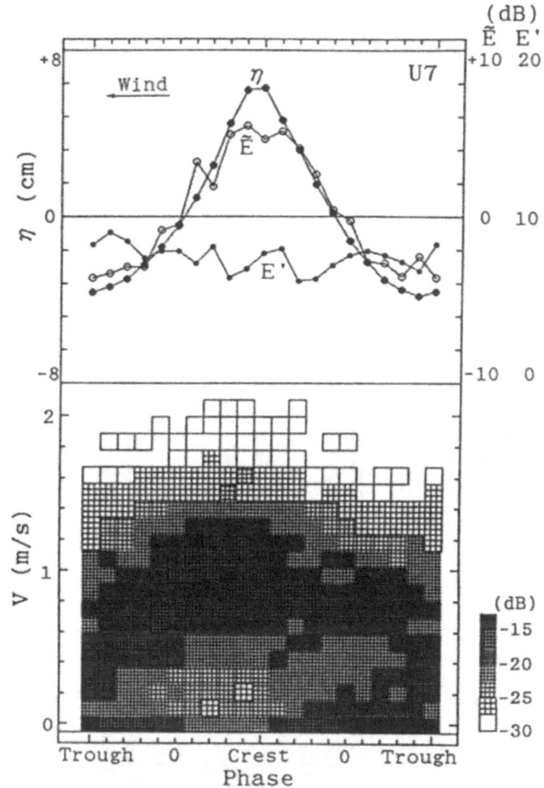

Fig.3. Phase average of backscattered signature (see the text).

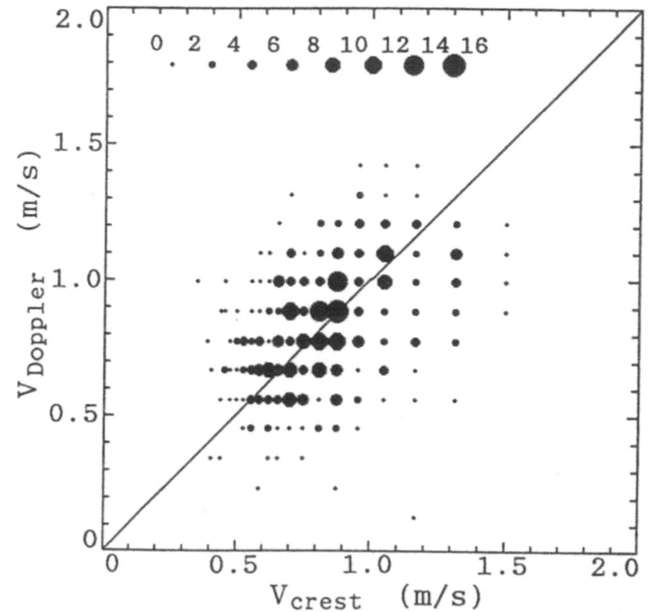

Fig.4. Comparison of the Doppler velocity at the crests of individual waves and the crest speed measured by the pair of wave gauges (see the text).

backscattering for the condition of this experiment. One possible interpretation is that, when the wave breaking with bubble entrainment is occurring, the wind-wave surface itself is already very rough, and that the breaking does not cause additional surface roughness, which could contribute to the microwave backscattering.

Table 2. Effect of wave breaking

	Breaking Wave		
	Non-	Partially	Fully
Number of Waves	18	18	12
Relative Power (dB)	-23.3	-23.1	-23.2
Doppler Velocity (m/s)	1.07	1.08	1.14

References

1. Toba, Y.: Wind waves and turbulence. in *Recent Studies on Turbulent Phenomena*. eds.: Tatsumi, T.; Maruo, H.; Takami, H.: Assoc. for Sci. Doc. Inform., Tokyo (1985) pp 277-296.

2. Ebuchi, N.; Kawamura, H.; Toba, Y.: Fine structure of laboratory wind-wave surfaces studied using an optical method. *Boundary-Layer Meteorol.* **39** (1987) 133-151.

3. Kawai, S.: Structure of air flow separation over wind-wave crests. *Boundary-Layer Meteorol.* **23** (1982) 503-521.

4. Okuda, K.: Internal flow structure of short wind waves. Part I. On the internal vorticity structure. *J. Oceanogr. Soc. Japan* **38** (1982) 28-42.

5. Kawamura, H.; Toba, Y.: Ordered motion in the turbulent boundary layer over wind waves. *J. Fluid Mech.* **197** (1989) 105-138.

6. Yoshikawa, I.; Kawamura, H.; Okuda, K.; Toba, Y.: Turbulent structure in water under laboratory wind waves. *J. Oceanogr. Soc. Japan* **44** (1988) 143-156.

7. Ebuchi, N.; Kawamura, H.; Toba, Y: Bursting phenomena in the turbulent boundary layer beneath the laboratory wind-wave surface. in *Natural Physical Sources of Underwater Sound*. ed.: Kerman, B.: Kluwer Acad. Pub., Dordrecht (1991, in press).

Acoustical Measurement of Breaking Surfaces Waves

DAVID M FARMER & LI DING

Institute of Ocean Sciences, P.O. Box 6000, Sidney, B.C.

Summary

Breaking surface waves can be detected acoustically in the open ocean with a small hydrophone array. We describe the technique used for this purpose and show a few results, including measurement of breaking wave propagation speed and the characteristics of the acoustical radiation.

Introduction

Surface wave breaking has been studied in some detail in the laboratory tank under controlled conditions, but observations in the open ocean remain sparse. The ocean surface is a difficult environment in which to obtain measurements, so that some form of remote sensing appears necessary for continuous observation. The fact that breaking waves radiate sound suggests the suitability of employing passive acoustical detection. Here we report some initial steps towards development of a passive detection sensor suitable for use in the open ocean.

Passive acoustical detection of breaking waves is not new. Time series measurements of ambient sound fluctuations thought to be caused by wave breaking events were identified by Farmer & Vagle [1] using moored WOTAN sensors. Some inferences were made about the repetitive nature of breaking and its dependence on the dispersion relationship (Donelan, Longuet-Higgins and Turner, [2]). Subsequent observations using simultaneous video and acoustical recording confirmed that nearby wave-breaking events create an increase in background sound (Farmer & Vagle [3]). Crowther (personal communication, 1990) has described the use of ambient sound to track the motion of individual events; he exploited the higher frequency radiation (50kHz and above) and employed narrow beam hydrophones that detected the wave as it

M. L. Banner, R. H. J. Grimshaw (Eds.)
Breaking Waves
IUTAM Symposium Sydney/Australia 1991

passed a surface patch towards which the hydrophones were oriented. His observations were made with a bottom mounted system connected by cable to shore.

Technical Approach

The measurements described here were obtained with a self-contained freely drifting instrument conceptually and technically related to earlier versions (Farmer & Vàgle [3], Zedel and Farmer [4]) used for studying ocean surface processes. A primary concern underlying the development of these instruments was the desirability of obtaining data in open ocean conditions without being tied to a particular location or a shallow costal environment. The difficulty and cost of deep ocean moorings favoured a drifting deployment with the instrument suspended by rubber cord some 20 - 40m below a small surface float. Use of a rubber cord and a high drag design for the instrument ensured rather efficient decoupling from the ocean surface. The residual instrument motion, which is monitored with inclinometers, an accelerometer, a pressure sensor and a compass, closely approximates the small orbital motion of the swell at the deployment depth; subsequent analysis includes a correction for this motion.

The instrument is equipped with a variety of sensors, but the passive array pertinent to the present discussion consists of 4 broad band hydrophones, each one mounted at the end of a motor driven arm (*Figure 1*). The horizontal span between opposite hydrophones is 9m; the arms are extended after deployment and retracted prior to recovery. The signal from each hydrophone is digitised at 11kHz (16 bit resolution) and recorded on video tape. The entire system has an endurance of 64h continuous operation, although this may be modified either by use of additional sensors or intermittent operation.

The basic principle employed is that of time delay estimation. Each hydrophone is omnidirectional. With only 4 hydrophones it is hardly useful to 'beam-form' in the traditional sense. The

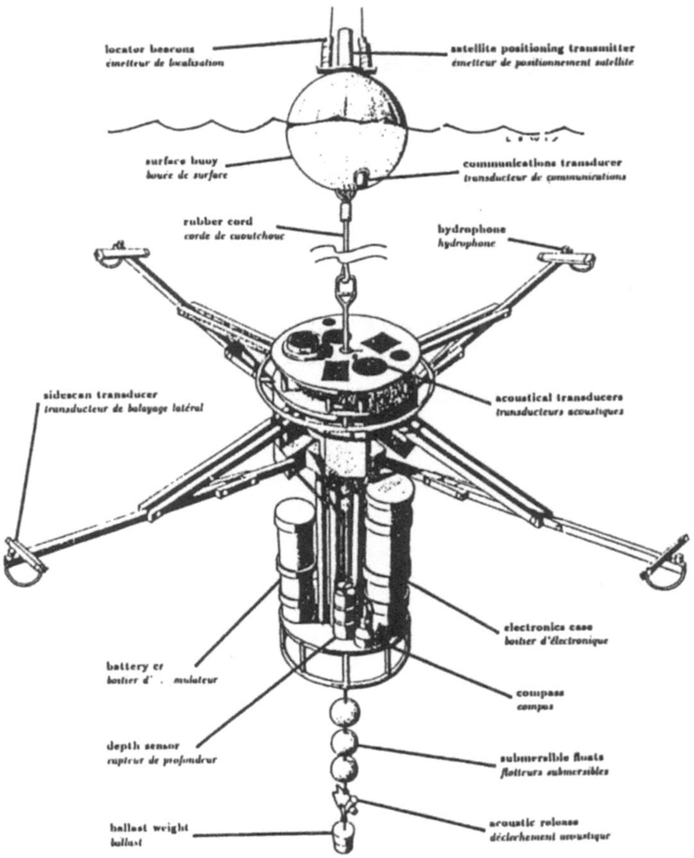

Fig. 1. Sketch showing acoustical instrument used for tracking breaking surface waves with naturally occurring sound. The instrument is suspended by rubber cord from a small (1m diameter) surface float. Hydrophones at the end of each arm detect the sound of individual breaking waves, the positions of which are subsequently found by time delay processing.

cross-correlation between signals detected at each of the 4 locations is determined as a function of time delay. A discrete sound source should then appear as a local maximum in the cross-correlation at a time-delay corresponding to the relative arrival time. A single hydrophone pair locates the source on the intersection of a plane with the ocean surface; a second pair, which need not be orthogonal constrains the source to a similar but different intersection. The common locus then fixes the source position in 2 dimensions, after taking due account of instrument depth and orientation. The calculations are performed at 85.6ms intervals leading to a sequence of measured positions as the breaking wave travels across the ocean surface.

Observations

We discuss an example of signals detected during a 12m s^{-1} wind in the Surface Wave Processes Project (SWAPP) in February 1990. Following cross-correlation between hydrophone pairs, and extraction of the cross-correlation peaks, both speed and position are found from a near-field calculation subject to the correction for instrument depth and orientation.

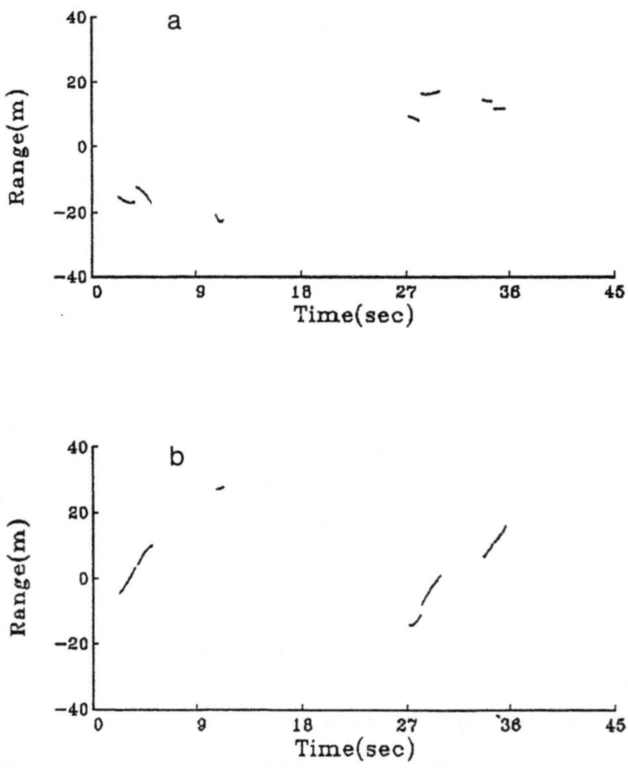

Fig. 2. Horizontal range in two orthogonal directions to breaking waves, found by cross-correlation analysis of acoustical signals from pairs of hydrophones.

Time series locations of breaking wave events for a 45s period are shown in Figure 2 where the vertical axis gives the horizontal ranges of breaking events relative to the hydrophone array in the downwind (b) and crosswind (a) direction. For the most part the breaking waves travel in the direction of the wind. There is some evidence of a grouping of the events. For example, event 2 follows event 1, with a displacement downwind of approximately 7m. Analysis of events 4, 5, 6 and 7 shows similar montonic downwind displacement.

A full analysis of the space-time distribution of breaking events would require knowledge of the directional properties of the wavefield which are not yet available. (Directional spectra were obtained by another investigator.) Nevertheless, it appears that data of this kind have the potential for examination of the dependence of wave breaking on the interaction of wind waves and the swell.

Figure 3a shows the relative acoustic power radiated from each breaking wave event, referenced to a range of 1m directly beneath the source, under the assumption of a dipole radiation pattern. The speed at which the breaking wave travels is shown in *Figure 3b*. For very brief events such as the third one, rapid changes in speed can occur but are probably not very significant. Although the speeds are scattered, they show a general clustering around the group speed of the dominant waves (5.25m s^{-1}).

Surface elevation of the longer period waves is shown in *Figure 3c*, inferred from the pressure signal. The elevation does not include allowance for the residual orbital motion of the instrument, and is probably an understatement of the amplitude. Moreover, wind waves do not show up on the pressure signal at the instrument depth of 24m.

The method of detection depends upon coherent acoustical energy being received at the separate hydrophones. For a broad band nearfield source to produce coherent sound signals at spaced hydrophones requires either that the source be compact or that it radiate coherently. Coherent radiation might occur if the

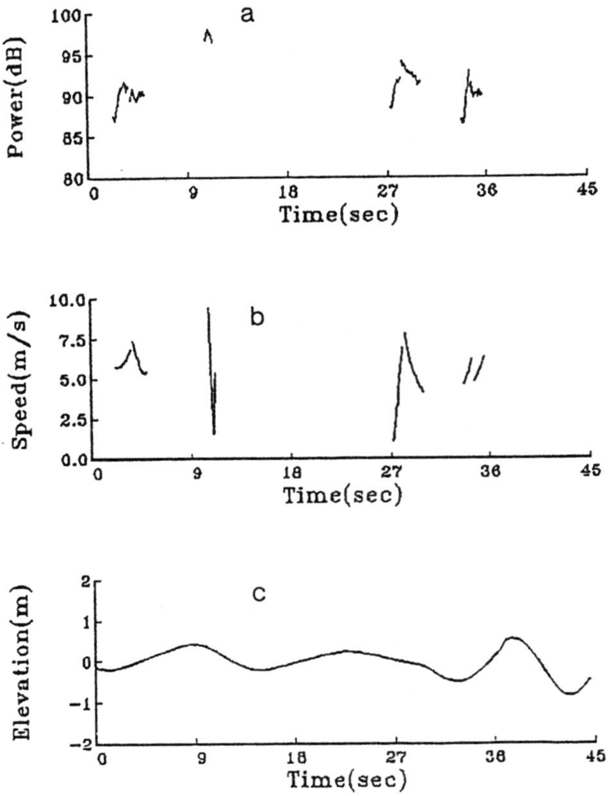

Fig. 3(a) Relative acoustical power, referenced to 1m below each wave breaking event. Fig. 3(b) Speed of breaking events calculated from breaker position data. Fig. 3(c) Surface elevation inferred from pressure sensor.

air-bubbles entrained in the breaking process oscillate collectively. Alternatively, if the breaking event consists of a distribution of incoherent sources, formed for example by the random entrainment and consequent radiation from bubbles at many different locations in the breaking event, then coherence will depend upon source dimensions. Since the breaking wave is typically anisotropic (nearer a line than a point or a disc) we would expect the coherence to be anisotropic also. Thus coherence of the detected signal should serve as an additional probe of breaking wave properties.

Figure 4(a) shows the power spectral density (PSD) for a breaking wave. The PSD is shown relative to the background sound field as measured just prior to the breaking event. Thus the vertical scale is in dB relative to the background power spectrum and shows the departure from this spectrum as the breaking event proceeds. Significant departures occur in the form of a sharp peak at 250Hz.

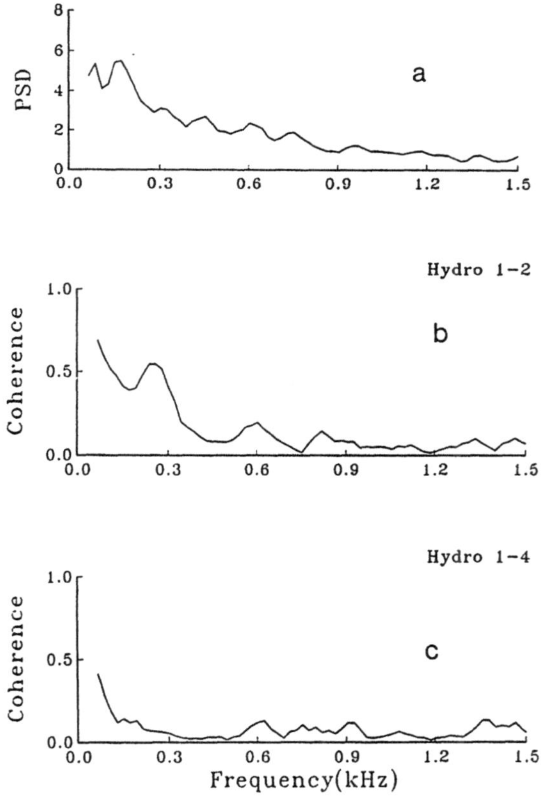

Fig. 4(a) Power Spectral Density (PSD) relative to background noise levels when a wave is breaking close to the instrument.

Fig. 4(b) Coherence between two hydrophones aligned orthogonally to the wave crest.

Fig. 4(c) Coherence between two hydrophones aligned with the wave crest.

The corresponding coherence is shown for the same wave breaking event for two orthogonal hydrophone pairs in *Figures 4b and 4c*. The breaking events are asymmetrical and tend to be aligned with the wave crest. Therefor the linear dimension of the source appears greater when viewed upwind or downwind than when it is viewed cross-wind. The resulting coherence is less when the breaking event is viewed head-on than when it is viewed from the side.

The coherence can be explained in terms of differences in the path length from the source to the two hydrophones. For a source to be compact over a bandwidth the least wavelength of which is λ, and thus to provide coherence between separated hydrophones of the array, the difference in path length from a radiating point to each hydrophone must be independent of location of that point in the source, to within a fraction of λ. Coherence thus depends upon acoustical wavelength λ and the size of the source as observed by the hydrophones.

Discussion

Consider first the well defined peak in the power spectrum at 250Hz. Two possibilities exist. Either the sound is generated by entrainment of individual bubbles resonant at that frequency, or there is a collective oscillation of the bubbles, each of which might individually have much higher resonant frequency. A single bubble oscillating in the breathing mode with resonant frequency 250Hz would have a radius of 1.4cm. Bubbles of this size could probably be formed during the breaking process, although they would be unlikely to last very long.

An alternative explanation is that bubbles in the breaking wave oscillate collectively. Recent laboratory observations (Lamarre and Melville [5] have shown the void fraction β can be 5-20%. Prosperetti [6] gives the resulting sound speed c_m as C^2_m = $p/\rho\beta(1-\beta)$, where p is the ambient pressure and ρ the water density. The lowest eigenfrequency for a void fraction field of linear dimension L is then c_m/L. This corresponds to a bubble layer of 10cm thickness for an eigenfrequency of 250Hz at β =

20%, or 18.4 cm thickness for β = 5%. Coherence plots are shown in *Figures 4b* and *c* for hydrophone pairs orthogonal (*4b*) and aligned (*4c*) with the breaking wave event. As expected, the coherence is generally higher and extends over a greater frequency range for the orthogonal hydrophone pair (*Figure 4b*) than for the aligned pair. If the sound radiation mechanism consists of a distribution of sources, for example randomly phased bubbles or perhaps randomly phased groups of bubbles in collective oscillation, then the upper frequency coherence bound will be determined by the effective width of the incoherent source as viewed by a given hydrophone pair.

In summary, it has been shown that a simple array of 4 omnidirectional hydrophones can be used to map the location and movement of breaking waves in the open ocean. A single event has been analyzed in detail. The spectral and coherence properties of radiated sound allow detection of additional features of the breaking events, including their geometrical evolution. There is also some evidence suggesting collective oscillation of bubbles is a source of radiated sound in the early stages of breaking.

It is worth noting here that the development and testing of a wave breaking model using open ocean data requires simultaneous directional wave measurements at the same location. While directional wave data were obtained from a nearby platform (FLIP) during the SWAPP, experiment, it would clearly be desirable to obtain such data from the same location as the hydrophone array. Recent developments in acoustic doppler measurement of surface wave fields will in future allow such combined observations from the same instrument.

Acknowledgement: This work received support from the U.S. Office Naval Research and the Canadian Panel of Energy Research and Development.

References

1. Farmer, D.M. and S. Vagle, 1988: On the Determination of Breaking Surface Wave Distributions, J. Geophys. Res. 93(C4), pp3591-3600.

2. Donelan, M., M.S. Longuet-Higgins, and J.S. Turner, 1972: Periodicity in Whitecaps, Nature, 239, pp449-451.

3. Farmer, D.M., and S. Vagle, 1989: Waveguide Propagation of Ambient Sound in the Ocean Surface Bubble Layer, J. Acoust. Soc. Am., 85(5), pp1897-1908.

4. Zedel, L., and Farmer, D.M., 1991: Organized Structures in Subsurface Bubble Clouds: Langmuir Circulation in the Open Ocean, J. Geophys. Res. 96(C5), pp8889-8900.

5. Lamarre, E., and W.K. Melville, 1991: Air Entrainment and Dissipation in Breaking Waves, Nature, 351, pp469-472.

6. Prosperetti, A., 1988: Bubble Related Ambient Noise in the Ocean, J. Acoust. Soc. Am., 84, pp1042-1054.

Dependence of Wave-Breaking Statistics on Wind Stress and Wave Development

Kristina B. Katsaros and Serhad S. Ataktürk

Department of Atmospheric Sciences, University of Washington, Seattle, Washington 98195.

Summary

Incidence of wave breaking for pure wind driven waves has been studied on Lake Washington at wind speeds up to 8 m s^{-1}. Video recordings were employed to identify and categorize the breaking events in terms of micro–scale, spilling and plunging breakers. These events were correlated with the magnitude of the wave spectrum measured with a resistance wire wave gauge and band pass filtered between 6 and 10 *Hz*. An equivalent percentage of breaking crests were found for spilling and plunging events.

Wave forcing as measured by wind stress (or friction velocity, u_*, squared) and by inverse wave age, u_*/C_p, where C_p is the phase velocity of the waves at the peak of the frequency spectrum, were found to be good predictors of percentage of breaking crests. When combined in a two parameter regression, those two variables gave small standard deviation and had a high correlation coefficient (66%). The combination of u_*^2 and u_*/C_p can be understood in physical terms. Furthermore, for the larger values of u_*^2 the dependence of wave breaking on wave age was stronger than at the low end of the values for u_*^2 and u_*/C_p. Thus, both the level of wave development as determined by inverse wave age, which we may term *relative wind effectiveness for wave forcing* and the wind forcing on the water surface determine the incidence of wave breaking.

Substituting $U_{10}^{3.75}$ (which is the dependence of whitecap cover found by Monahan and coworkers) an equivalent correlation was found to the prediction by u_*^2. Slightly better standard deviation value and higher correlation coefficient were found by using a Reynolds number as predictor. A two–parameter regression involving u_*^2 and a Reynold's number proposed by Toba and his colleagues (e.g., Toba and Koga, 1986) which relates u_*^2 and peak wave frequency, improves the correlation even more but is less easy to interpret in physical terms.

The equivalent percentage of breaking crests obtained in our previous study (Weissman et al., 1984) was reported at 8.6% for a short record obtained at U_{10N} of about 6 *ms*$^{-1}$. Typical values in the current study for similar conditions are 6%, which is consistent with the previous study in view of the scatter. In that study we did not have a video recording system, so the observed breaking may include more of the micro–scale breaking events, and the value, 8.6%, is well within the range of highly probable sampling variability.

INTRODUCTION

Wave breaking can be a dramatic change in sea surface characteristics and occurs on many scales. It plays a significant role in air–sea exchanges of momentum, energy and mass in various ways (e.g., Donelan, 1990; Banner, 1990). In particular, it is the dominant mechanism responsible for wave dissipation; it removes momentum and energy from the wave field and transfers the momentum to surface currents and the energy into both turbulence and currents. The dissipation of wave energy, due

M. L. Banner, R. H. J. Grimshaw (Eds.)
Breaking Waves
IUTAM Symposium Sydney/Australia 1991

to breaking, is one of the key parameters in wave prediction models which are based on the energy transfer equation:

$$\frac{\partial E}{\partial t} + V \cdot \nabla E = S_{in} + S_{nl} + S_{diss} \qquad (1)$$

where E is the wave spectral energy density, V is velocity at which the wave energy propagates (i.e., group velocity plus currents), S_{in} is source function due to wind input, S_{nl} is source function due to nonlinear wave–wave interactions, and S_{diss} is source term due to dissipation. In the state of the art, third generation ocean wave prediction model, the WAM model (the WAMDI Group, 1988), S_{in} based on field measurements has been adopted from that of Snyder et al. (1981), S_{nl} derived on theoretical grounds has been taken from Hasselmann et al. (1985) and, S_{diss} inferred from the residual term in numerical energy balance experiments has the form proposed by Komen et al. (1984). Despite its importance, S_{diss} is the least known of these three source terms.

Whitecaps, which are related to the wave breaking but also includes the more persistent visible foam patch, have been studied extensively by Monahan and coworkers using statistical analysis of photographs and video recordings (e.g., Monahan, 1968, 1971; Monahan and O'Muircheartaigh, 1980; Monahan et al., 1983; O'Muircheartaigh and Monahan, 1986; O'Muircheartaigh et al., 1991). In these studies areal whitecap coverage has been found to correlate with the 10 m neutral wind speed raised to the power 3.75, $U_{10N}^{3.75}$ (also see, Wu, 1979, 1986).

Because of the increased significance of small scale perturbations on the sea surface for the return of incident radar signals and for microwave emissions, renewed interest in defining wave breaking and relating its occurrence to other measurable quantities has developed (e.g., Banner and Fooks, 1985). Jessup et al. (1990, 1991a) and Bush et al. (1991) have related the *sea spikes* in radar return at X, C and Ku bands to wave breaking observed from video records. The latter studies found that sea spikes often occurred without visible breaking events, but Jessup et al. (1991b) comment that the largest spikes in their study were all directly relatable to readily visible plunging breakers (i.e., dramatically breaking wave crests).

The Bragg scatter is the dominant mechanism responsible for the variations in the normalized radar backscatter cross section, σ_o, as a function of wind speed or stress. It is currently recognized that breaking events may modify this relationship, especially since it is unlikely that occurrence of wave breaking has the same dependence on wind speed or stress as the part of the wave spectrum corresponding to Bragg scattering water waves. It has also been observed that wave breaking is related not only to the wind forcing but that details of the underlying sea state must also be considered. Numerous studies have considered the hydrodynamic instability of the water surface itself, i.e., in situations with no atmospheric forcing at all (see other articles in this volume). On theoretical grounds

Phillips (1988) predicted that contributions to σ_o by breaking waves should be proportional to u_*^3, where u_* is friction velocity defined through $\tau_o/\varrho = u_*^2$. τ_o is wind stress at the sea surface and ϱ is air density. Jessup et al. (1990, 1991b) verified this relationship.

Thus, the variability of processes responsible for wave breaking events may also be responsible for some of the scatter found between measured radar cross sections and wind speed (or wind stress). So far as the relationship to wind stress is concerned, it is quite possible that the intensity of wave breaking affects the wind stress in a manner which more closely parallels the effect on radar backscatter than the relationship between 10 m neutral wind speed, U_{10N}, and σ_o. It is customary to use U_{10N} rather than the true wind speed in this relationship. U_{10N} is convertible to a wind stress by defining a relationship between the neutral drag coefficient, C_{DN}, and U_{10N} viz:

$$\tau_o = \varrho\, C_{DN}\, U_{10N}^2 \,. \tag{2}$$

This relationship represents an average condition if typical values for C_{DN} are used (e.g., Smith, 1980; Liu et al., 1979). For anomalous wave breaking a different C_{DN} value should probably be employed and σ_o may be a very good measure of the effective τ_o.

In this study we extend the work of Weissman et al. (1984), who developed a technique for identifying breaking waves from the record of a resistance wire wave gauge. That study was based on a limited time series and only visual observations of the conditions at the wave wire. In this study video recordings were employed.

A preliminary study on sea spikes and wave breaking by Bush et al. (1991) with data obtained on our Lake Washington site has been carried out. Further work on the relationship between wind and wave parameters, frequency of occurrence of wave breaking and radar cross–sections at X, C and Ku bands will be reported elsewhere in the near future.

EXPERIMENTAL SET–UP AND DATA SETS ANALYZED

This work was carried out at our field station on Lake Washington where wind and waves arrive after an over water fetch of 7 *km* (Figure 1.a). A complete suite of environmental measurements was collected, which includes mean wind speed, temperature and humidity at 2 *m* and 8 *m* heights, turbulent fluxes of momentum, heat and water vapor at 8 *m* and wave height from a resistance wire wave gauge hanging from a boom extending 2 *m* upwind of the supporting platform (Figure 1.b). The wave wire was made of stainless steel, 100 μm in diameter.

For the purpose of this study, 10 data sets were analyzed. The length of each data set is 2 hours. The experimental conditions during these data sets are summarized in Table I where for each run the

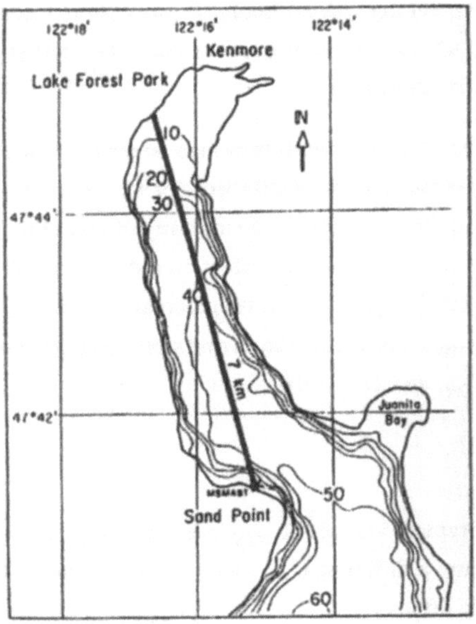

Figure 1a: The location of our field station (MSMAST) on Lake Washington. The tower is 15 *m* offshore and at a depth of 4 *m*. The contours show the water depth in meters.

Figure 1b: The Lake Washington tower with meteorological instrumentation, wire wave gauge (at the left most tip of the low level boom) and the video camera (at the lower section of the platform).

mean values (the first line) and the standard deviations (the second line) of various environmental variables are provided.

ANALYSIS

Weissman et al. (1984) defined three types of breaking events; micro–scale, spilling and plunging. The three types are fairly easily separated. Typical examples are seen in Figure 2. Two independent observers of the video record agreed on the classification of the breaking events into these three categories with insignificant deviations. For the event to be counted in the video record it had to occur within a 0.1 *m* radius of the penetration of the wave wire through the mean water level.

For the wave breaking analysis, wave wire data were processed to obtain a time series of the spectral energy density, E_{6-10}, in the frequency band 6–10 *Hz* with a sampling interval of 1/8 second (see Ataktürk, 1991). In our previous study (Weissman et al., 1984) we had devised a technique to detect the breaking events from such a time series by using the breaking criteria: (i) the spectral energy in a high frequency band exceeds a threshold, and (ii) the data point is in the crest region (i.e., in the vicinity of a local maximum). Since the threshold, i.e., the absolute value of the measured band energy varies with wind speed (stress), gustiness, underlying long waves, currents etc., it had to be determined individually for each run. In the current study, this difficulty was greatly reduced by defining a new detection parameter, N_σ, based on the fluctuating component of band energy normalized by its root–mean–square value;

$$N_\sigma = \frac{E_{6-10} - \overline{E}_{6-10}}{E_{\sigma,6-10}} \tag{3}$$

Table I: General description of the data sets analyzed. Duration of each run is two hours. For each run, the first row indicates the mean values and the second row the standard deviations.

Run#	Date Y/M/D	Time	U_{10N} (m/s)	u_* (m/s)	T_{s-10} (°C)	T_s (°C)	z/L	f_p (Hz)	H_s (m)	ak	$\frac{U_{10N}}{C_p}$
1100	86/8/25	15:32	4.35	0.16	−1.15	23.26	−0.27	0.67	0.13	0.09	1.88
			0.38	0.02	0.22	0.03	0.04	0.04	0.02	0.01	0.23
1101	86/8/25	17:36	4.69	0.17	−1.18	23.13	−0.24	0.70	0.12	0.09	2.09
			0.93	0.04	0.31	0.11	0.10	0.05	0.03	0.01	0.38
1102	86/8/26	11:20	4.21	0.15	−1.83	23.09	−0.21	0.75	0.09	0.08	2.01
			0.34	0.01	0.24	0.12	0.03	0.09	0.01	0.01	0.32
1103	86/8/26	15:00	5.18	0.18	−2.70	23.78	−0.08	0.64	0.13	0.10	2.13
			0.32	0.01	0.42	0.06	0.03	0.03	0.00	0.00	0.20
1104	86/8/27	12:03	3.36	0.12	−2.60	23.26	−0.25	0.70	0.10	0.07	1.51
			0.83	0.03	0.62	0.06	0.21	0.05	0.02	0.01	0.37
1106	86/9/11	18:35	5.84	0.22	4.21	20.50	−0.44	0.55	0.18	0.12	2.05
			1.01	0.04	0.33	0.05	0.29	0.02	0.02	0.01	0.31
1107	86/9/12	14:41	4.31	0.16	1.82	21.04	−0.56	0.66	0.13	0.10	1.81
			0.42	0.02	0.32	0.05	0.13	0.07	0.01	0.01	0.27
1013	87/5/15	12:30	5.85	0.22	3.63	16.29	−0.34	0.55	0.19	0.12	2.05
			0.92	0.04	0.11	0.01	0.16	0.02	0.02	0.01	0.32
1014	87/5/15	14:45	5.23	0.19	2.90	16.41	−0.44	0.59	0.17	0.11	1.96
			1.12	0.05	0.26	0.05	0.25	0.04	0.03	0.01	0.35
89.10	89/7/19	17:05	6.83	0.26	1.09	22.06	−0.28	0.57	0.21	0.13	2.47
			0.55	0.02	0.18	0.18	0.05	0.06	0.03	0.01	0.16

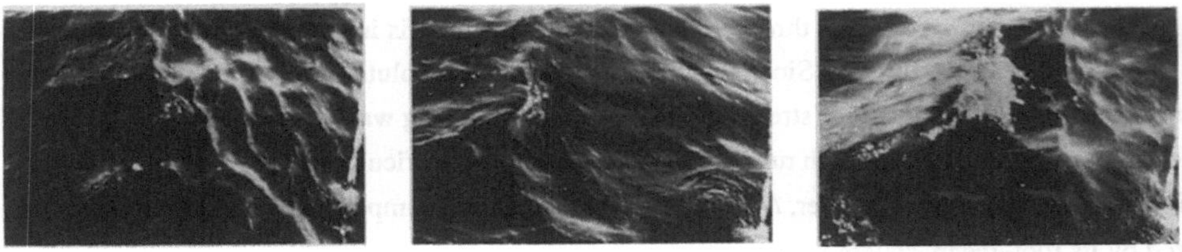

(a) micro–scale (b) spilling (c) plunging

Figure 2: The three types of wave breaking characterized from our video records.

where $\overline{E}_{6-10}$ and $E_{\sigma,6-10}$ respectively, are the mean and standard deviation of E_{6-10} calculated from a record typically 17 minutes in length. (Out of 66 records, two records were short – 8.5 minute in duration. The statistics from these runs were included by simply doubling the number of breaking events.) Also, data points within ±90° phase of the local maximum were considered to be *on the crest*.

Long wave spectra were calculated from the 17 min wave height records. An equivalent number of wave crests, N_{crest}, in the record was defined as the duration, T, times the frequency of the waves at the peak of the spectrum, f_p:

$$N_{crest} = T f_p \ . \qquad (4)$$

From video records, N_{break}, number of crests on which breaking occurred, was found. Percentage of crests with breaking,

$$\%B = \frac{N_{break}}{N_{crest}} \times 100 \qquad (5)$$

was initially calculated separately for the events of plunging, spilling and micro–scale breakers. Since the frequencies of occurrence of spilling and plunging breakers were about the same and the micro–scale breaking turned out to be ubiquitous and not well correlated with any turbulence or sea state measure, presentation of results has been done in terms of the sum of percentages of crests with breaking of type plungers and spillers, $\%B_{s+p}$.

In a separate study, Ataktürk (1991) has analyzed the wind stress measurements of the current study and additional data runs, and obtained the relationship between the drag coefficient and neutral 10 *m* mean wind speed valid for the fetch and water surface conditions at our Lake Washington site;

$$C_{DN} = (0.75 + 0.10U_{10N}) \times 10^{-3} \quad (6)$$

This formula was used with the Liu et al. (1979) formulation for calculating turbulent fluxes from mean atmospheric measurements. The calculation of wind stress or equivalently the friction velocity from mean wind speed with this numerical scheme includes corrections for the influence of atmospheric stratification.

We also determined the inverse wave age, U_{10N}/C_p or u_*/C_p, for each run, where $C_p = g/\omega_p$ is the phase speed of the dominant waves corresponding to the angular frequency, $\omega_p = 2\pi f_p$, of the spectral peak and, g is the acceleration due to gravity.

$\%B_{s+p}$ was correlated with various measures of the atmospheric turbulence and the *relative wind forcing* (identically equivalent to the inverse wave age). Both single and dual parameter linear regressions (Lapin, 1983) were calculated for several variables using the statistical program package *Quattro–Pro*. A curvilinear fit was also tried when it was noted that a linear fit for a certain variable was not optimal.

RESULTS

Figure 3a is a plot of $\%B$ determined from video records and separated into micro–scale, spilling and plunging types, versus U_{10N}. Micro–scale breaking occurs to about the same extent at all wind speeds, while spilling and plunging type breakers display a similar behaviour in frequency of occurrence which increases with increasing U_{10N}. Therefore, further analyses were carried only for the combined effects of spilling and plunging breakers.

Comparisons of simultaneous video and wave wire records showed that, in general, micro–scale breaking is associated with events for which $N_\sigma < 8$ and, spilling or plunging breaking may be characterized by $N_\sigma \geq 8$. Events with $N_\sigma \geq 12$ were always due to plungers. In Figure 3b, percentage of crests with spilling or plunging breaking, $\%B_{s+p}$, determined from video records and percentage of crests with events $N_\sigma \geq 8$ obtained from wave wire records are plotted as a function of U_{10N}. Although the distributions of the data points are similar, some differences in their magnitudes are visible: values from video records are smaller at low winds, and larger at high winds. These differences may result from, respectively, (i) sensitivity of wave wire to small scale features which may not be visible on video records, (ii) insensitivity of wave wire to events that occur nearby but do not directly touch the sensor. The latter may be improved by using an array of wave wires which would allow spatial averaging.

In the studies by Jessup et al. (1990, 1991a) and Bush et al. (1991) some spikes in the radar backscatter were found not to correspond to any visible breaking events. Those events may be carried by the

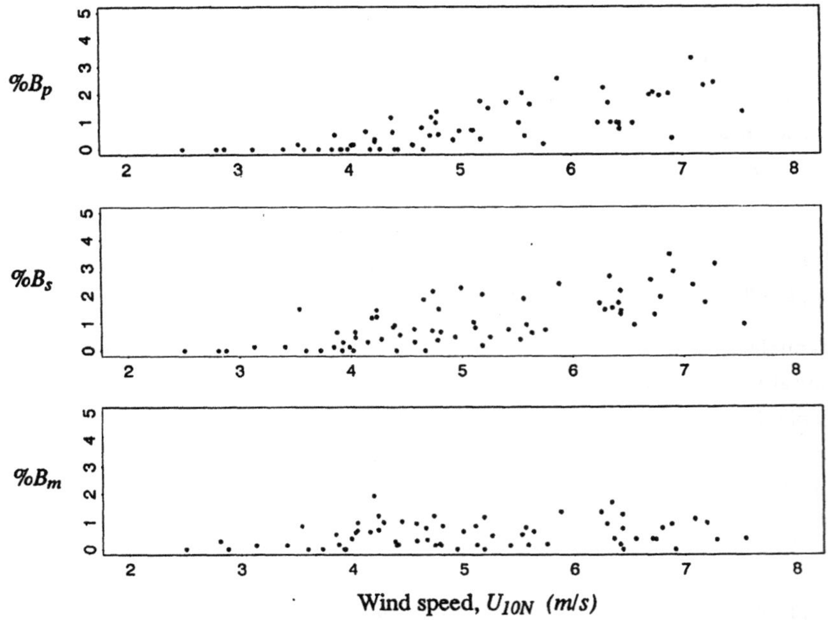

Figure 3a: Percentages of crests with micro–scale, %B_m, spilling, %B_s, or plunging, %B_p, breaking versus wind speed, U_{10N}, determined from video records.

same features (for instance sharp corners) that cause N_σ to be large (but still $N_\sigma < 12$). This indicates that the radars are quite sensitive to such features, even if they do not cover the whole radar footprint.

Some of our $8 < N_\sigma < 12$ events (or perhaps even a few of the $N_\sigma > 12$ cases) correspond to strong Doppler shifts causing the amplitudes of the signals at 6–10 *Hz* to be increased (being that they are really due to surface features of much lower intrinsic frequency. See, for instance, Ataktürk and Katsaros (1987) for a discussion of Doppler shifts in wave wire data.).

Figure 4 shows the dependence of %B_{s+p}, on *Re*, $U_{10}^{3.75}$, u_*/C_p and u_*^2. Reynolds number, *Re*, was defined as (Toba and Koga, 1986):

$$\mathrm{Re} = \frac{u_*^2}{\nu \omega_p} \tag{7}$$

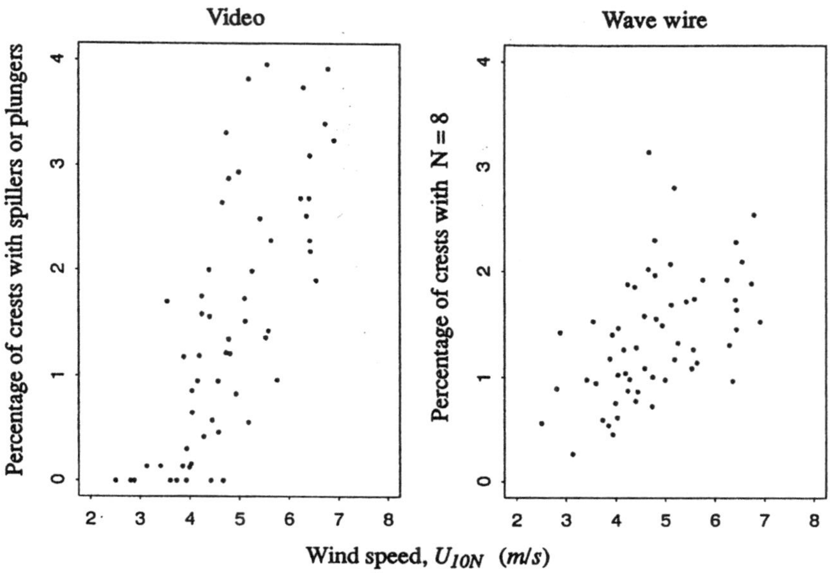

Figure 3b: Statistics of breaking events obtained from video and wave wire records versus wind speed, U_{10N}.

where ν is the kinematic viscosity of air. All these parameters are about equally good predictors of $\%B_{s+p}$. However, from the statistics of the regression analysis provided in Table II it is seen that u_*^2 which explains 62% of the variations in $\%B_{s+p}$, is the best one among the individual variables used. (Although $u_*^{1.6}$ gives a slightly better correlation, for the limited data set the difference is not significant.)

This point is further illustrated in Figure 5 of $\%B_{s+p}$ versus u_*/C_p where the data points have been stratified according to the values of u_*^2. There is obviously a dependence on both the value of u_*^2 and u_*/C_p. Even though the two measures are not independent (since both contain u_*), it makes sense to examine for a certain wave age the variability in the $\%B_{s+p}$ as a function of wind stress (here calculated from U_{10N}). The regression lines show that for strong relative wind forcing (large inverse wave age), the increase in $\%B_{s+p}$ as a function of u_*/C_p is greater for large values of u_*^2. Also, from Table II it is seen that the pair, u_*^2 and u_*/C_p, explains a higher proportion of the variance of $\%B_{s+p}$ than u_*^2 alone with a proportional reduction of 11% in the previously unexplained variation.

Multiple regressions using (u_*^2, Re) or $(u_*^2, Re, u_*/C_p)$ produced still better statistical results by explaining, respectively, 69% and 72% of the variance of $\%B_{s+p}$. However, physical interpretations of

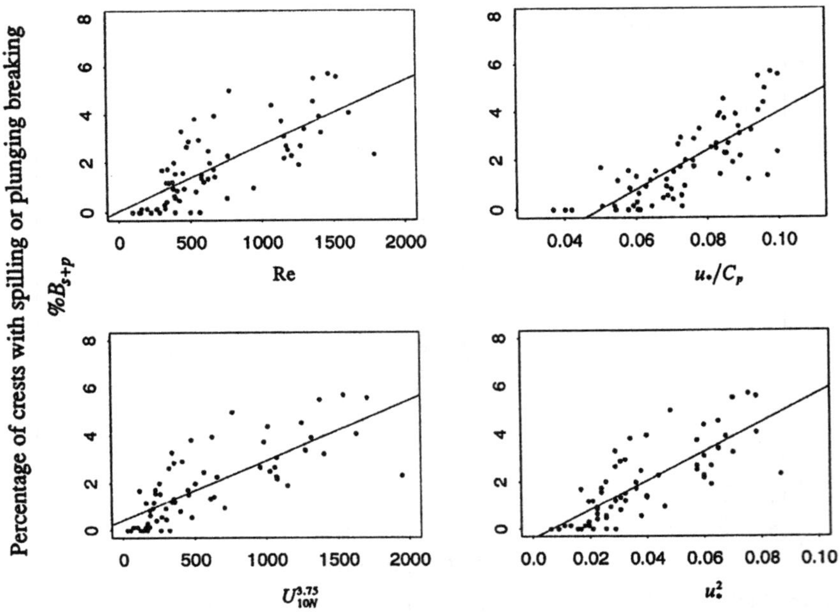

Figure 4: Linear regressions of %B_{s+p} for various predictors. A summary of the statistics obtained from regression analysis is provided in Table II.

Table II: Summary of the statistics obtained from regression analysis. The analyses were carried out following the definitions in Lapin (1983). Square–root of the *3rd* column is the correlation coefficient.

Regression	Standard Error of Estimate	Sample Coefficient of Determination
$\%B_{s+p} = 0.03 + 2.68 \times 10^{-3} Re$	1.07	0.54
$\%B_{s+p} = 0.44 + 2.52 \times 10^{-3} U_{10N}^{3.75}$	1.02	0.58
$\%B_{s+p} = -3.97 + 79.48\, u_*/C_p$	0.99	0.60
$\%B_{s+p} = -0.41 + 61.09\, u_*^2$	0.99	0.62
$\%B_{s+p} = -2.47 + 32.87\, u_*^2 + 42.37 u_*/C_p$	0.93	0.66

such combinations of variables are not as clear to us (but see Toba and Koga (1986)). Also, in view of the large scatter, it does not make sense to try very complicated functional forms until a greater data base has been formed or a convincing physical relationship has been proposed. Therefore, these results are not included in Table II.

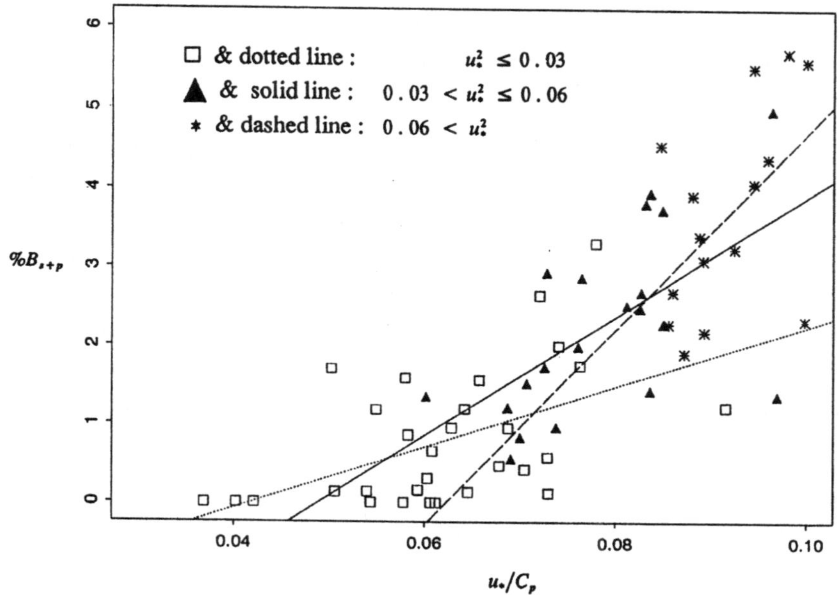

Figure 5: Percentage of crests with spilling or plunging breakers, $\%B_{s+p}$, versus u_*/C_p. For three ranges of u_*^2 separate symbols and linear fits have been used, as indicated on the figure.

DISCUSSIONS

Predictability of the percentage of crests with spilling or plunging breakers, $\%B_{s+p}$, were investigated using a set of variables, Re, $U_{10}^{3.75}$, u_*/C_p and u_*^2. These variables were chosen because such correlations make sense from a physical point of view. (For interpretations of the first two variables, see Toba and Koga (1986) and Wu (1979), respectively). In Table II we note that the highest sample coefficient of determination and the minimum standard error are obtained for the regression of $\%B_{s+p}$ as a function of u_*^2 and u_*/C_p. This result implies that for wave breaking both the magnitude of the wind stress and the degree of evolution of the surface waves are important.

It is easy to understand that young waves which are being strongly forced by a large wind stress should be breaking more frequently. Whether young waves, which are also short (i.e., the wavelength at the peak of the spectrum is short) break more frequently than those that have a wider spectral range through which to distribute the wind input by non–linear interactions – we intend to investigate further. We hypothesize that the absolute width of the wave spectrum (or the value of the wave number at the peak of the spectrum) could be considered as another parameter for improving the prediction of the incidence of wave breaking.

Jessup et al. (1990) suggested that the information content of the frequency of spikes in the radar cross section may also be used to study the variability of wave breaking. We suggest that the band passed data from an array of simple wave gauge wires may serve the same purpose. Since we have seen that $N_\sigma \gg 12$ consistently corresponds to visible breaking in all our wind wave records, this characteristic can now be used to count breaking waves. Enough cases can then be added by objective analysis to establish statistically significant information about the dissipation of surface waves under various forcing conditions as determined by wind stress and relative wind stress forcing. We are currently adding more cases on the breaking of pure wind waves on Lake Washington from two more summer seasons.

Once the interrelationships have been established for the simple case of wind driven waves, the role and importance of swell (of variable amplitude and relative direction to the wind waves) as well as surface currents in wave breaking can then be studied more effectively.

Acknowledgements: We would like to thank Mr. Ralph C. Monis who spent countless hours watching the video records for detection and classification of breaking events and assisted in data processing. This study was supported by the *National Aeronautics and Space Administration* under Grant *NAGW–1322*.

REFERENCES

Ataktürk , S.S., 1991: Characterization of roughness elements on a water surface. Ph.D. Dissertation, Department of Atmospheric Sciences, University of Washington, Seattle, WA, 98195, 196 pp.

Ataktürk , S.S. and K.B. Katsaros, 1987: Intrinsic frequency spectra of short gravity–capillary waves obtained from temporal measurements of wave heights on a lake. *J. Geophys. Res.*, **92**, 5131–5141.

Banner, M.L., 1990: The influence of wave breaking on the surface pressure distribution in wind–wave interactions. *J. Fluid Mech.*, **211**, 463–495.

Banner, M.L. and E.H. Fooks, 1985: On the microwave reflectivity of small–scale breaking water waves. *Proc. R. Soc. Lond.*, **A 211**, 93–109.

Bush, D.A., S.P. Gogineni, R.K. Moore, K.B. Katsaros, and S.S. Ataktürk, 1991: A video–aided study of sea spikes in radar backscatter at moderate incidence. (Manuscript submitted to *IEEE Trans.*)

Hasselmann, S., K. Hasselman, L.H. Allender, and T.P. Barnett, 1985: Computations and parameterizations of the nonlinear energy transfer in a gravity–wave spectrum. Part II: Parameterizations of the nonlinear transfer for applications in wave models. *J. Phys. Oceanogr.*, **15**, 1378–1391.

Jessup, A.T., W.C. Keller, and W.K. Melville, 1990: Measurements of sea spikes in microwave backscatter at moderate incidence. *J. Geophys. Res.*, **95**, 9679–9688.

Jessup, A.T., W.K. Melville, and W.C. Keller, 1991a: Breaking waves producing sea spikes in microwave backscatter. Part II: Dependence on wind and wave conditions. (To appear in *J. Geophys. Res.*)

Jessup, A.T., W.K. Melville, and W.C. Keller, 1991b: Breaking waves producing sea spikes in microwave backscatter. Part I: Detection and visual verification. (To appear in *J. Geophys. Res.*)

Komen, G.J., S. Hasselmann, and K. Hasselmann, 1984: On the existence of a fully developed wind–sea spectrum. *J. Phys. Oceanogr.*, **14**, 1271–1285.

Lapin, L.L., 1983: *Probability and Statistics for Modern Engineering*. PWS Publishers, B/C Engineering Division, Boston, MA, 624 pp.

Liu, W.T., K.B. Katsaros and J.A. Businger, 1979: Bulk parameterization of air–sea exchanges of heat and water vapor including the molecular constraints at the interface. *J. Atmos. Sci.*, **36**, 1722–1735.

Monahan, E.C., 1968: Sea spray as a function of low elevation wind speed. *J. Geophys. Res.*, **73**, 1127–1137.

Monahan, E.C., 1971: Oceanic whitecaps. *J. Phys. Oceanogr.*, **1**, 139–144.

Monahan, E.C. and I.G. O'Muircheartaigh, 1980: Optimal power–law description of oceanic whitecap coverage dependence on wind speed. *J. Phys. Oceanogr.*, **10**, 2094–2099.

Monahan, E.C., D.E. Spiel and K.L. Davidson, 1983: Model of Marine Aerosol Generation via Whitecap and Wave Disruption. *Ninth Conference Aerospace and Aeronautical Meteorol., Am. Meteorol. Soc.*, pp. 147–152.

O'Muircheartaigh, I.G. and E.C. Monahan, 1986: Statistical aspects of the relationship between oceanic whitecap coverage wind speed and other environmental factors. In *Oceanic Whitecaps*, E.C. Monahan and G.M. Niocaill, Eds., D. Reidelberg Publishing Company, pp. 125–128.

O'Muircheartaigh, I.G., M. Claffey, and E.C. Monahan, 1991: Modeling the wind–dependence of whitecapping using hierarchical models and parametric empirical Bayes methodology. Presented in *XX General Assembly of the IUGG*, 11–24 August, 1991, Vienna, Austria.

Phillips, O.M., 1988: Radar returns from the sea surface–Bragg scattering and breaking waves. *J. Phys. Oceanogr.*, **18**, 1065–1074.

Smith, S.D., 1980: Wind stress and heat flux over the ocean in gale–force winds. *J. Phys. Oceanogr.*, **10**, 709–726.

Snyder, R.L., F.W. Dobson, J.A. Elliott, and R.B. Long, 1981: Array measurements of atmospheric pressure fluctuations above surface gravity waves. *J. Fluid Mech.*, **102**, 1–59.

The WAMDI Group; S. Hasselmann, K. Hasselmann, E. Bauer, P.A.E.M. Janssen, G.J. Komen, L. Bertotti, P. Lionello, A. Guillaume, V.C. Cardone, J.A. Greenwood, M. Reistadt, L. Zambresky, and J.A. Ewing, 1988: The WAM model–A third generation ocean wave prediction model. *J. Phys. Oceanogr.*, **18**, 1775–1810.

Toba, Y. and M. Koga, 1986: A parameter describing overall conditions of wave breaking, whitecapping, sea–spray production and wind stress. In *Oceanic Whitecaps*, E.C. Monahan and G.M. Niocaill, Eds., D. Reidelberg Publishing Company, pp. 37–47.

Weissman, M. A., K. B. Katsaros, and S.S. Ataktürk, 1984: Detection of breaking events in a wind–generated wave field. *J. Phys. Oceanogr.*, **14**, 1608–1619.

Wu, J., 1979: Oceanic whitecaps and sea state. *J. Phys. Oceanogr.*, **7**, 1064–1068.

Wu, J., 1986: Whitecaps, bubbles, and spray. In *Oceanic Whitecaps*, E.C. Monahan and G.M. Niocaill, Eds., D. Reidelberg Publishing Company, pp. 113–124.

Effect of Breaking Waves on the Transport of Heat and Vapor Fluxes from the Oceans

S. C. Ling

The Catholic University of America
Washington, D.C. 20064, U.S.A.

ABSTRACT

One way of studying the complex subject of atmospheric surface-layer over the ocean during high-sea states is the consideration of the waves as moving drag objects for the wind. The resultant transport equations for the momentum, temperature, humidity, and micro-water droplets fields are treated as inter-coupled fields. In this study, we have found that the micro droplets produced by breaking waves is a major source of atmospheric moisture and latent heat from the ocean. Field data reduced by the present model gives highly consistent results. During high-sea states, the vertical energy flux from the ocean was found to be more than four times larger than the existing formulation, in which the effect of water droplets was not taken into consideration.

INTRODUCTION

The world's climate is strongly influenced by the interaction of sea and air during high-sea states. The mechanics of the physical process is not yet fully understood due to its complex nature. It is the purpose of this paper to address this problem through the study of micro air-bubbles and water-droplets produced by breaking waves. Our present study has indicated that the micro water-droplets created by breaking waves play a dominant role on the vertical transport of moisture and heat from the ocean to the atmosphere. The evaporation of droplets provides both a major source of water vapor and latent heat to the atmosphere. From data taken during the Joint Air-Sea Interaction Project, JASIN 1978, and from the corresponding analytical modeling of the coupled set of momentum, temperature, humidity, and micro-droplet fields, we have found a simple empirical expression for the vertical heat and moisture fluxes that are much higher than previously expected. This permits one to model the cooling of the ocean along the path of a storm, and to simulate the driving source of energy for a storm.

Based on the measured data, we have now determined that the production rate of micro water-droplets from breaking waves and wind can be best expressed as

$$q_{s2} = 1.2 \times 10^4 U_{10}^2, \qquad \text{for} \quad U_{10} > 3 \text{ m/s}, \tag{1}$$

where q_{s2} is the surface flux of the 20-60 μm size group of water droplets in number of drops/m^{2}s, and U_{10} is the 10 m wind in m/s. The flux for other size groups can be

M. L. Banner, R. H. J. Grimshaw (Eds.)
Breaking Waves
IUTAM Symposium Sydney/Australia 1991

related to this reference group through previously determined universal spectrum of droplet size distribution[1]. More recent measurements of the concentration of micro air-bubbles produced by breaking waves[2] were found to provide an independent confirmation for the result indicated by Eq. 1. Several hundred micro droplets are produced by the popping of each entrained air bubble. It should be noted that both the temperature and humidity fields near the sea surface are strongly dependent on the magnitude of the water-droplet flux. This dependency is due to the large latent heat sink and water vapor source produce by the evaporating water droplets; i.e., an over estimation of Eq. 1 will cause the simulated temperature to be too low and the humidity too high. The constant in Eq. 1 was originally evaluated for the pre-JASIN trial experiment[3]. It was subsequently found to fit equally well without a single exception for all cases measured during the JASIN experiment. With the successful parameterization of the water droplet flux, the analytical model[3] can be used to predict the profiles for the four stated field quantities under any atmospheric and oceanic conditions. From these information, the net vertical fluxes for both water and heat can be determined with confidence.

Twenty sets of selected data collected from August 2nd to September 10th, 1978, in the North Atlantic Ocean at approximately 59°N and 12°W, were analyzed. The corresponding vertical fluxes for water, latent and sensible heat were calculated. The data were evaluated at 6 wave heights above the mean wave trough or 5.5 wave heights above the mean sea level. Except during sea fog and drizzle conditions, most droplets were found to have evaporated above this level. With the successful parameterization of the atmospheric surface layer, the integration of the vertical fluxes over a vast area of the ocean under all sea states is now possible.

RESULTS AND DISCUSSIONS

Two samples of the soundings are presented in Figs. 1 and 2. The analytically simulated profiles[3] are plotted as solid curves, and the experimentally measured data are indicated as solid dots. In the figures, $\overline{z}$ is the vertical distance z above the mean wave trough normalized by the significant wave-height, $\overline{T} = (T - T_o)/|T_w - T_o|$, T the air temperature, T_w sea-surface temperature, $\overline{U}$ wind speed normalized by the 10 m wind U_{10}, H* relative humidity, $\overline{K}$ eddy diffusivity normalized by U_{10} and the significant wave-height, and $\overline{Q}_n$ concentration of the n size group of water droplets Q_n normalized by q_{s2}/U_{10}. The subscript n represents the 1- 5 droplet size groups of 1-20μm, 20-60μm, 60-100μm, 100-180μm, and 180-420μm droplet sizes, respectively. Note that in some of the droplet concentration profiles, the $\overline{Q}_n$ values at the wave trough $\overline{z} = 0$ are beyond the plotting range of the graph; their values are given on the right-hand border of the figure. Also on the right-hand side of each normalized data plot is the corresponding dimensional plot of the dew point and dry bulb temperature soundings. This data is provided to give some physical connection between the upper and sea surface air. The height of the analytical simulation z_o is indicated in each figure. Subscript o indicates conditions at the z_o level. The exact upper air data for most cases were not available, hence they were estimated from various weather data taken near the area and

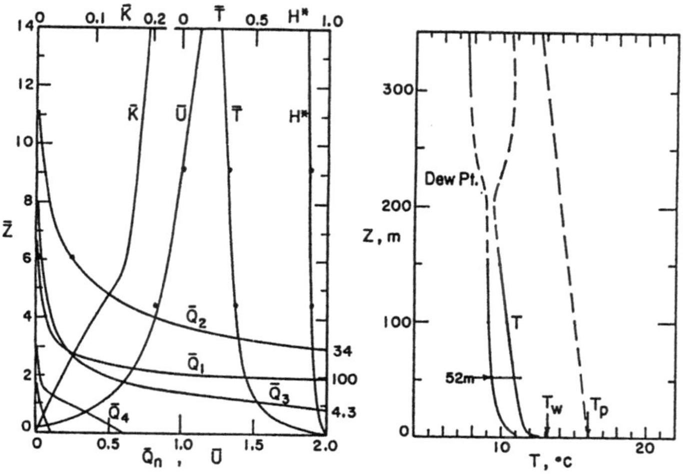

Fig. 1. Sounding August 7th, 18:18; U_{10}=6.3 m/s, 350°; z_o=52 m; T_o=10.8°C; H_o^*= 0.78; T_w =13.2°C; q_w =5.4 x 10^{-5} kg/m^{2}s; h_l=32 cal/m^{2}s; Bowen ratio=0.23 at $\bar{z}$=6.

indicated as dashed curves. The estimated dry adiabatic line for the upper air, at approximately 1000 m level, is shown in each figure. The estimated potential temperature of this air is indicated as T_p. Because of unknown accuracy, it is only used as a rough reference.

Figure 1 shows the typical atmospheric conditions before a storm. There is always a surface layer of air which is cooler than the sea by 2° to 3°C. The air is basically unstable and humid. The rising moist air condenses to form low level clouds varying from 150 m to 220 m in height. Due to the latent heating of condensing cloud layers, the upper air is generally warmer than the sea as indicated by the potential temperature T_p. The most important feature to be noted is that there is a constant unstable cooler surface layer of air sandwiched in between a warmer sea and a warmer upper air. Although the present analysis did not include the mechanics of the upper air, the very physical existence of the cold surface-layer indicated that this is due to the latent heat sink of the evaporating water droplets. There is no other known radiative or convective heat sink that is capable of maintaining such a layer. Hence the unstable cold surface-layer is an efficient means for extracting heat from the ocean even when the upper air is warmer than the sea. This type of enhanced convection is very common over the ocean.

The average vertical water flux during this period was q_w = 5.4x10^{-5} kg/m^{2}s or 0.46 cm/day. The corresponding latent heat-flux was h_l = 32 cal/m^{2}s, and the ratio of sensible to latent heat-flux (the Bowen ratio) evaluated at $\bar{z}$ = 6 was 0.23 indicating a total energy flux of 39 cal/m^{2}s from the ocean.

Figure 2 represents data taken near the eastern edge of the storm's eye. Wind was from the south, 190°. The upper air was dry and warm, and the sky was clear. There

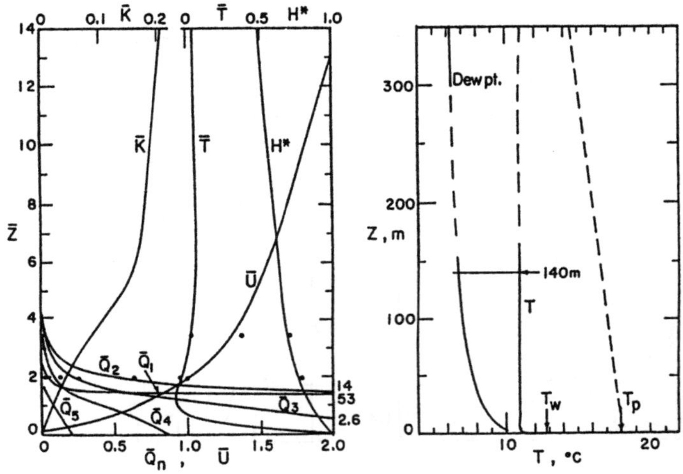

Fig. 2. Sounding August 20th, 15:45; U_{10}=15.6 m/s, 190°; z_0= 140 m; T_0=11.0°C; H_0^* =0.40; T_W =12.8°C; q_W =4.7 x 10^{-4} kg/m^{2}s; h_l=280 cal/m^{2}s; Bowen ratio= -0.2 at $\bar{z}$=6; storm.

was very strong mixing of the upper and surface air. The vertical water flux was q_W = 4.7x10^{-4} kg/m^{2}s or 4.1 cm/day. The corresponding latent heat-flux was h_l = 280 cal/m^{2}s. In this case, the analytical result correlated well with the measured droplet concentration. Note that the present theory gives the correct temperature and humidity profiles without the need for a large artificial increase in the eddy diffusivity for the heat and moisture fields. The stiff temperature profile is clearly due to the strong latent heat-sink produced by the evaporating droplets. The corresponding profiles of sensible heat-flux h_s and latent heat-flux h_l are plotted in Fig. 3. A positive heat-flux indicates heat going up from the ocean, and a negative for heat coming down from the upper air. It is interesting to note that the sensible heat h_s below the wave crest $\bar{z} < 1.2$ is positive, and becomes negative and constant at higher levels. There is a negative flux of 55 cal/m^{2}s from the upper air. At $\bar{z} > 6$ the latent heat-flux h_l is a maximum and constant with height; i.e., all droplets have evaporated at this level. It has a positive value of 280 cal/m^{2}s. Hence, the net heat flux from the ocean is $h_l - h_s$ = 225 cal/m^{2}s. This is the energy flux that cools the ocean and at the same time fuels the storm.

Based on the present analyses of a wide range of sea states, a simple empirical equation for the net vertical water-flux over the ocean can be expressed as

$$q_w = 6.54\times10^{-7}(e_{10s} - e_{10})U_{10}^2 + 5.58\times10^{-7}(e_{ws} - e_{10})U_{10}, \qquad (2)$$

where q_w is the vertical water flux in kg/m^{2}s, e_{10s} the saturated vapor pressure at 10 m in mbar , e_{ws} the saturated vapor pressure at the sea surface, e_{10} the vapor pressure at 10 m, and U_{10} the 10 m wind in m/s. For low sea states the present result is very close

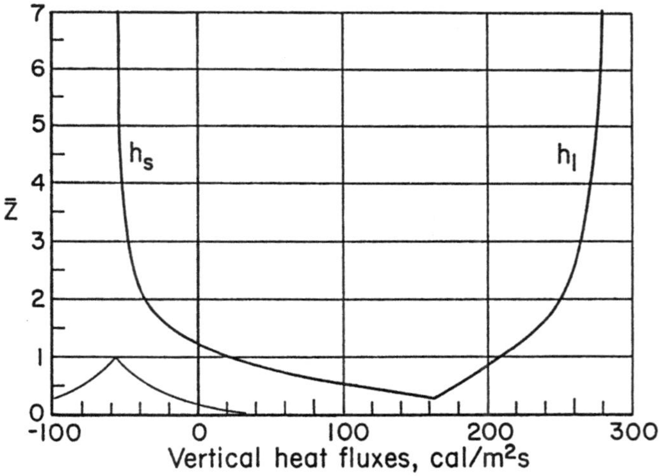

Fig. 3. Profiles of sensible heat h_s and latent heat h_l for soundings shown in Fig. 2.

to those obtained in the AMTEX 1975 experiment[4]. For high sea-states, water droplets from breaking waves greatly increase the evaporation rate.

CONCLUSIONS

What we have clearly proven thus far is the significance of the latent heat-sink caused by the evaporation of micro water-droplets. Failure to take this into consideration is the primary reason for all the previous problems relating the transfer of energy from the ocean to the atmosphere. From this study, it is also clear that the system is too complex to admit simple universal profile functions for any one of the four field quantities. However, it is fortunate that a simple expression for the vertical water-flux does exist, and that it can be easily applied to provide a more realistic simulation for the dynamics of sea-air interaction.

REFERENCES

1. Ling, S. C.,Saad, A. I., and Kao, T. W., 1978, Mechanics of multiphase fluxes over the ocean, Turbulent Fluxes through the Sea Surface, Wave Dynamics and Prediction, Ed. Faver and Hasselmann, Plenum Pub., N.Y., p. 185-197.

2. Ling, S.C.,and Pao, H. P., 1988 Study of micro-bubbles in the North Sea, Sea Surface Sound, Ed. Kerman, Kluwer Academic Publishers B.V., Netherlands, p. 197-210.

3. Ling, S. C., Kao, T. W., and Saad, A. I., 1980, Micro-droplets and transportation of moisture from ocean, J. of Engr. Mech. Div., ASCE, Vol. 106, p. 1327-1339.

4. Francey, R. J. and Garratt, J. R., 1978, Boundary-layer Meteorology, Vol. 14, p. 153-166.

Sound Production and Air Entrainment by Breaking Waves: A Review of Recent Laboratory Experiments

W.K. Melville, M.R. Loewen and Eric Lamarre
R.M. Parsons Laboratory
Department of Civil Engineering
Massachusetts Institute of Technology
Cambridge, MA 02139, U.S.A.

ABSTRACT. We review a series of laboratory experiments on the sound production and air entrainment by breaking waves. The acoustic radiation and microwave backscatter from individual breaking events are found to correlate with the important kinematic and dynamic properties of the breaking waves. Air entrainment measurements show that the bubble plume characteristics evolve as simple functions of time. A large fraction (30% to 50%) of the surface wave energy dissipated is found to be expended in entraining air against buoyancy forces.

1. INTRODUCTION

There has been much interest over the past five years in natural sources of sound at the ocean surface [1]. Several recent papers have reported on direct measurements of the sound generated by breaking waves. Banner and Cato [2] avoided the difficulties associated with a large bubble number density and studied quasi-steady gently breaking waves. Correlating the sound with high speed movies, they found that the dominant noise bursts were associated with the air entrainment at the leading edge of the break, and with the coalescence and splitting of bubbles. In a series of recent papers Medwin and his colleagues [3,4] have reported on laboratory experiments in which breaking results from intrinsic hydrodynamic instabilities of a uniform wave field. On the basis of their hydrophone measurements they have concluded that the laboratory analogue of the Knudsen wind-noise spectrum, including the characteristic 5dB/octave slope, results from the incoherent sum of the noise from individual resonant bubbles. The work of Medwin and his colleagues has clearly shown that for gently breaking waves, the dominant source of sound is associated with the relaxation of the bubble to its spherical shape following generation at the surface.

For more vigorously breaking waves, it is not clear that individual bubble resonances can be simply isolated and identified, nor is it clear that the bubble dynamics are uncoupled. Both Prosperetti [5] and Carey & Browning [6] addressed the role of collective bubble oscillations as the source of sound at frequencies less than 1kHz. The significance of collective bubble oscillations in acoustic radiation from breaking waves is of considerable interest, and evidence of the phenomenon has been observed in the acoustic signals radiated by a bubble column in the laboratory [7].

In our research on the sound production and air entrainment by breaking waves, we have proceeded by first doing the simplest acoustic experiments: measurement of the sound generated by breaking waves and its correlation with the dynamical wave variables. In addition, measurement of the microwave scattering by breaking waves in both the laboratory and the field were conducted. We then went on to directly quantify the air entrainment in breaking through a series of air volume fraction (void-fraction) measurements. In this paper we review these recent laboratory measurements.

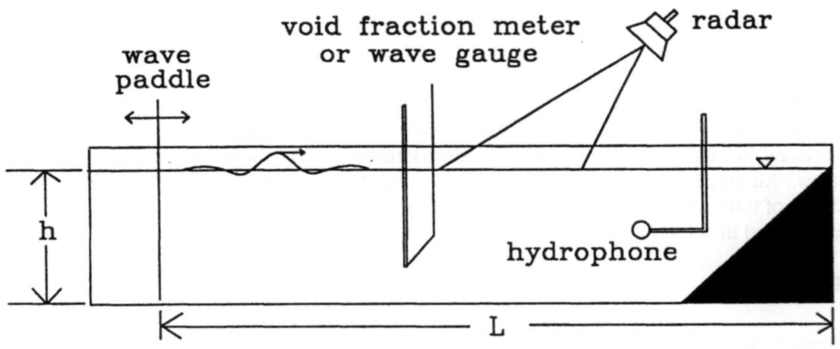

Dimensions

48–422	48–110
L = 21m	L = 22m
h = 0.38m	h = 0.60m
w = 0.38m	w = 0.76m

Figure 1: A schematic of the experimental setup. The channels labeled 48-422 and 48-110 correspond to channel A and B in the text.

2. LABORATORY FACILITIES

The experiments reported here were conducted in two glass-walled steel-framed channels in the R.M. Parsons Laboratory, M.I.T. (figure 1). Breaking wave packets were generated using a chirp pulse, which leads to a focussing of the waves at the break point down the channel. The surface displacement is measured with fine resistance wire wave gauges. These measurements are then used to calculate the surface wave energy dissipated by breaking. Details can be found in Rapp & Melville [8].

Acoustic measurements were made with two omnidirectional hydrophones and the microwave backscatter was measured with an X-band CW Doppler scatterometer. Measurements of the volume fraction of air in water (or void-fraction, given in percent hereafter) were made with a three-wire conductivity probe. Details on the void-fraction instrumentation is given in Lamarre & Melville [9].

3. SURFACE WAVE DISSIPATION AND ACOUSTIC RADIATION

The typical experimental arrangement for this series of experiments in Channel A is shown in figure 1. The primary aim of this series of experiments was to determine whether the acoustic radiation and microwave scattering correlate with the energy dissipated. Figure 2 shows a set of radar time-series at different locations along the channel and a hydrophone time series along with photographs of the breaking wave at times marked on the figure [10]. A number of features are evident from this figure. First, the departure of the acoustic signal above the background noise level marks the impact of the surface (i.e. the onset of breaking). The duration of the hydrophone signal is comparable to the wave period and the sound terminates at the same time as the air entrainment at the leading edge of the breaker.

The simultaneous termination of sound generation and active air entrainment strongly supports the hypothesis that the dominant source of sound is associated with bubble generation. The dominant microwave signal occurs prior to wave breaking. That is, the dominant microwave scattering is not from the whitecap but from the steepening unbroken wave just prior to impact.

We undertook to investigate whether the sound radiated and the microwave energy scattered by the breaking waves correlated with the dynamical variables describing the wave field [10,11]. The primary parameter of interest for the hydrodynamicist is the dissipation due to breaking. The fractional dissipation D is measured by taking difference of the mean square surface displacement upstream and downstream of the break. In all our work we have found that D correlates closely with the characteristic slope of the wave packet S; where S=Nak, and N is the number of Fourier components in the wave packet and (ak) is the slope of each component. Figure 3 shows the correlation between the dimensionless mean square acoustic pressure P_d and D for a range of wave frequencies. The dimensionless mean square acoustic pressure is defined by

$$Pd = \frac{\overline{p^2(x)}}{(\rho c_c^2)^2} \tag{1}$$

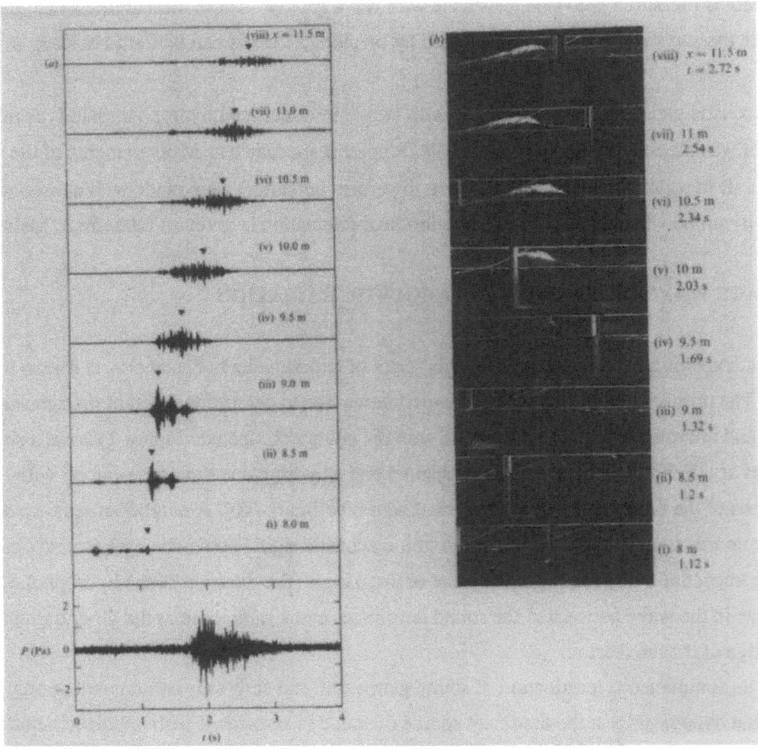

Figure 2: a) Bottom trace is a hydrophone time-series bandpass filtered in the range 500Hz to 10kHz, upper traces are radar time series at various locations down the channel bandpass filtered in the range 50 to 250Hz. b) Photographs of a breaking wave matching the data in figure 2a. Each photograph corresponds to one radar time series and the time the photograph was taken is marked in a) with an arrow above the matching radar time-series. I - x = 8m t = 1.12s, II - x=8.5m t=1.2s, III - x=9.0m t=1.32s, IV - x=9.5m t=1.69s, V - x=10m t=2.03s, VI - x=10.5m t=2.34s, VII - x=11m t=2.54s, VIII - x=11.5m t=2.72s [10].

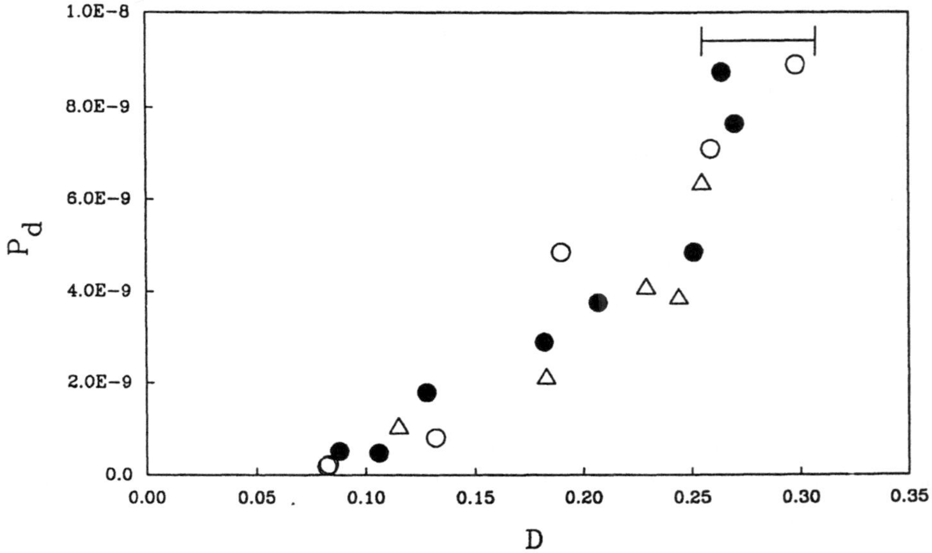

Figure 3: Dimensionless mean square acoustic pressure P_d as a function of the fractional dissipation, D , for three different wave packet center frequencies: O - 0.88Hz, ● - 1.08Hz, Δ - 1.28Hz [10].

where c_c is the characteristic phase speed of the water wave. We conclude from this data that the acoustic energy radiated by breaking waves is approximately proportional to the mechanical energy dissipated.

4. AIR ENTRAINMENT AND DISSIPATION

In view of the fact that the acoustic energy radiated is only a very minor contributor to the total energy budget of the breaking process, it is at first surprising that the sound radiated should correlate well with the wave energy dissipated. However, if the process of air entrainment is a dynamically significant part of breaking then it would be less surprising that the acoustic processes associated with the entrained air (e.g. bubble oscillations) correlated with the wave dynamics.

The initial aim of our experimental program to measure air entrainment by breaking waves has been to measure the evolution of the entrained air and to determine whether the moments of the void-fraction field are simply related to the wave variables [12]. Our earlier measurements of mixing due to breaking suggested that this would be the case [8].

The two-dimensional void-fraction field under breaking waves generated in channel B was mapped out on a grid with a spacing of 5cm (or less). For each location of the void-fraction meter at least three repeats of the experiment were conducted and the time series averaged. These time-series were then used to construct time-dependent two-dimensional maps of the void-fraction field. The signal to noise ratio of the instrument permitted us to measure the void-fraction down to a threshold of 0.3% and to within 3cm of the surface. (As a reference, the characteristic wave length of the wave packet is 2m). For small times after breaking void-fractions in excess of 20% were sustained for up to half a wave period.

Moments of the void-fraction field provide a useful measure of the integral properties of the mixture as they evolve in time. We have computed the total volume of air, V, normalized by the initial volume of air entrained, V_o. The latter was determined from video images of the cylinder of air formed as the surface impacted on itself. The area of the bubble plume, A, within the 0.3% threshold was measured and the mean void-fraction calculated by dividing the total volume of air by the area. Work is done against buoyancy forces by the water in mixing down the air. This work is defined by

$$E_b = \rho g b \int_A \alpha z \, dA \tag{2}$$

where α is the void fraction, z is the distance to the water surface, and ρ, g, b are water density, gravity and channel width. Normalizing E_b by the total energy lost from the wave field during breaking (E_d) gives the fraction of the energy dissipated by breaking which goes into entraining the air. These variables are shown in figure 4 where they have been fitted by simple power law and exponential functions of time normalized by the characteristic period of the surface waves. We see that only a few percent of the initial air entrained remains in the water column after one wave period. Much of the air has been lost due to the larger bubbles rising back to the surface. In the same time the mean void-fraction is reduced from O(10%) to O(1%). The area of the bubble plume increases rapidly at first and then decreases as the bubbles are lost. However, the area of the bubble plume is dependent on the threshold of the void-fraction measurement. We anticipate that as the threshold is reduced the rate of decrease of the area will decline and in the limit as the threshold goes to zero we would expect to see the area of the plume continue to increase as was found in the dye experiments of Rapp & Melville [8]. The conclusion to be drawn from figure 4 is that the lower moments of the void-fraction field are simply related to the prebreaking wave variables and time [12]. Of particular importance is the fact that up to 50% of the energy lost from the surface wave field is expended in entraining the air. The subsequent decrease in the normalized potential energy corresponds to the rise of the larger bubbles back to the surface. Thus the potential energy associated with the bubble position acts as a buoyant source of kinetic energy in the water column. This kinetic energy is ultimately dissipated by viscosity.

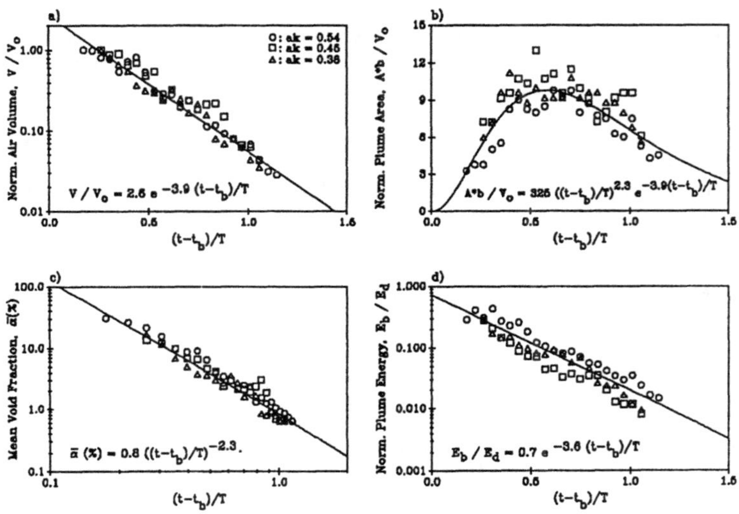

Figure 4: a) Air volume V normalized by V_o. The data for the three waves collapse onto a simple decaying exponential for V/V_o. b) Bubble plume cross-sectional area A normalized by V_o/b. The data was fitted by using the functional form for V/V_o divided by that for $b\bar{\alpha}$. (This procedure was used because extrapolation of the curves for V and $\bar{\alpha}$ is evidently better conditioned than extrapolation of the data for A directly.) c) Mean void-fraction $\bar{\alpha}$. The data is well correlated by a power law with $\bar{\alpha} = 100\%$ near $t = 0$ as one would expect for the initial air pocket. d) Bubble plume potential energy E_b normalized by the total energy dissipated by breaking E_d. Data is fitted with a decaying exponential. Time for all four plots is normalized by the wave period T. The center frequency and bandwidth of the packet were both 0.88Hz [12].

5. DISCUSSION

The work reviewed here presents the beginning of our attempt to investigate the coupling between the hydrodynamics and acoustics of wave breaking. Our finding that the sound radiated by breaking waves increases with the mechanical wave energy dissipated is supported by our subsequent finding that the entrained air is dynamically significant [10,11,12]. For, if the sound is due to oscillations of individual bubbles which break off from the initial volume of air entrained, then we expect that the integral properties of the sound field must be related to the initial volume of air entrained, which is in turn related to the energy dissipated through the buoyancy forces.

ACKNOWLEDGEMENTS

This work was supported by ONR (Surface Reverberation Special Research Program) and NSF (Physical Oceanography).

REFERENCES

1. KERMAN, B.R. (ed.), (1988) Sea Surface Sound, Kluwer Academic Publishers, Dordrecht.
2. BANNER, M.L. & CATO, D.H., (1988) 'Physical Mechanisms of Noise Generation by Breaking Waves - a Laboratory Study', in B.R. Kerman (ed.), Sea Surface Sound, Kluwer Academic Publishers, Dordrecht, 429-436.
3. MEDWIN, H. & BEAKY, M.M. (1989) 'Bubble Sources of the Knudsen Sea Noise Spectra', J. Acoust. Soc. Am., 86, 1124-1130.
4. MEDWIN, H. & DANIEL JR., A.C., (1990) 'Acoustical Measurements of Bubble Production by Spilling Breakers', J. Acoust. Soc. Am., 88, 408-412.
5. PROSPERETTI, A. (1988) 'Bubble Dynamics in Oceanic Ambient Noise', in B.R. Kerman (ed.), Sea Surface Sound, Kluwer Academic Publishers, Dordrecht, 151-171.
6. CAREY, W.M. & BROWNING, D. (1988) 'Low Frequency Ocean Ambient Noise: Measurements and Theory', in B.R. Kerman (ed.), Sea Surface Sound, Kluwer Academic Publishers, Dordrecht, 361-376.
7. YOON, S.W., CRUM, L.A., PROSPERETTI, A. & LU, N.Q. (1991) 'An Investigation of the Collective Oscillations of a Bubble Cloud', J. Acoust. Soc. Am., 89, 700-706.
8. RAPP, R.J., & MELVILLE, W.K. (1990) 'Laboratory Measurements of Deep Water Breaking Waves', Phil. Trans. R. Soc. Lond., A331, 735-800.
9. LAMARRE, E. & MELVILLE, W.K. (1992) 'Instrumentation For the Measurement of Void-fraction in Breaking Waves: Laboratory and Field Results', J. Ocean. Eng., (to be published).
10. LOEWEN, M.R. & MELVILLE, W.K. (1991) 'Microwave Backscatter and Acoustic Radiation from Breaking Waves', J. Fluid Mech., 224, 601-623.
11. MELVILLE, W.K., LOEWEN, M.R., FELIZARDO, F.C., JESSUP, A.T. & BUCKINGHAM, M.J. (1988) 'Acoustic and Microwave Signatures of Breaking Waves', Nature, 336, 54-59.
12. LAMARRE E. & MELVILLE, W.K. (1991) 'Air Entrainment and Dissipation in Breaking Waves', Nature, 351, 469-472.

Wave Breaking in the Presence of Wind Drift and Opposed Swell

H. Mitsuyasu

Research Institute for Applied Mechanics
Kyushu University
Kasuga 816, Japan

Summary
Results of two our recent studies are combined to show interesting phenomena in the ocean surface where the wind waves are co-existing with a swell propagating against the wind. It is shown that the growth of the wind waves is much intensified by swell propagating against the wind, when the swell steepness increases. This phenomenon shows clear contrast to the attenuation of the wind waves by swell propagating in the direction of the wind. Another interesting phenomenon is that the drift current is much intensified by swell propagating against the wind when the swell steepness increases. The increase of the drift current velocity is much larger than the theoretical prediction of the mechanism proposed by Phillips and Banner (1974).

Introduction

Many years ago Mitsuyasu [1] found an interesting phenomenon that swell propagating in the direction of the wind suppresses wind waves considerably if the swell steepness increases. Phillips and Banner [2] obtained a similar result in their experiment and presented an attenuation mechanism of the wind waves by the swell, which explained quite well the experimental result.

Quite recently Mitsuyasu and Yoshida [3] studied experimentally the growth of wind waves co-existing with swell propagating against the wind. A little later Cheng and Mitsuyasu [4] studied experimentally the surface drift current induced by the wind and swell. The results of these two experiments are combined to show very interesting phenomena that both of the wind waves and surface drift current are intensified by the swell propagating against the wind, when the swell steepness increases. According to the Phillips and Banner's mechanism [2], wind waves attenuate even when the swell is propagating

M. L. Banner, R. H. J. Grimshaw (Eds.)
Breaking Waves
IUTAM Symposium Sydney/Australia 1991

against the wind. Particularly the intensification of the surface drift current by the opposed swell should accelerate the breaking of wind waves. Therefore, the phenomena found in these two studies are quite contradictory to the theoretical prediction, although new mechanisms for explaining the phenomena are left for future studies.

Experiments

The following two laboratory experiments have been conducted;

Experiment I (Mitsuyasu and Yoshida [3])

The wind and wind waves in a laboratory tank were measured under the existence of swell propagating against the wind. The laboratory tank used in the experiment is 0.8m high (water depth 0.353m), 0.6m wide and with a usual test section 15m long. The arrangement of equipment is shown schematically in Fig.1. A centrifugal fan (left outside of the figure) sends the wind from the left to the right in the flume. The reference wind speed U_r, which is roughly the same to the cross sectional mean wind speed in the flume, was changed stepwise as U_r=5.0, 7.5, 10.0, 12.5 (m/s). A flap-type wave generator at the downwind side generates a swell (regular

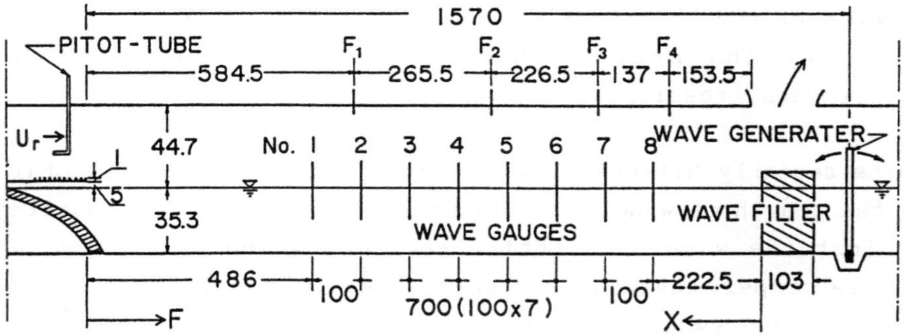

Fig.1. Schematic diagram of wind-wave flume (units in cm).

oscillatory waves). Waves were measured simultaneously at eight stations (No.1-No.8) by using resistance-type wave gauges and recorded on a digital data recorder with a sampling frequency of 200 Hz. Vertical wind profiles over the water surface were measured with Pitot-tubes at fetches, F_1, F_2, F_3 and F_4 in the flume (Fig. 1) to determine the friction velocity of the wind.

Experiment II (Cheng and Mitsuyasu [4])
The wind-induced surface drift current was measured in a laboratory tank under the existence of swell propagating against the wind. The wind and waves were also measured. The laboratory tank used in the Experiment II is the same as that used in the Experiment I. The arrangement of the equipment and the measurements were almost the same as those in the Experiment I except for the following points; the waves were measured only at the stations No.3, No.4 and No.5. The wind profile over the water surface was measured only at the station No.4. The wind speed used in the experiment was U_r = 3, 5, 7 (m/s). The surface drift current was determined by measuring, with a digital recording stopwatch, the time required for waxed paper punchings (diameter 0.5 cm) to travel a distance 2m between the stations No.3 and No.5.

More detailed description of the both experiments are referred to the previous two papers, [3] and [4].

Results

The effect of the opposed swell on the growth of wind waves.
Figure 2 shows some examples of the wave spectra of co-existent system of wind waves and a swell propagating against the wind. For comparison the spectra of pure wind waves measured under almost the same conditions are shown in the same figure. It can be seen clearly that the growth of the wind wave spectra are intensified by the opposed swell. Such

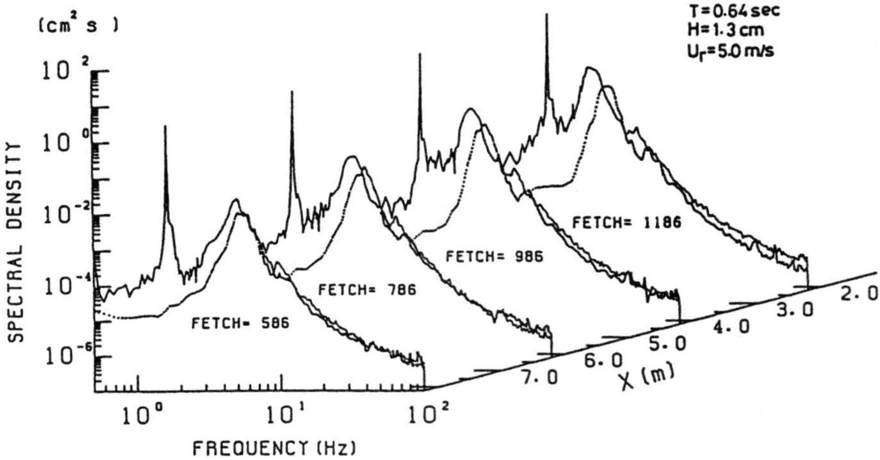

Fig.2. Comparison of the wave spectra of the co-existent system (solid line) and those of pure wind waves (dotted line). swell, T=0.64sec, H=1.3cm; wind speed, U_r=5m/s.

phenomenon is much different from the phenomenon observed when the swell is propagating in the direction of the wind. In the latter case wind waves are suppressed by the swell if the swell steepness increases [1] and [2]. In order to show the change of the wind wave energy quantitatively, the ratio of wind wave energy of the co-existent system to that of the pure wind waves is shown in Fig.3 as a function of wave steepness of the swell H_1/λ, where H_1 and λ are the wave height and length of the swell under the action of the wind.

For comparison, Fig.3 also shows, in the lower part, the results of our previous study [1] where the swell is propagating in the direction of the wind. Although the present data scatter considerably, the relative wind wave energy is not much affected by the swell when the swell steepness is small, but it increases with the increase of the swell steepness. The present results, in association with the previous results [1] and [2], show that the effects of the swell on the growth of wind waves are much different depending

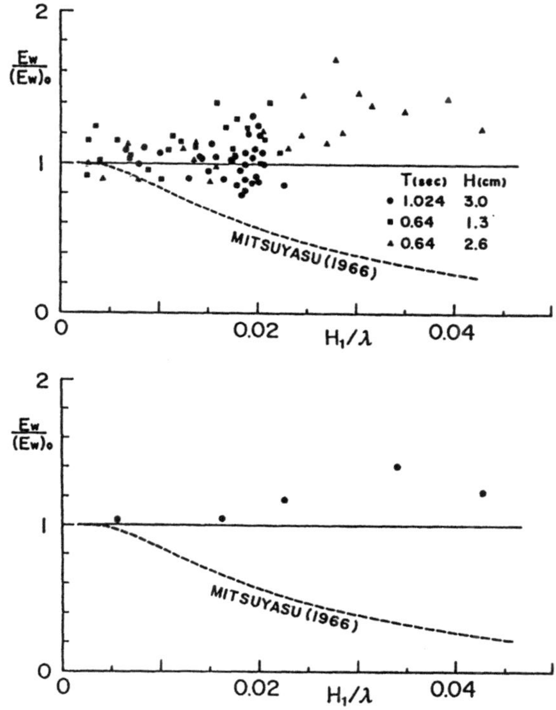

Fig.3. The ratio of the energy of wind waves co-existing with the opposed swell, Ew to that of the pure wind waves, (Ew)o. (a) original data, (b) sectional mean values. The previous result for the swell propagating in the direction of the wind [1] is shown with a broken line.

on the direction of the swell relative to the wind waves. Wind waves are suppressed by the swell of the same propagation direction, while the wind waves are amplified by the swell of the opposite propagation direction.

The effect of the opposed swell on the drift current.

Figure 4 shows the effect of the opposed swell on the surface drift current. In fig.4 the drift current velocity u_s normalized by the wind friction velocity u_* is plotted against the swell steepness H_0/λ by taking the reference wind speed

U_r as a parameter. Here H_0 is the wave height of swell without wind action. It can be seen from Fig.4 that u_s/u_* decreases with the increase of the wind speed and increases with the increase of the swell steepness though it tends to saturate for very large swell steepness.

In addition the experiment for the opposed swell, similar experiment has been made for the swell in the same direction of the wind. In this experiment, however, the change of the surface drift current by the swell was much smaller than that for the case of the opposed swell.

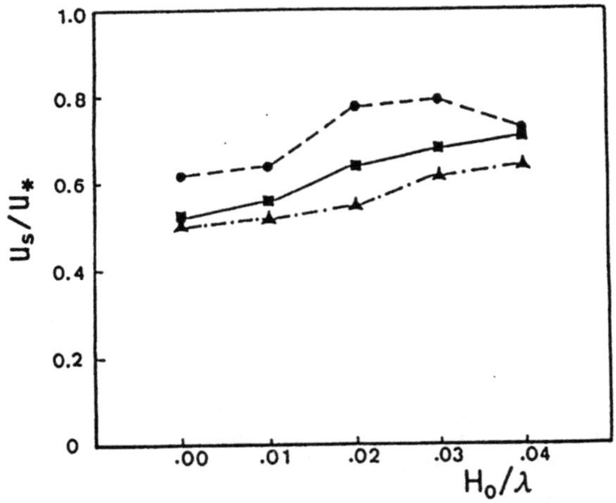

Fig.4. Dimensionless surface drift speed versus the steepness of the swell propagating against the wind. ●, U_r=3m/s; ■, U_r=5m/s; ▲, U_r=7m/s.

Discussions

Phillips and Banner [2] presented a theoretical model on the change of wind drift by the swell in relation to the wave breaking in the presence of surface drift current. According to their theory, when the swell is propagating against the wind, The distribution of the surface drift velocity u_s with respect to the phase of the swell is given by

$$u_s = -(C-u) + \{(C-u)^2 + u_{so}(2C + u_{so})\}^{1/2} \quad (1)$$

where C is the phase velocity of the swell, u is the orbital velocity of the swell and u_{so} is the surface drift current without the swell. This equation (1) shows the augmentation of the surface drift at the crest of the swell. However, the drift current estimated by (1) for the limiting case of our present experiment, i.e., for the maximum wind speed, U_r= 7m/s, and for the maximum swell steepness H_0/λ = 0.04, shows that the increase of the drift current by the opposed swell is at most 2% of the drift current unaffected by the swell. Therefore, the large increase of the surface drift current by the opposed swell can not be explained by the theoretical model. Another contradiction to the theory is that, according to the theory, the large augmentation of the drift current by the opposed swell should accelerate the breaking of the wind waves and attenuates them, while the actual phenomenon show an increased growth of wind waves. Further studies are needed to clarify the augmentation of the drift current and wind waves by the opposed swell.

References

1. Mitsuyasu, H.: Interaction between water waves and wind (1). Rep. Res. Inst. Appl. Mech., Kyushu Univ., 14, 1966, 67-88.
2. Phillips, O.M.; Banner, M.L.: Wave breaking in the presence of wind drift and swell. J. Fluid Mech., 66, 1974, 625-640.
3. Mitsuyasu, H.; Yoshida, Y.: The effect of swell on the growth of wind waves. Proc. JECSS V Workshop 1989, (in Press, Elsevier).
4. Cheng, Z.; Mitsuyasu, H.: Laboratory studies on the surface drift current induced by the wind and swell. (Submitted to J. Fluid Mech.).

The Sideband Instability and the Onset of Wave Breaking

J.S.Reid
CSIRO
Division of Oceanography
Hobart
Tasmania

Abstract
Packets of surface gravity waves of constant frequency and similar raised cosine envelopes were generated with a variety of amplitudes by means of a hydraulic paddle. The behaviour of each packet was observed with two capacitance probes, one near the paddle and the other some distance along the tank. In the time domain each packet was observed to evolve into a single envelope soliton and a "tail" as it travelled down the tank. In the frequency domain the the initial spectral peak was observed to split into two peaks at frequencies on either side of the initial frequency in accordance with the well known "sideband" instability. In addition, for packets in excess of a particular threshold, wave breaking also occurred. The sideband instability was observed at amplitudes both above and below the threshold of breaking. At amplitudes below the threshold, energy was partitioned almost equally between the upper and lower sidebands but with the onset of breaking a smaller fraction of the initial energy was transferred to the upper sideband. The spectral behaviour of these evolving wave packets shows a number of features in common with that of wind waves observed in the open sea.

Introduction

Much of the work on instabilities of surface gravity waves has ignored the phenomenon of wave breaking. The aim of this study was to set up a very simple tank experiment involving a nonlinear instability so as to examine the effect of breaking on the instability. Rather than set up a continuous periodic wave train, only isolated packets of waves were used in the belief that periodicity introduces an unnecessary constraint on a dynamical system, one not commonly encountered in nature.

Experimental Setup

These experiments were conducted in the wave tank at the Australian Maritime College in Launceston Tasmania. The tank was 60m long, 3.5m wide and 1.5m deep. At one end was a hydraulically powered paddle hinged at the bottom of the tank

M. L. Banner, R. H. J. Grimshaw (Eds.)
Breaking Waves
IUTAM Symposium Sydney/Australia 1991

and driven by the output signal from a digital to analogue converter. At the other end was a sloping beach to help suppress reflections. Two capacitance probes, (termed "near" and "far"), were set up in the tank, at distances of 4.4m and 34.4m from the paddle respectively.

The computer driving the digital to analogue converter was programmed to generate a wave packet, V(t), of the form

$$V(t) \quad = \quad 0.5\ [\ 1 - \cos(\omega_e t)]\ \sin(\omega_1 t) \qquad (1)$$

in the range $0 \le t \le t_e$ where $t_e = 2\pi/\omega_e$ was the period of the modulating envelope and $t_1 = 2\pi/\omega_1$ was the period of the "carrier" waves themselves. For all of the trials t_e was set to 15 seconds and t_1 to one second. In practice (1) was programmed so as to maximize the range of the digital to analogue converter and the amplitude of the analogue signal was controlled seperately by means of a graduated potentiometer. The experiment entailed passing the same waveform to the wave maker for different "gain" settings of the potentiometer and

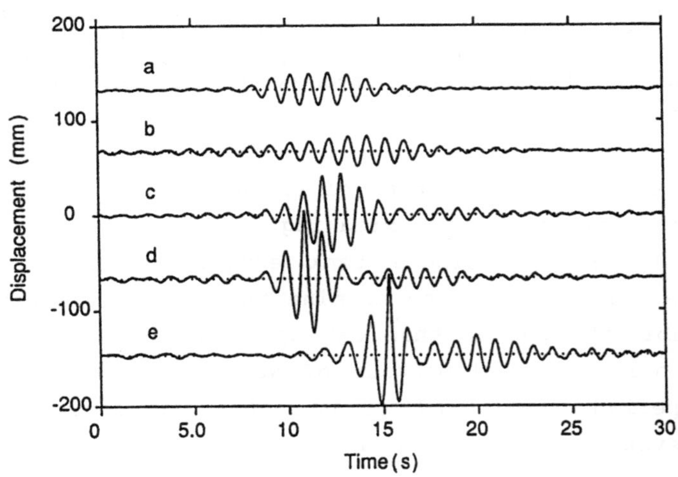

Fig.1. Time series of surface displacement:
(a) near probe, ak = 0.07, (b) far probe, ak = 0.13,
(c) far probe, ak = 0.18, (d) far probe, ak = 0.21,
(e) far probe, ak = 0.24.

recording the surface height as a function of time as the wave disturbance passed each probe.

Results

Examples of time series recorded by the probes are shown plotted in Figure 1. For a small initial amplitude (b) the packet becomes more dispersed while for greater amplitudes (c,d,e) the disturbance forms itself into an envelope soliton followed by a dispersed tail. There is no obvious difference between (d) and (e) despite the fact that spontaneous wave breaking was observed to occur in case (e) but not in case (d).

The variance spectra of the same five time series are shown in Figure 2. The effect of the sideband instability can be clearly seen in cases (d) and (e) whereby the initial carrier frequency has been replaced by two or more sidebands. There is a noticeable difference between (d) and (e) in that there is proportionately less variance in the upper sidebands in the case, (e), for which spontaneous wave breaking had occurred.

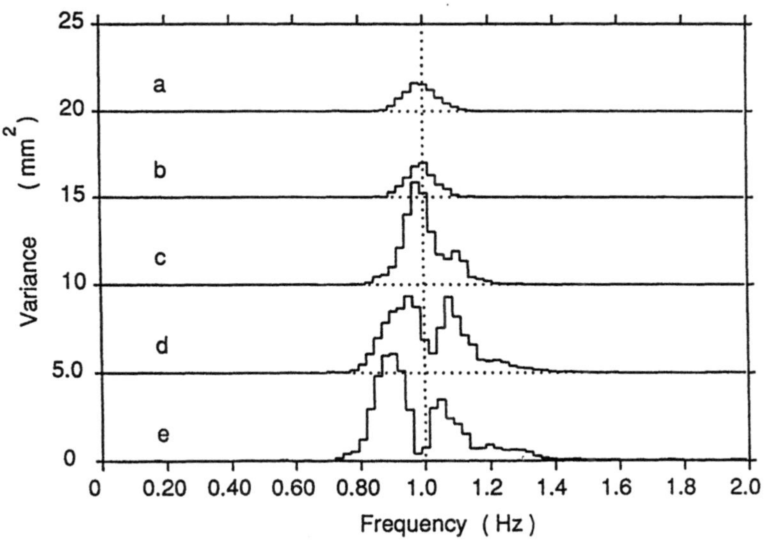

Fig.2. Variance spectra of the time series displayed in Fig.1. The zeroes are displaced for display purposes.

The results for nine different runs with varying initial steepnesses are summarised in Figure 3. A nominal initial steepness, ak, was defined as $a_m \omega_1^2 / g$, where a_m was the maximum excursion of the water surface measured at the first probe. The arrow indicates the interval in which spontaneous breaking commences when the initial steepness is gradually increased. Breaking was observed during all of the runs plotted to the right of the arrow. The figure shows that there was a nett loss in variance with increasing initial steepness and that this variance loss was due to losses from the upper sideband in the range of steepnesses for which breaking occurred.

Conclusions

To first order the variance of surface height is proportional to the energy density of the wave disturbance and the missing variance can be accounted for in terms of the transfer of energy by wave breaking from the wave field into turbulence.

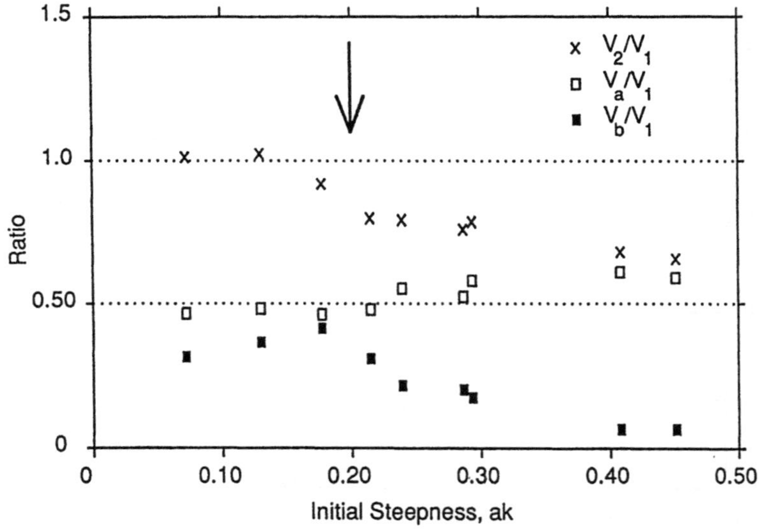

Fig.3. Ratios of far probe variances to near probe variance, V_1. V_2 is the total far probe variance, V_a is the far probe variance summed over frequencies less than 1 Hz and V_b is the far probe variance summed over frequencies greater than 1 Hz.

The preferential removal of energy from the upper sidebands by this process results in a nett downward shift in the average frequency of the spectrum.

Frequency downshifting is a common phenomenon of wind seas in the open ocean. Figure 4 shows spectra (a) and (e) of Figure 2 replotted using logarithmic scales. The upper spectrum (e) shows some other features in common with wind sea spectra, viz a steep low frequency roll-off, a power law high frequency roll-off and a peak enhancement above the power law trend. This suggests that the broad features of wind sea spectra may be accounted for in terms of the sideband instability or some similar process coupled with wave breaking. The fourth order wave-wave interactions usually invoked to describe open sea spectra are certainly not applicable to the present experiment because of its two dimensional nature.

Acknowledgement

The author wishes to thank Dr Martin Renilson and the Australian Maritime College for use of the wave tank.

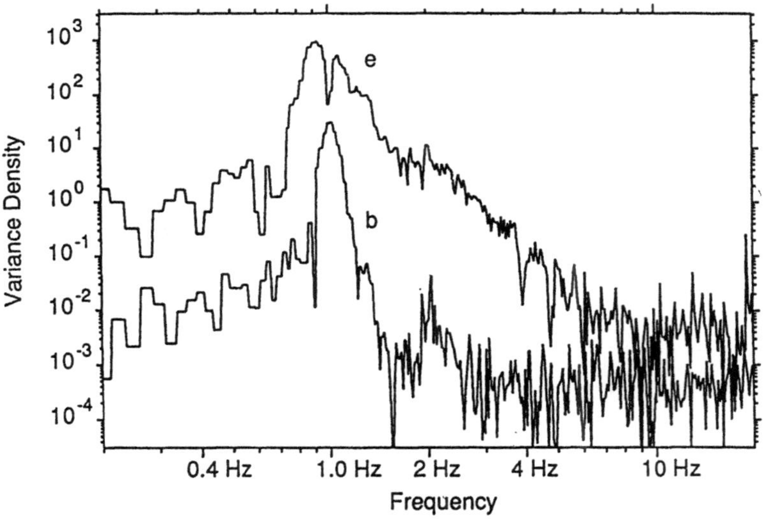

Fig. 4. The variance spectra (b) and (e) of Fig. 2 replotted with logarithmic scales. The upper curve, (e), is raised by a factor of ten for display purposes.

Breaking Wave Statistics Obtained During 'Swade'

MING-YANG SU
Naval Oceanographic and Atmospheric Research Laboratory
Stennis Space Center, Mississippi, U.S.A. 39529-5004

JOHN CARTMILL
Planning Systems Incorporated
Slidell, Louisiana, U.S.A. 70458

Summary

Void fraction meters are very simple, but effective automatic sensors for measuring air entrainment due to wave breaking. The maximum void fraction value measured during SWADE (Weller, et. al, 1991) with the wind speed up to 15 m/s and gusts to 18 m/s is about 60%. The duration for the void fraction event is about, on average, 0.5 sec at the depth of 12" for the average wave period of about 5.5 sec. Much less void fraction is observed below 24". The void fraction statistics turn out to be a very logical and quantitative way for describing the breaking wave statistics. All of the previous five published field measurements of breaking wave statistics, using different techniques, can be encompassed nicely by the present void fraction statistics in a more meaningful and orderly way. The frequency, or number fraction, of breaking waves levels off as the wind speed increases.

Introduction

One of the most obvious and characteristic features of wave breaking in the open sea is the entrainment of air by the plunging of, and/or creeping near, the crest of a steep wave, which generates in turn, a series of bubble plumes beneath and behind the breaking wave. Our interest in this investigation is concerned equally with the frequency of occurrence of breaking waves and the amount of air entrained due to such wave breaking in the moderate to high sea states. To accomplish simultaneously these dual, but related, objectives of our research interest, we decided to detect the presence of breaking waves by direct measurement of near surface void fraction, which is expressed as percentage of air content in the water, by means of a floating vertical array tethered to a R/V, on which several void fraction meters are installed at various depths. A very brief summary of the results of this field measurement on both the void fraction statistics and breaking wave frequency statistics under moderate to strong wind are presented here.

M. L. Banner, R. H. J. Grimshaw (Eds.)
Breaking Waves
IUTAM Symposium Sydney/Australia 1991

Results and Comparison

The cumulative average in percentages for the four cases, A-D, during SWADE are plotted in Figure 1 with the five ranges of void fraction denoted by letters *a, b, c, d,* and *e,* respectively, and with the mean wind speeds at 10 and 15 m/s.

Each "void fraction event" may be interpreted as one "breaking wave event", if we define a breaking wave to be a wave which entrains air into the water that becomes a plume of bubbles with a measurable void fraction to a depth of 12" below the surface. This definition will preclude all the micro-breaking by the capillary waves and the gravity-capillary waves with wave length shorter than about 2 m. We shall next compare our field measurements with several published field measurements on breaking wave statistics by various investigators employing different techniques and definitions on what constitutes a "breaking wave".

In the paper by Holthuijsen and Herbers (1986), the observed occurrence percentage (fraction) of breaking waves from five different field measurements as a function of wind speed (U_{10}) are compared. Their Figure 7 is reproduced here as Figure 1 for comparison with our results.

Toba et. al (1971) and Holthuijsen and Herbers (1986) use the similar technique of coupling visual identification of white capping with wave height measurement. Even though this method might be the most positive way to say a wave is in the process of breaking, some degree of subjective judgement is still involved, particularly when the breaking area is small. Thorpe and Humphries (1980) use telephotography in place of direct visual observation. This method could miss counting smaller breaking waves in contrast to the former method. As such, the observed number of breaking waves should be less in the latter case, as exhibited in Figure 1. Weissman et. al (1984) employed the method of detecting the presence of high frequency components (5-50 Hz) in the waves as the definition of breaking waves. Their measurement at a low wind speed of about 6 m/s involved only very small breaking waves, which should agree better with those by direct visual observation (such as Toba, et. al). Longuet-Higgins and Smith (1983) employed a "jump meter" to measure the slope of large and/or dominant waves near a North Sea tower. Waves with the steep slope beyond some chosen threshold values are then considered to be breaking waves. Since the "jump meter" technique is designed for detecting larger waves, many small breaking waves are ignored. It is reasonable to suggest that small breaking waves occur more often than large breaking waves. Hence, Longuet-Higgins and Smith's count would be much lower than all of the other methods described earlier.

It is very interesting to note that curves *a* and *e* encompass all the earlier data points, and curves *b*, *c*, and *d* provide some dividing boundaries to data points from different techniques. The fact that *e* and *c* with 1% > VF > 0.30% agrees more with Toba et. al (1971) might imply that Toba and his colleagues have been more careful than the others to pick up many more smaller breaking waves. The observed data by Holthquijsen and Herbers (1986) are bounded nicely by curves *d* and *c* with 2% > VF > 1%. The data obtained by Thorpe and Humphries (1980) fall more into the bounds defined by curves *b* and *c* with 5% > VF 2%. The data presented by Longuet-Higgins and Smith (1983) seem to lie more below curve *b* with VF > 5% and above curve *a* with VF > 10%.

The surprisingly good overall match between the void fraction event statistics, interpreted as the equivalent breaking wave statistics, from the five previous field measurements may strongly suggest that the quantitative measure of void fraction is physically better than other techniques for detecting breaking waves and their relative scale and intensity in the open ocean.

A much more detailed and complete description of the bubble and void fraction measurement during 1991 SWADE shall be published elsewhere in the near future.

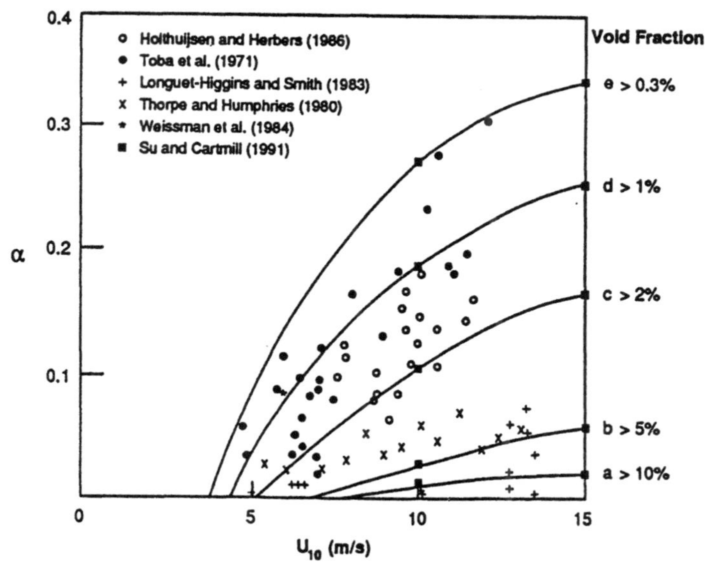

Figure 1. Observed fraction of breaking waves (α) and void fraction as a function of wind speed (U_{10}).

References

1. Holthuijsen, L.H. and T.H.C. Herbers. 1986. Statistics of Breaking Waves Observed as Whitecaps in the Open Sea. J. Phys. Oceano. 16. pp. 290-297.

2. Longuet-Higgins, M.S. and N.D. Smith. 1983. Measurement of Breaking Waves by a Surface Jump Meter. J. Geophys. Res. 88-C14. pp. 9823-9831.

3. Thorpe, S.A. and P.N. Humphries. 1980. Bubbles and Breaking Waves. Nature 283. pp. 643-665.

4. Toba, Y., H. Kunishi, K. Hishi, S. Kawai, Y. Shimada and N. Shibota. 1971. Study on Air-Sea Boundary Processes at the Shirahama Oceanographic Tower Station. Disaster Prevention Research Institute. Kyoto University Annals. 148. pp. 519-531. (In Japanese with English abstract)

5. Weissman, M.A., S.S. Atakturk and K.B. Katsaros. 1984. Detection of Breaking Events in a Wind-Generated Wave Field. J. Phys. Oceano. 14. pp. 1608-1619.

6. Weller, R.A., M.A. Donelan, M.G. Briscoe and N.E. Huang. 1991. Riding the Crest - A Tale of Two Wave Experiments. Bulletin Am. Met. Soc. 72(2). pp. 163-183.

Strong Coupling of Wind and Wind Waves

Y. TOBA, H. KAWAMURA and N. EBUCHI

Department of Geophysics, Faculty of Science, Tohoku University, Sendai 980, Japan

Abstract

Observational evidence is presented that the wind and the wind waves are in strong coupling in their fluctuations. In this strong coupling the wave breaking should play an essential role. A 4.5-hour time series of wind profile and wind-wave statistics, obtained at an oceanographic tower station under a winter monsoon wind, were reanalyzed. The main part of the equilibrium range of wind-wave spectra always follows the σ^{-4}-slope, though the energy level fluctuates following the fluctuating wind speed. The time scale of the adjustment of the energy level to the fluctuation of the wind is of the order of ten minutes, and within this time scale over- and under-saturation of the energy level occurs. During this time scale, there is significant energy transfer between the waves in the equilibrium range and those at the spectral peak, involving both upward and downward cascading of the wave energy. The variation of z_0 directly reflects the degree of over- and under-saturation of the level of the equilibrium range.

Introduction

For many years we have demonstrated by wind-wave tank experiments that the wind waves are phenomena which are strongly coupled with the wind, through the stress distribution along the phase of wind waves, the existence of a highly vortical layer under the crest which is frequently accompanied by wave breaking with or without air entrainment, the air-flow separation and ordered motions above and below the wind-wave surfaces (e.g., [1],[2],[3],[4]). The physics governing the phenomena is considered continuous from laboratory wind waves to wind waves over the ocean though geometrical similarity breaks down; e.g., the steepness of wind waves gradually changes from steeper laboratory waves to less steep field waves [5].

In this paper, a 4.5-hour time series of wind and wind-wave statistics obtained at an oceanographic tower station is presented to demonstrate evidence of coupled fluctuations of wind, wind waves and the sea surface roughness length. Detailed description is referred to [6] and [7].

M. L. Banner, R. H. J. Grimshaw (Eds.)
Breaking Waves
IUTAM Symposium Sydney/Australia 1991

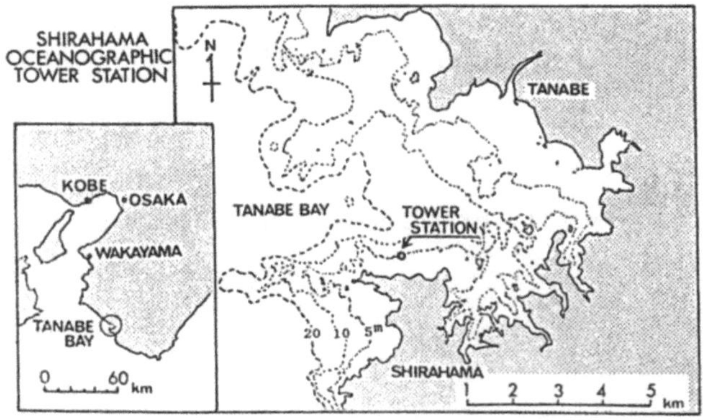

Fig.1. The location of the Shirahama Oceanographic Tower Station of Kyoto University.

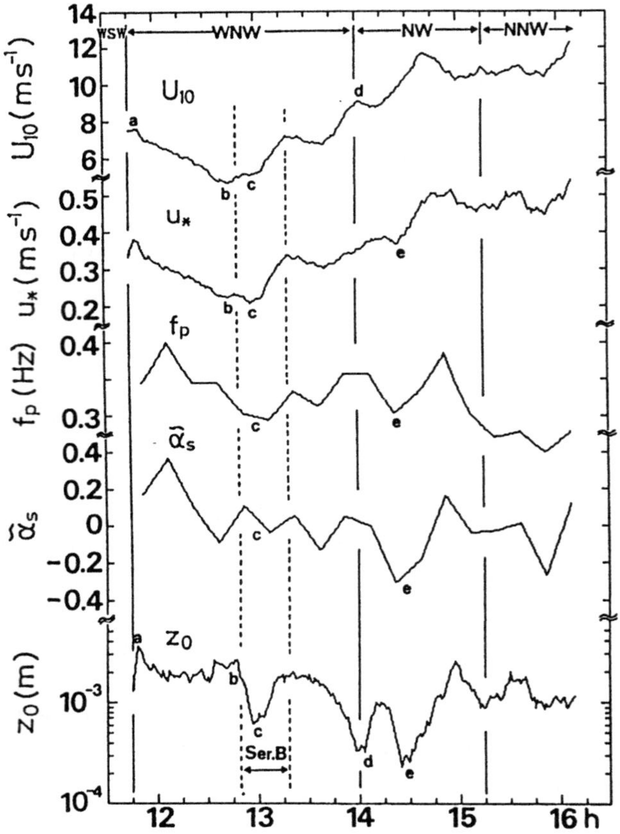

Fig.2. The 4-hr. time series of U_{10}, u_*, the wind-wave peak frequency f_p, the degree of under-saturation $\tilde{\alpha}_s$ defined by Eq.(1), and the roughness length z_0. The data of U_{10}, u_* and z_0 are from the 15-min. moving average wind speeds.

Data

The data were obtained at the Shirahama Oceanographic Tower Station of Kyoto University in November 1969 under a condition of the winter monsoon wind with slightly fluctuating wind speed and steadily growing wind-wave conditions. Detailed description of the observation conditions is referred to [8].

Detailed investigation of the time series

Figure 2 shows time series of the 10-m wind speed U_{10}, the air friction velocity u_*, the wind-wave peak frequency f_p, the degree of under-saturation of the equilibrium range of wind-wave spectra $\tilde{\alpha}_s$ and the roughness length z_0 which is in one-to-one correspondence with the drag coefficient C_D under neutral stratification. The $\tilde{\alpha}_s$ is defined by

$$\tilde{\alpha}_s = 1 - \alpha_s / \overline{\alpha_s} \tag{1}$$

where $\overline{\alpha_s}$ is the average value of the coefficient of energy level α_s of the equilibrium range of wind waves given by

$$\phi(\sigma) = \alpha_s g u_* \sigma^{-4} \tag{2}$$

where $\phi(\sigma)$ is the spectral density as a function of the angular frequency σ. The values of α_s are the same as reported in [6], and $\overline{\alpha_s}$ is 0.062. The fluctuation in $\tilde{\alpha}_s$ is caused by the time delay for the adjustment of the energy level in the equilibrium range to the variation of wind, as already discussed in [6]. It is seen that the variation of z_0 is related to the variation of U_{10}, passages of small fronts, and also the variation of f_p and $\tilde{\alpha}_s$. In particular, it is noted that the general trend of z_0 is very similar to that of $\tilde{\alpha}_s$.

Figure 3 shows a similar time series but for more detailed fluctuations, for the period indicated as Ser.B in Fig.2. Many observations similar to Fig.2 are recognized here extending down to the time scale of several minutes.

These observations lead us to the following physical interpretation. The variation of U_{10} on the time scales of several minutes to one hour causes the variation of f_p and $\tilde{\alpha}_s$. The larger $\tilde{\alpha}_s$ corresponds to the situation where the level of the wave equilibrium range is rapidly increasing. This condition is associated with larger z_0. Probably the wave components of

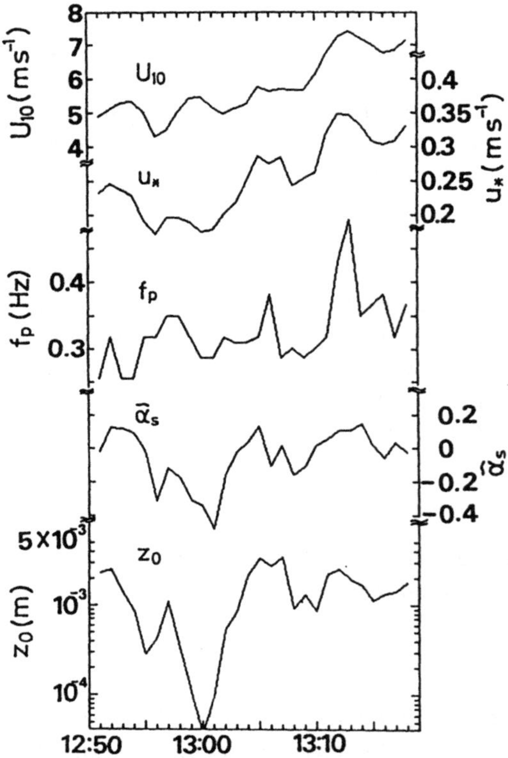

Fig.3. Detailed time series for the period indicated as Ser.B in Fig.2. The data of U_{10}, u_* and z_0 are here from the 3-min. moving average wind speeds, and the data of f_p and $\tilde{\alpha}_s$ are here in 1-min. series.

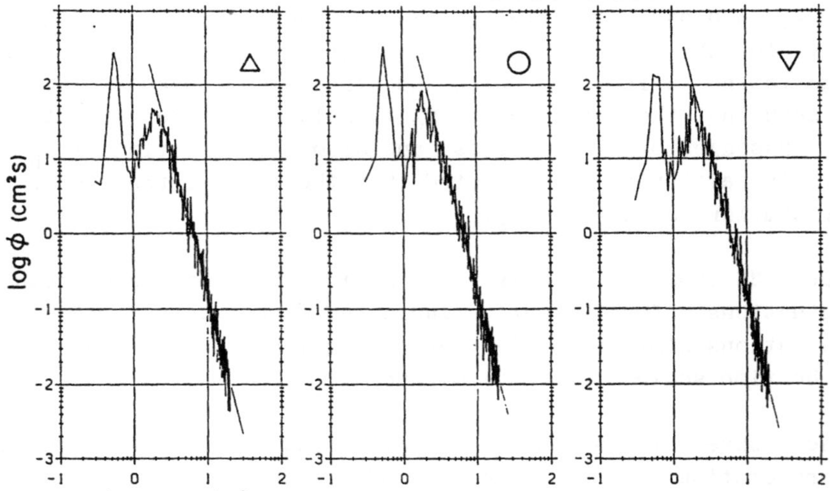

Fig.4. Ensemble averages of ten of the wind-wave spectra for increasing-wind case (left panel), constant-wind case (middle), and decreasing wind case (right). The wind-wave spectra were from Ser.B period indicated in Figure 2.

the very high-frequency part of the wave spectra outside the equilibrium range are steeper in this situation, and small-scale processes related to viscosity, surface tension and wave breaking will be more important in the momentum flux.

The variation in the wave spectral form corresponding to the variation of z_0 and $\tilde{\alpha}_s$ in Fig.3 is shown in Fig.4. While $\tilde{\alpha}_s$ is changing, the shape of the spectra of the equilibrium range always has a slope very close to σ^{-4}. In other words, it seems that there is a strong constraint for the spectra to be kept to the statistical σ^{-4} slope presumably caused by wave interactions which may include wave breaking adjustment. Consequently, the air momentum entering the waves due to the increased roughness at the very high frequency part of the spectrum is quickly transferred to the entire equilibrium range to feed the wave energy in the equilibrium range near to f_p. Also the shape of the spectral peak is flatter in positive $\tilde{\alpha}_s$ periods (left panel of Fig.4), and steeper in negative $\tilde{\alpha}_s$ periods (right panel). In the former case the wave energy near the spectral peak should have been cascading downward, and the latter cascading upward. The interpretation is as follows. The adjustment of the σ^{-4} slope occurs almost instantaneously, and the adjustment of the energy level of the equilibrium range occurs within several minutes. The development of the peak-frequency wave cannot follow such a quick response of the energy level, and a positive correlation of f_p with U_{10} or u_* occurs in the time scales of 15 minutes as seen in Fig.2. These observations indicate that the coupling of wind and wind waves is much stronger than usually considered.

Figure 5 shows the time series of $\sigma_p u_*/g$ and gz_0/u_*^2, calculated from the time series of u_* and z_0 from the 3-min. moving average wind data. Figure 6(a) is the gz_0/u_*^2 - $\sigma_p u_*/g$ diagram (the nondimensional roughness-wave age diagram) plotted from the time series of Fig.5. Figure 6(b) is the same except that the values were calculated from the 15-min. moving average wind data.

As seen in Figs.6(a) and (b), in the first half of the period when the wind felt no land effect, the averaged variation in this diagram is along the line of $gz_0/u_*^2 = \gamma(\sigma_p u_*/g)^m$ with m = -1 and, say, γ = 0.015. This corresponds to the evident averaged variation of gz_0/u_*^2 which has an inverse correlation with $\sigma_p u_*/g$ as seen in Fig.5. However, it is evident that there is a

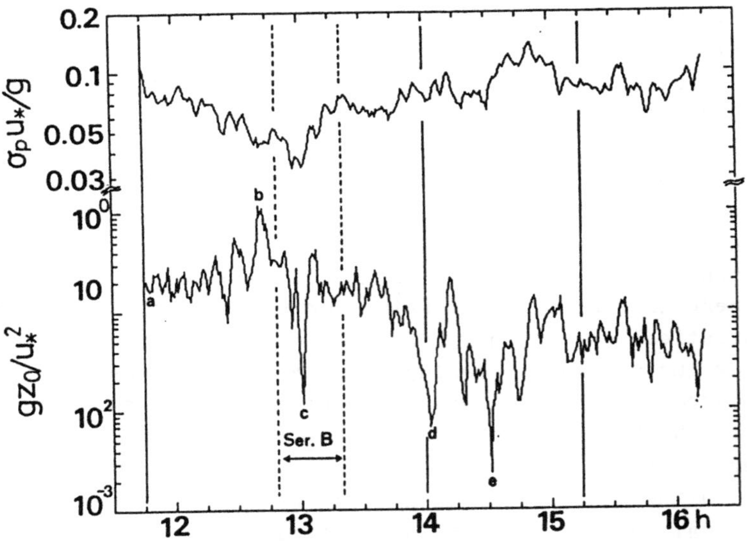

Fig.5. Time series of $\sigma_p u_*/g$ and gz_0/u_*^2, calculated from the time series of u_* and z_0 from the 3-min. moving average wind data.

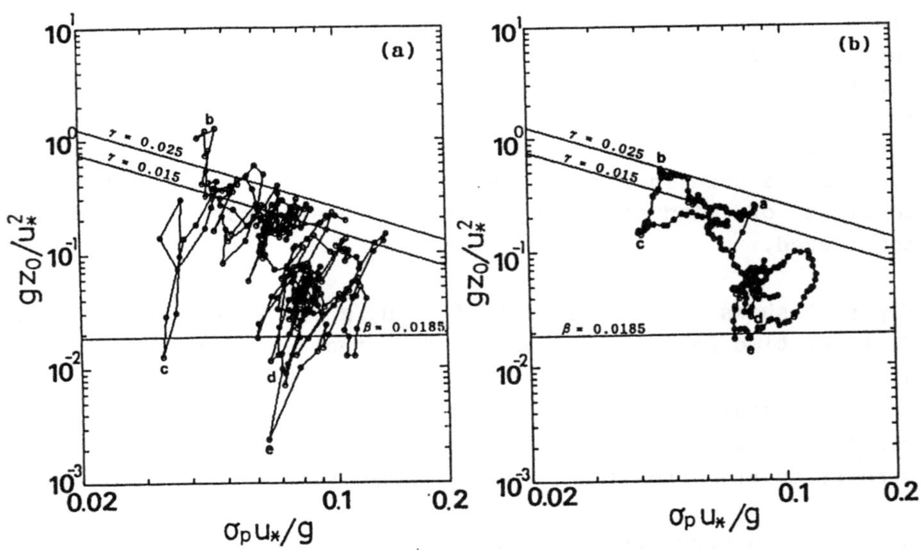

Fig.6. (a) The nondimensional roughness-wave age diagram plotted from the time series of Figure 4. (b) The same as (a) except that the values were calculated from the 15-min. moving average wind data.

fluctuation of the data points which is transverse to this line, i.e., in the direction of positive m. This transverse fluctuation corresponds to the in-phase fluctuation of gz_0/u_*^2 and $\sigma_p u_*/g$, as seen clearly in Fig.5 at shorter time scales, and this corresponds to the above described strong coupling of wind and wind waves as seen in the good correlation of z_0 and f_p or $\tilde{\alpha}_s$ shown in Fig.2.

References

1. Toba, Y.: Wind waves and turbulence. in Recent Studies on Turbulent Phenomena, eds.: Tatsumi T.; Maruo, H.,; Takai, H.: Assoc. for Sci. Doc. Inform., Tokyo (1985) pp 277-296.

2. Kawamura, H.; Toba Y.: Ordered motion in the turbulent boundary layer over wind waves. J. Fluid Mech. **197** (1988) 105-138.

3. Yoshikawa, I.; Kawamura, H.; Okuda, K.; Toba, Y.: Turbulent structure in water under laboratory wind waves. J. Oceanogr. Soc. Japan **44** (1988) 143-156.

4. Ebuchi, N.; Kawamura, H.; Toba, Y.: Bursting phenomena in the turbulent boundary layer beneath the laboratory wind-wave surface. in "Natural Physical sources of Underwater Sound" ed.: Kerman, B.; Kluwer Acad. Pub., Dordrecht (1991, in press).

5. Bailey, R.J.; Jones, I.S.F.; Toba, Y.: The steepness and shape of wind waves. J. Oceanogr. Soc. Japan **47** (1991, to be published).

6. Toba, Y.; Okada, K.; Jones, I.S.F.: The response of wind-wave spectra to changing winds. Part I: Increasing winds J. Phys. Oceanogr. **18** (1988) 1231-1240.

7. Toba, Y.; Ebuchi, N.: Sea-surface roughness length fluctuating in concert with wind and waves. J. Oceanogr. Soc. Japan **47** (1991, in press).

8. Kawai, S.; Okada, K.; Toba, Y.: Field data support of three-seconds power law and $gu_* \sigma^{-4}$-spectral form for growing wind waves. J. Oceanogr. Soc. Japan **33** (1977) 137-150.

Theoretical and Numerical Results

Possible Mechanisms for Wave Breaking

L. Cavaleri and P. Lionello
Istituto Studio Dinamica Grandi Masse-CNR
San Polo 1364, 30125 Venice, Italy

Abstract

We discuss possible mechanisms for wave breaking focusing in particular on the modulation of the short waves, associated with the tail of the spectrum, by the dominant ones. We seek confirmation for this in the analysis of accurate wave records taken from an oceanographic platform in the Adriatic. Consistently with the forementioned mechanism, we find evidence of grouping of short waves just ahead of the dominant crests in the active wind cases, while no grouping is found in the case of swell.

Time and space evolution of a wave field depend on three basic processes: generation by wind, conservative energy exchange due to nonlinear wave-wave interactions, dissipation by breaking. Other various mechanisms can be present in shallow water, but they are not considered in the present general discussion, that applies therefore to deep water conditions. Of the three processes mentioned above, generation by wind is relatively well understood and accurate experiments have provided usable quantitative estimates of the energy budget involved. The nonlinear wave-wave interactions are mathematically well-defined, and the limit to the accuracy of their evaluation depends only on the available computer power. On the contrary, we know little of the physics of breaking, and we, consequently, face a poor mathematical description of the involved energy exchanges.

The traditional engineering approach is to consider the sea surface as the random superposition of a sufficiently large number of wave components. There follows, at least under certain conditions, a Gaussian distribution for the elevation and a Rayleigh distribution for the single wave heights, defined as the difference between the maximum crest and the minimum trough enclosed between two sequential up-zero crossings. Each wave, as defined, has a wave length that, related to its corresponding height, establishes the possible breaking of the wave according to some accepted criterion. Several arguments point against this approach, the basic one being that a random linear superposition implies an up-down symmetry (no curtosis) of the sea surface profile. This is not supported by the experimental evidence.

A more sophisticated approach, based on the assumption that the breaking is a strong sparse process, weak on the mean, leads to a

M. L. Banner, R. H. J. Grimshaw (Eds.)
Breaking Waves
IUTAM Symposium Sydney/Australia 1991

statistical quantification of the loss of energy at each spectral frequency [1]. The loss depends on some mean parameters of the spectrum and, calibrated on the experimental data of the growth curve of an active sea and on the dynamical equilibrium of the fully developed conditions, leads to quite acceptable results in a large number of cases. However, the approach is open to criticism for two obvious facts: the relative direction of wind and waves is not considered, which leads, when applied to cross wind-wave cases, to unrealistic results; and wind, a basic forcing element in the process of breaking, is neglected too, while the experimental evidence shows that, for a given spectrum, a sudden change of wind speed leads to an immediate response of the intensity of breaking.

We have explored the possibility that breaking depends substantially on the high frequency part of the spectrum. Spilling is considered as a consequence of the modulation of the short, steep waves of the spectral tail by the dominant waves in the spectrum. Several interaction phenomena act in this sense - a) the orbital velocity of the dominant wave is felt by the short waves as a periodic variable current, with local convergence and divergence. The consequent strain enhances the steepness of the short-waves on the front of the large ones; b) the vertical acceleration, again due to the large wave's orbital velocity, effectively changes the dynamical behaviour of the short ones, increasing and decreasing their phase velocity, hence wavelength, hence steepness. These two phenomena lead to a concentration of steep wavelets on the front of the large waves. This, in turn, increases the local drag due to wind and, while locally producing breaking and consequently loss, can locally increase the input from wind to the large waves.

To prove the general lack of firm knowledge of the process it is enough to consider that those who regard the breaking from the atmospheric point of view have a quite opposite attitude. The breaking, and all its related quantities, are parametrized simply as a function of local wind speed (see e.g. [2] [3]). It is our opinion that much of the scatter of the data that is found with this approach derives, besides from the undeniable difficulties of the experiment, from the lack of consideration of the wave spectrum that underlies the process.

We strongly believe that both the wave spectrum and the wind speed have to be taken into account. Possibly the explicit dependence on the wind speed could be neglected if proper consideration was given to the high frequency tail of the wave spectrum. Under the hypothesis that most of the deep sea breaking is due to spilling by short waves riding on the crest of the large and long ones, and if, as it is believed, the tail of the spectrum adapts within a very short time to any wind change, then the wave spectrum carries in itself the whole information required for a proper evaluation of breaking.

We have explored the above phenomenology by analyzing accurate wave records taken from an oceanographic tower in the Northern Adriatic Sea [4] under variable sea conditions, ranging from very active generation to pure swell. We have considered in particular the statistical distribution of the short-waves with respect to the dominant ones in the spectrum. Our still preliminary results are consistent with the above description of breaking as spilling. Fig. 1a shows a section of a record taken during active generation conditions. The original record, obtained with a resistence wave gauge with a sampling frequency of 4 hz, has been first filtered with a ± 2 points cosine function obtaining the continuous line in the figure. Its envelope represents the distribution of short waves along the record, and its has been spectrally and cross-spectrally analyzed versus the original record. Fig. 1b shows the spectra of the original record, of the envelope and of the filtered record (the last two amplified 100 times). The envelope spectrum shows three peaks at low frequency, likely to be associated with wave grouping, and an evident peak at the peak frequency of the wave spectrum. The phase, not shown, places the envelope peak about 30° ahead of the wave peak, i.e. it locates the short waves just ahead of the wave crests, consistently with the spilling hypothesis.

By contrast we can consider the record in Fig. 2a, taken under swell conditions. Note the long period and the modulation of the signal. Note also, with respect to Fig. 1a, the limited amplitude of the envelope. The corresponding spectra are in Fig. 2b, and we see that the envelope spectrum (amplified here 1000 times) is much smaller than in the generation case. The phase does not show any tendency with respect to the crests of the wave.

If spilling takes place with the very short waves, the sampling frequency of the above records, 4 hz, is likely not to be sufficiently high to reveal the details of the phenomenon. However, the fact that, notwithstanding this limitation, the results are consistent with this hypothesis is in strong support of it and an indication of its relevance.

References

[1] Hasselmann, K., 1974, "On the spectral dissipation of ocean waves due to white capping", Boundary Layer Meteorol., 6, pp. 107-127.

[2] Wu, J., 1986, "Whitecaps, bubbles and spray", In: the Proceedings of Oceanic Whitecaps and Their Role in Air-Sea Exchange Processes, Galway, Reidel Publishing Company, pp. 113-124.

[3] Omuircheartaigh, I.G. and E.C. Monahan, 1986, "Statistical aspects of the relationship between oceanic whitecap coverage, wind speed and other environmental factors", In: the Proceedings of Oceanic Whitecaps and Their Role in Air-Sea Exchange Processes, Galway, Reidel Publishing Company, pp. 125-128.

[4] Cavaleri, L., 1979, "An instrumental system for detailed wind wave study", Il Nuovo Cimento, Series 1, Vol. 2C, pp. 288-304.

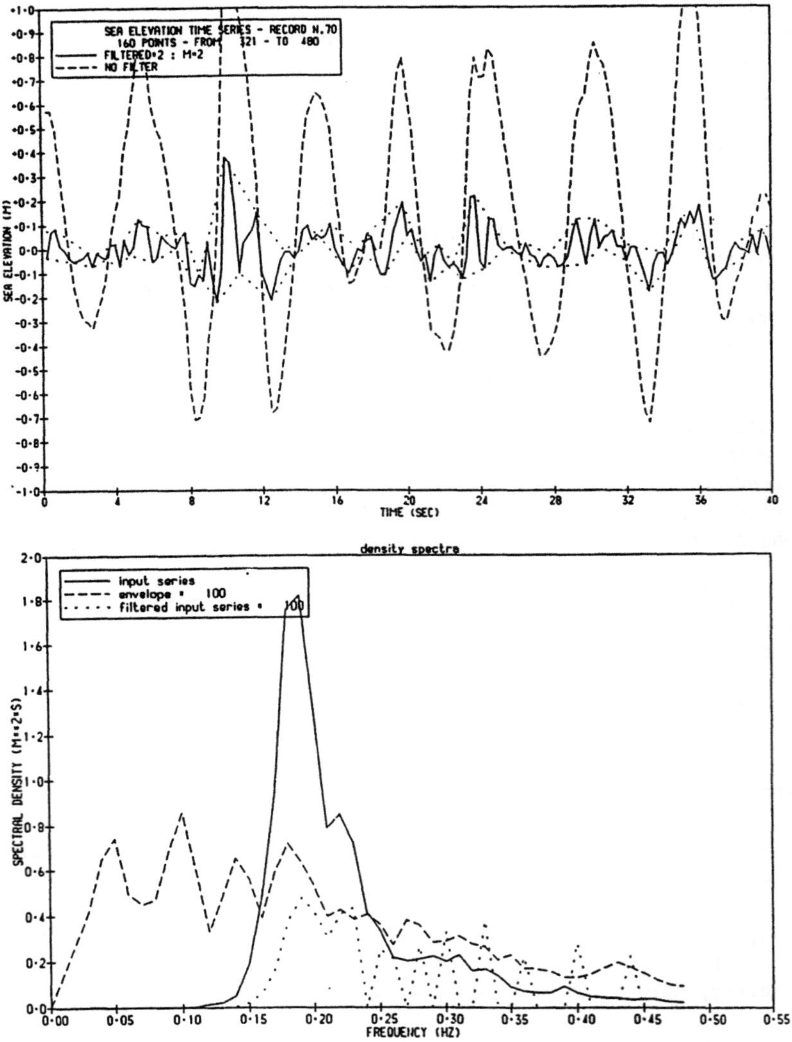

Fig. 1. a) Active generation conditions. Dash line = original record, continuous = filtered record, dotted = envelope of filtered record.

b) Spectra of the series in fig. 1a. The spectra of the envelope and filtered series have been amplified 100 times.

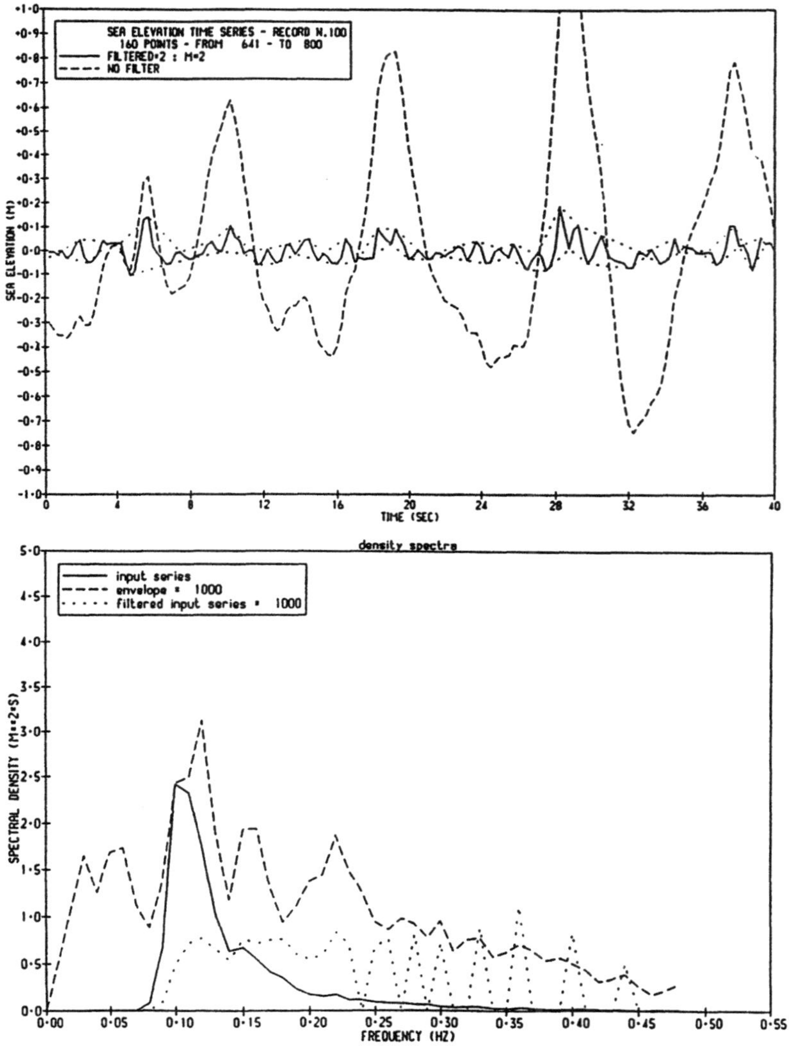

Fig. 2. As fig. 1, but for swell conditions. In fig. 2b the amplification factor is now 1000.

The Influence of Wave-Breaking Bubbles on Low-Frequency Underwater Ambient Noise Formation

I. N. DIDENKULOV

Institute of Applied Physics USSR Academy of Sciences
46, Ulyanov str., Nizhny Novgorod,
603600, USSR

Summary

The possible mechanisms of underwater ambient noise generation and propagation in the ocean associated with the presence of breaking-wave bubbles are discussed. The subsurface bubble layer can promote the transformation of the pressure fluctuations in the atmosphere into acoustic noise in the ocean due to scattering at the sound speed fluctuation in the layer.

Introduction

It is not a new idea that bubbles in the ocean can play a significant role in the processes of underwater noise generation [1-3]. Here we will consider a mechanism of this influence which seems to fall out of the previous considerations and, need experimental verification; mechanism refers to generation of low-frequency sound in the ocean due to turbulent pressure fluctuations in the near-water atmospheric layer and related to Isakovitch-Kur'yanov mechanism [4] updating. As is known, the pressure fluctuation propagation speed in the atmosphere can not exceed the sound speed and, therefore, such waves can not practically be radiated into water due to the fact that the sound propagation speed in the sea is much greater. The presence of the gas bubble layer makes it possible, in principle, to radiate noise into the sea under turbulent pressure pulsations. The mechanism of such a radiation consists in scattering at the sound speed inhomogeneities in a subsurface layer. These inhomogeneities emerge as a result of inhomogeneous spatial distribution of bubbles in the sea.

M. L. Banner, R. H. J. Grimshaw (Eds.)
Breaking Waves
IUTAM Symposium Sydney/Australia 1991

Wind-turbulence ambient noise generation in the presence of breaking-wave bubbles

As it follows from the work by Isakovitch and Kur'yanov [4] who were the first to study theoretically the process of underwater noise generation due to the turbulent pressure pulsations above the ocean surface the noise radiation into water occurs due to spatial harmonic components of the pressure pulsations which have the phase speed of on-surface travel in excess of the sound speed in water. Inasmuch as the basic energy-carrying components of the wind turbulence belong to quite another range of spatial-temporal frequencies this radiation proceeds very inefficiently and depends, to a great extent, on a particular form of the corresponding correlation function. Nowadays there is a lack of complete experimental data concerning the natural sea environment which could provide an exhaustive analysis of this mechanism. It is supposed to be interesting to assess the role of the subsurface bubble layer in the process of underwater noise generation due to wind turbulence. At frequencies under consideration, up to several tens of Hz, the bubble influence leads mostly to the sound speed fluctuations due to the bubble concentration fluctuations. An acoustic field arising under the ocean surface under the action of the wind turbulence will scatter at the sound speed inhomogeneities. Therefore, the spatial spectrum of the scattered field will contain combinational spatial frequencies and as a result of it part of the energy of inhomogeneous plane sound waves (whose amplitude falls off exponentially from the surface and into the depth) is transformed into the energy of the plane waves propagating depthward. Let us consider this model. Later on we shall adhere to the order set forth by Isakovitch and Kur'yanov in [4]. Suppose that the surface of the ocean appearing as a semiunbounded medium is under the action of a statistically stationary and homogeneous pressure field $\pi(\vec{\xi},t)$ (where $\vec{\xi}=(x,y)$ is a vector in horizontal plane) which can be expressed in the form of a Fourier transform in term of the spatial-temporal harmonics with δ-correlated amplitudes:

$$\pi(\vec{\xi},t) = \iint \Pi(\vec{x},\omega)\exp[i(\vec{x}\vec{\xi}-\omega t)]\,d\vec{x}d\omega \qquad (1)$$

$$\langle \Pi(\vec{x},\omega)\Pi^*(\vec{x}',\omega')\rangle = P(\vec{x},\omega)\delta(\vec{x}-\vec{x}')\delta(\omega-\omega'), \quad (2)$$

where $\langle\ldots\rangle$ means statistical averaging, $*$ - stands for complex conjugation. The sound speed in the ocean will be given as

$$c = c + \Delta c(\vec{r}) \quad (3)$$

where $\Delta c(\vec{r})$ are random fluctuations of the sound speed in a subsurface layer due to the presence of bubbles ($|\Delta c| \ll c_\circ$). In order to find the Fourier-components of the acoustic field we shall proceed from the Helmholtz equation

$$\Delta p' + k^2(\vec{r})p' = 0 \quad (4)$$

and the boundary condition $\quad p'_{|z=0} = \pi(\vec{\xi},t) \quad (5)$

Taking (3) into account the wave number $k^2 = k_\circ^2(1 + \varepsilon_\circ(\vec{r}))$, where $\varepsilon_\circ = -2\Delta c/c_\circ$. Solution of the problem (4),(5) will be sought using the perturbation method setting

$$p' = p_\circ + p_1 \quad (6)$$

where $p_\circ$ is unperturbed field (for $\varepsilon_\circ = 0$), which can be can be presented in the form of a set of plane waves with amplitudes $\Pi(\vec{x},\omega)$ [4]

$$p_\circ(\vec{r}) = \int \Pi(\vec{x},\omega)\ \exp[i(\vec{x}\vec{\xi} + z\sqrt{k_\circ^2 - x^2})]\,d\vec{x} \quad (7)$$

The solution of (4)-(6) can be written as

$$p_1 = -\frac{k_\circ^2}{4\pi}\int \varepsilon_\circ(x',y',z')\ p_\circ(x,y,z)\ G(\vec{r}',\vec{r})\ d\vec{r}$$

where $G(r',r)$ is Green's function. Assuming the Green's function in the form of the expansion in terms of plane waves and the function $\varepsilon_\circ$ as a Fourier transform

$$\varepsilon_\circ(x,y,z) = \int E(\vec{q},z)\ [e^{i(\vec{q}\vec{\xi})} + e^{-i(\vec{q}\vec{\xi})}]\,d\vec{q}$$

$$\langle E(\vec{q},z)\cdot E(\vec{q}',z')\rangle = \mu(z)\Psi(\vec{q})\delta(\vec{q}-\vec{q}')\delta(z-z')$$

where $\mu(z)$ describes the intensities of the sound speed fluctuation distributed in deph, the spectral density of intensity $\langle|p_1|^2\rangle$ for the scattered field can be obtained. After the transformations one can write

$$\langle|p_1|^2\rangle = A\iint_{k^2>(\vec{x}\pm\vec{q})^2} P(\vec{x},\omega)\Psi(\vec{q})\,d\vec{x}d\vec{q} \ +$$

$$+ \; A \iint_{k_\circ^2 < (\vec{x} \pm \vec{q})^2} P(\vec{x},\omega)\Psi(\vec{q})\exp(-2z\sqrt{(\vec{x}\pm\vec{q})^2 - k_\circ^2})\, d\vec{x}\, d\vec{q} \qquad (8)$$

$$A = 4\pi k_\circ^2 \cdot \int z^2 \mu(z)\, dz \qquad (9)$$

Expression (8) differs from that for $\langle |p_\circ|^2 \rangle$ deduced by Isakovitch and Kur'yanov in only changing the range of integration and emerging of an additional integral over the power spectrum of the sound speed fluctuations

$$\langle |p_\circ|^2 \rangle = \int_{k_\circ^2 > x^2} P(\vec{x},\omega)\, d\vec{x} + \int_{k_\circ^2 < x^2} P(\vec{x},\omega)\exp(-2z\sqrt{x^2 - k_\circ^2})\, d\vec{x} \qquad (10)$$

In order to find a particular form of the scattered field $\langle |p_1|^2 \rangle$ one needs to have data on $P(\vec{x},\omega)$ and $\Psi(\vec{q})$ which are lacking in current literature. Therefore, when estimating the efficiency of the mechanism considered we shall make use of an approach due to Isakovitch and Kur'yanov [4] and Wilson [5] who related the spectrum $P(\vec{x},\omega)$ with that of surface seas. As regards the inhomogeneity spectrum, we shall assume it to be relatively narrow and unifomly distributed over depth in a subsurface layer with thickness h which is in agreement with some experimental data [6]

$$\Psi(\vec{q}) \simeq \Psi_\circ \delta(\vec{q} - \vec{q}_\circ)$$

$$\mu(z) = \begin{cases} 1/h, & 0 \le z \le h \\ 0, & z > h \end{cases} \qquad (11)$$

Comparing (8) and (10) and allowing for (11) one can use final expressions for $\langle |p_1|^2 \rangle$ deduced in [4], to gain the purpose of the present investigation, performing the following replacement in them $k_\circ \to k_\circ \pm q_\circ$. It appears, therefore, that spectral space-time harmonics of the pressure pulsations which do not radiate sound into water in the absence of inhomogeneities as their phase speed is less than that of sound in water ($k_\circ < x < k_\circ + q_\circ$) make a contribution into the noise field in the case of a scattered field. Performing the necessary calculations and taking into account that bubble layer locates near the ocean surface ($k_\circ h \ll q_\circ h \ll 1$), where field $p_\circ$ is, mainly, inhomogeneous and described by the second integral in (10), one obtains

$$\langle |p_1|^2 \rangle \simeq \Psi_\circ \cdot (k_\circ h)^2 (1 + q_\circ / k_\circ)^2 \langle |p_\circ|^2 \rangle_{|z=h} \simeq$$

$$\approx \Psi_\circ \cdot (q_\circ h)^2 \langle |p_\circ|^2 \rangle_{|z=h}$$

It is known from [4] that inhomogeneous part of field $p_{\circ|z=h}$ differs from homogeneous one (described by the first integral in (10)) by the multiplier $1/[2(k_\circ h)^2]$. This yeilds the relation between the propagating (homogeneous) parts of the fields p_1 and $p_\circ$:

$$\frac{\langle |p_1|^2 \rangle}{\langle |p_1|^2 \rangle} \approx (\Psi_\circ/2) \cdot (q_\circ/k_\circ)^2 \qquad (12)$$

Let us make the numerical estimation from (19). From [6] it is clear that the value of $\sqrt{\langle |\Delta c|^2 \rangle}$ in a subsurface sea layer at relatively small wind speeds attains ≈15 m/s for the average layer depth h≈10 m. To estimate the values of $q_\circ$ we shall use the measured quantities of bubble concentration by nonlinear sound scattering performed in Arabian sea using the equipment described in [7]. Distinct bubble layers are observed such that under the ship drift speed amounting to 0.2-0.3 knots characteristic distance between the layers adds up to ≃ 100 m. Substitution this value into (12) yields $\langle |p_1|^2 \rangle / \langle |p_\circ|^2 \rangle \simeq 1$ at the frequency 0.2 Hz.

Conclusion

A possible mechanism of influence of a subsurface bubble layer on the ambient noise generation at low frequences was considered above. From the estimations of their efficiencies it is followed, that the scattering of the turbulent wind pressure pulsations by the horizontal inhomogeneities of sound speed in a subsuface bublle layer can play a significant role at very low frequency region (f≤1 Hz). However it is nessesary to note that the absence of data on temporal-spatial spectra of turbulent wind pressure pulsations near the ocean surface does not allow to make more reliable estimations.

References

1. Knudsen, V.O.;Alford, R.S.;Emling, J.W.: Underwater ambient noise. J.Marine Res. 7 (1948) 410-429.

2. Wenz, G.: Acoustic ambient noise in the ocean: spectra and sources. J. Acoust. Soc. Am. 34 (1962) 1936-1956.
3. Furduev, A.V.: The underwater cavitation as a source of noise in the ocean.: Fiz. atm. i okeana. 11 (1966) 523-533.
4. Isakovitch, M.A.; Kur'yanov, B.F.: To the theory of low-frequency noise in the ocean. Akust. Zh.: 16 (1970) 62-74.
5. Wilson, J.H.: Very low frequency (VLF) wind-generated noise produced by turbulent pressure fluctuations in the atmosphere near the ocean surface: J. Acoust. Soc. Am. 66 (1979) 1499-1507.
6. Clay C.S., Medwin H. (1977) Acoustic oceanography: principles and applications, John Wiley and sons, New York.
7. Ostrovsky, L.A.; Sutin, A.M.: Nonlinear acoustic methods in diagnostics of bubbles in fluid. In: Grechova, N. (ed.) Ultrasound diagnostics. Gorky: Inst. of Applied Physics 1983. P. 139-150.

Do Stokes' Double Series Converge for Large Amplitude Waves?

W.M. Drennan[1], W.H. Hui and G. Tenti

Department of Applied Mathematics,
University of Waterloo,
Waterloo, Ontario (N2L 3G1) Canada

[1] National Water Research Institute,
Burlington, Ontario (L7R 4A6) Canada

Summary

This paper reports on a study of the behaviour of Stokes' series for large amplitude waves, in which a combination of analytical work (using symbolic computation) and accurate numerical calculations performed on a supercomputer was used. Clear numerical evidence is presented showing that the Stokes double series appear to diverge for large amplitude waves.

Introduction

The pioneering work of Stokes on two–dimensional, finite amplitude gravity waves [1,2] was brought into the computer age over a century later by Schwartz [3] who developed a nonlinear recursive relation for the Stokes coefficients and then computed them to 117th order in wave height h (half the distance from crest to trough). This limit was imposed by numerical roundoff errors resulting from the nonlinearity of the algorithm and available computer precision. Schwartz then employed Padé approximants to sum the resulting series with the hope of both extending the radius of convergence and accelerating the rate of convergence. A similar procedure, though with a different perturbation parameter, was followed by Cokelet [4].

As is well known, Stokes' method of calculating the Stokes wave consists of first expanding the coordinates x and y as Fourier series, followed by perturbation expansion in some small parameter ϵ, e.g. the wave height h, for the calculation of the Fourier coefficients, resulting in a solution in the form of a double series. This method has been followed by researchers for over a century under the assumption that it yields convergent results. It is therefore important to note that this does not, in fact, appear to be the case: for large amplitude waves, the Stokes series are seen to diverge.

M. L. Banner, R. H. J. Grimshaw (Eds.)
Breaking Waves
IUTAM Symposium Sydney/Australia 1991

On the convergence of the Stokes double series

An investigation of the convergence of the Stokes series is perhaps best carried out by investigating the surface profile:

$$x(\phi,0) = -\phi - \sum_{n=1}^{\infty}\sum_{i=0}^{\infty} \epsilon^{n+2i}\frac{\alpha_n^{(i)}}{n}\sin n\phi \qquad (1)$$

$$y(\phi,0) = \sum_{n=1}^{\infty}\sum_{i=0}^{\infty} \epsilon^{n+2i}\frac{\alpha_n^{(i)}}{n}\cos n\phi. \qquad (2)$$

For given values of ϵ and ϕ these are double series and, in particular, (2) is of the type

$$\sum_{n=1}^{\infty}\sum_{i=0}^{\infty}\beta_{ni} \qquad (3)$$

where the entries in the double infinite array β_{ni} are real numbers. Questions about the absolute convergence of such series may be formulated in a manner analogous to the case of a single infinite series, except of course for the fact that there is no longer a unique way of constructing the sequence of partial sums. Of particular interest to us is the result (analogous to those of Tonelli and Fubini for iterated integrals) that a necessary condition for the series (2) to be absolutely convergent is that its value be independent of the particular way in which the sequence of partial sums is formed [5].

In the method employed by Schwartz [3], and by the researchers who subsequently employed his algorithm, the partial sums are computed by keeping the 'row index' n much lower than the 'column index' i; in particular, from his 117th order solution, Schwartz employed only the first thirty harmonics (i.e. $n \leq 30$) but retained all terms up to $O(\epsilon^{117})$ in the perturbation series for the coefficients of these harmonic components. Ad hoc methods such as Padé approximants were then used to obtain 'converged results'.

However, the double series (1) and (2) may equally be looked upon as perturbation series in ϵ, and it is therefore natural to construct sums having an $O(\epsilon^M)$ accuracy for a given M. This implies that all terms proportional to $\alpha_n^{(i)}$ with $n + 2i \leq M$ must be included. In particular, for an O(117) solution, the terms proportional to $\alpha_{31}^{(i)}(i = 0,1,2,\ldots,43)$, $\alpha_{32}^{(i)}(i = 0,1,2,\ldots,42),\ldots,\alpha_{117}^{(0)}$ must be retained as they are of the same or lower order as terms already included by Schwartz. By the theorem mentioned above, and under the assumption of absolute convergence, this sequence of partial sums should have the same limit as the one used by Schwartz. Accurate numerical evidence will now be presented casting serious doubts that this is indeed the case for Stokes' series (1) and (2) for large amplitude waves.

If the perturbation solution is carried out consistently to order M, i.e. $n+2i \leq M$, the surface profile is given by

$$x(\phi,0) = -\phi - \sum_{n=1}^{M} \frac{a_n(M,h)}{n} \sin n\phi, \tag{4}$$

$$y(\phi,0) = \sum_{n=1}^{M} \frac{a_n(M,h)}{n} \cos n\phi \tag{5}$$

where

$$a_n(M,h) = \sum_{i=0}^{[(M-n)/2]} h^{n+2i} \alpha_n^{(i)} \tag{6}$$

with the symbol $[p]$ representing the closest integer not greater than the rational number p. Prior to discussing the wave profile itself, we focus our attention on the behaviour of the partial sums and on the coefficients a_n. We present here detailed numerical results, based on Schwartz' algorithm as implemented on a Cray X-MP computer, for the partial sum (5) of the y-series at selected values of h and ϕ; the x-series behaves in a similar manner. For small to intermediate wave heights ($h < 0.3$) the partial sums appear to be well converged as M is increased; however, already by $h = 0.35$ the values of $y(\phi,0)$ do not converge with increasing M, and this behaviour worsens with increasing wave height. A clear illustration of this phenomenon is given in Table 1 for $h = 0.4$: If one limits the number of harmonics to, say, $n = 10$ and lets $i = 0,1,\ldots,45$, (hence $M = 100$) then the partial sums appear to converge, as can be seen by reading the table across a row. On the other hand, by including all the harmonics contributing to the same $O(\epsilon^M)$ the sequence appears to diverge as is evident by reading the table down a column or diagonally. This behaviour is, of course, also visible in the values taken by the Fourier coefficients (6) for the corresponding wave heights exhibited in Table 2.

Turning now to the wave profile itself, we give in Fig. 1 an example (again, with $h = 0.4$) which illustrates graphically the convergence problems associated with Stokes' method, represented by the dotted line. The solid line represents the results of the $y(x,\psi)$ method of [6] and [7]. We note that this method yields smooth profiles for all values of x and all wave heights – without the use of artificial or ad hoc summation techniques. In contrast to the Stokes method, the relevant series converge regardless of how the summation is carried out. This remains true for amplitudes up to h= 0.443, the highest Stokes wave. Clearly the Stokes double series are incapable of producing smooth profiles (i.e. converged results) at larger amplitudes. We have confirmed this behaviour of the wave profile by repeating the calculation with the perturbation parameter of Cokelet [4].

We emphasize that our results above were obtained by series summation – without the use of approximation techniques. On the other hand, as pointed out above, the results presented by Schwartz [3] and Cokelet [4] were obtained by using Padé approximants (PAs) to estimate the series sum. Although this semi-empirical technique has yielded the smooth results presented by the authors, it is necessary to raise the question as to whether these Padé sums are equal to the series sum. Unfortunately at this point, little is known about the convergence of Padé approximants of arbitrary series, with the important exception of series of the Stieltjes type. It is, however, readily shown (see Theorem 2.5 of Vrscay and Cizek [8]) that the Stokes series are *not* of this type.

Without a solid framework for establishing the convergence of PAs, researchers have instead relied on rules of thumb such as 'when the Padé table results appear to converge, they are correct'. Unfortunately, this empirical approach does not always result in the correct conclusion. It is entirely possible for the PAs to have the appearance of being converged, at some finite order, to a number different from the actual series sum. A vivid illustration of this is given in Figure 2, reproduced from [9] for the convenience of the reader, where the lower order Padé approximants appear to converge to something quite different from the higher order ones. We note here that the phase speed results reported by Schwartz ([3], Table 3) appear to correspond to the [16,16] Padé approximant.

In conclusion, the non-convergent results obtained by the method of Stokes, and its computer extensions, are not phenomena resulting from roundoff error. By using an independent method, (c.f. Hui and Tenti [10], Drennan, Hui and Tenti [9]) a linear recursion formula for the Stokes series coefficients was derived allowing them to be calculated *exactly* employing symbolic computation. We have carried out such a calculation to order thirty-three, and the non-convergent results for large h are already evident at that order. Rather, the phenomenon is the result of the increasing relative importance of the higher order harmonics as the wave height increases, and of the incapacity of Stokes' method of producing converged results for them. In particular, $a_M(M,h)/M$ does not tend to zero, violating the necessary condition for the convergence of the Fourier series (5).

References

1. Stokes, G.G.; On the theory of oscillatory waves. Mathematical and physical papers 1 (1847) 197-229, Cambridge University Press.
2. Stokes, G.G.; A supplement to a paper on the theory of oscillatory waves, Mathematical and physical papers 1 (1880) 314-326, Cambridge University Press.
3. Schwartz, L.W.; Computer extension and analytic continuation of Stokes' expansion for gravity waves. J. Fluid Mech. 62 (1974) 533-578.
4. Cokelet, E.D.; Steep gravity waves in water of arbitrary uniform depth. Phil. Trans. R. Soc. London A286 (1977) 183-230.
5. Goursat, E.; A course in mathematical analysis, Vol. I, Ch. VIII, translated by E.R. Hendrick. Dover, 1959.
6. Hui, W.H. and Tenti, G.; A new approach to steady flows with free surfaces. ZAMP 33 (1982) 569-589.
7. Drennan, W.M., Hui, W.H. and Tenti, G.; Accurate calculations of Stokes water waves of large amplitude. Submitted to ZAMP.
8. Vrscay, E. and Cizek, J.; Continued fractions and Rayleigh- Schrödinger perturbation theory at large order. J. Math. Phys. 27 (1986) 185-201.
9. Drennan, W.M., Hui, W.H. and Tenti, G.; Accurate calculations of the Stokes water wave, in Continuum mechanics and its applications, (Eds. G. Graham and S. Malik). Hemisphere, (1989) 624-634.
10. Hui, W.H. and Tenti, G.; Nonlinear wave theory via pressure formulation, in The ocean surface, (Eds. Y. Toba and H. Mitsuyasu). Reidel, Dordrecht, (1985) 17-24.

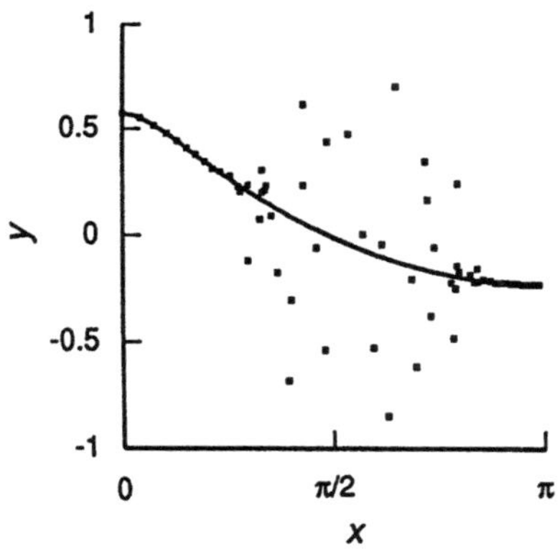

Figure 1 — Comparison of wave profiles using Schwartz's algorithm (▪) (without Padé approximants) and $y(x,\psi)$ formulation (—) of Refs. [6], [7].

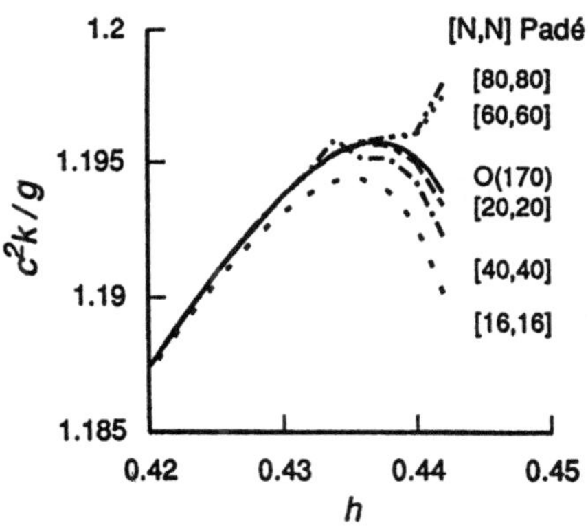

Figure 2 — Behaviour of Padé approximants applied to the Stokes series for phase speed.

Table 1 - Partial sums of y-series, eqn. (5) using Schwartz' algorithm
n harmonic components, perturbation to order M with $h = 0.4$

n	$M=5$	$M=15$	$M=30$	$M=50$	$M=100$
1	0.021474	0.021022	0.020984	0.020981	0.020981
2	-0.073565	-0.077022	-0.076624	-0.076599	-0.076597
3	-0.082649	-0.087634	-0.087180	-0.087148	-0.087147
4	-0.017102	-0.058231	-0.058023	-0.058011	-0.058010
5	0.001386	-0.051134	-0.050981	-0.050975	-0.050975
6		-0.064718	-0.064157	-0.064137	-0.064136
7		-0.070101	-0.069259	-0.069242	-0.069240
8		-0.065225	-0.062312	-0.062296	-0.062295
9		-0.063470	-0.058458	-0.058449	-0.058449
10		-0.076757	-0.062398	-0.062381	-0.062381
11		-0.090107	-0.065361	-0.065337	-0.065336
12		-0.104900	-0.063069	-0.063050	-0.063049
13		-0.124988	-0.060741	-0.060755	-0.060755
14		-0.138374	-0.062103	-0.062081	-0.062080
15		-0.159308	-0.063594	-0.063871	-0.063869
16			-0.063085	-0.063130	-0.063129
17			-0.059337	-0.061733	-0.061733
18			-0.060774	-0.062110	-0.062109
19			-0.050202	-0.063198	-0.063197
20			-0.052516	-0.063048	-0.063047
21			-0.013643	-0.062203	-0.062204
22			-0.014286	-0.062213	-0.062213
23			0.065202	-0.062854	-0.062864
24			0.077273	-0.062925	-0.062937
25			0.182781	-0.062325	-0.062440
26			0.215058	-0.062160	-0.062321
27			0.292515	-0.061751	-0.062696
28			0.327975	-0.061378	-0.062838
29			0.353546	-0.056535	-0.062558
30			0.369217	-0.052426	-0.062410
31				-0.032450	-0.062615
32				-0.010379	-0.062759
33				0.057485	-0.062613
34				0.150796	-0.062478
35				0.321359	-0.062579
36				0.616583	-0.062701
37				0.932921	-0.062635
38		$\psi = 0$		1.640987	-0.062527
39		$\phi = -1.5$		2.054349	-0.062567
40				3.327277	-0.062660
41				3.674659	-0.062639
42				5.361256	-0.062560
43				5.497919	-0.062567
44				7.091694	-0.062633
45				7.050813	-0.062636
46				8.065718	-0.062582
47				7.996751	-0.062573
48				8.386401	-0.062617
49				8.363829	-0.062629
50				8.432073	-0.062602

Table 1 - cont'd

n	$M=100$
51	-0.062565
52	-0.062660
53	-0.062453
54	-0.062920
55	-0.061008
56	-0.063971
57	-0.050427
58	-0.065487
59	0.017244
60	-0.046609
61	0.386727
62	0.171385
63	2.129488
64	1.614178
65	9.262042
66	8.838714
67	34.67606
68	38.22851
69	113.6734
70	138.1432
71	328.1393
72	425.7388
73	836.8420
74	1130.801
75	1890.672
76	2605.595
77	3795.235
78	5234.547
79	6792.233
80	9213.615
81	10886.07
82	14295.70
83	15718.93
84	19723.74
85	20618.22
86	24511.17
87	24844.44
88	27937.32
89	27906.44
90	29878.15
91	29735.35
92	30717.25
93	30611.84
94	30978.70
95	30935.43
96	31031.78
97	31021.64
98	31037.43
99	31036.36
100	31037.52

Table 2 - Fourier coefficients $a_n(M,h)/n$, eqn. (6)
via Schwartz' algorithm, $h = 0.4$

n	$M=5$	$M=15$	$M=30$	$M=50$	$M=100$
1	0.303573	0.297180	0.296649	0.296606	0.296605
2	0.096000	0.099035	0.098595	0.098566	0.098565
3	0.043093	0.050341	0.050076	0.050048	0.050046
4	0.068267	0.030622	0.030367	0.030347	0.030345
5	0.053333	0.020474	0.020316	0.020296	0.020295
6		0.014909	0.014461	0.014446	0.014445
7		0.011320	0.010750	0.010734	0.010733
8		0.005778	0.008244	0.008231	0.008230
9		0.002950	0.006479	0.006466	0.006465
10		0.017490	0.005187	0.005176	0.005175
11		0.019006	0.004218	0.004208	0.004207
12		-0.022403	0.003470	0.003465	0.003464
13		-0.025241	0.002925	0.002883	0.002882
14		0.024440	0.002486	0.002421	0.002420
15		0.023971	0.001708	0.002049	0.002048
16			0.001200	0.001746	0.001745
17			0.004016	0.001497	0.001496
18			0.004917	0.001290	0.001289
19			-0.010846	0.001117	0.001116
20			-0.015003	0.000971	0.000970
21			0.039012	0.000848	0.000847
22			0.048465	0.000743	0.000742
23			-0.079620	0.000642	0.000651
24			-0.094329	0.000557	0.000574
25			0.107635	0.000612	0.000507
26			0.121048	0.000621	0.000449
27			-0.082181	-0.000434	0.000398
28			-0.088655	-0.000931	0.000354
29			0.028863	0.005467	0.000316
30			0.029830	0.007823	0.000282
31				-0.024612	0.000252
32				-0.034478	0.000226
33				0.094119	0.000203
34				0.125730	0.000182
35				-0.276865	0.000164
36				-0.355987	0.000147
37				0.634293	0.000133
38				0.786856	0.000120
39				-1.112928	0.000108
40				-1.336529	0.000098
41				1.467755	0.000089
42				1.710724	0.000080
43				-1.406051	0.000073
44				-1.594336	0.000066
45				0.924327	0.000060
46				1.021658	0.000055
47				-0.373261	0.000050
48				-0.402842	0.000046
49				0.070185	0.000036
50				0.074038	0.000029

Table 2 - cont'd

n	$M=100$
51	0.000082
52	0.000111
53	-0.000361
54	-0.000601
55	0.002799
56	0.004358
57	-0.017377
58	-0.026433
59	0.096191
60	0.142505
61	-0.469257
62	-0.678396
63	2.022080
64	2.856000
65	-7.694903
66	-10.63081
67	25.84974
68	34.97003
69	-76.58226
70	-101.5516
71	199.7731
72	259.9123
73	-457.8321
74	-584.9417
75	919.0096
76	1153.994
77	-1609.381
78	-1987.732
79	2446.309
80	2974.003
81	-3206.629
82	-3839.806
83	3594.575
84	4242.452
85	-3408.967
86	-3967.922
87	2696.609
88	3097.246
89	-1745.575
90	-1979.450
91	900.3089
92	1008.468
93	-355.6937
94	-393.7442
95	101.0613
96	110.6059
97	-18.38179
98	-19.89806
99	1.607624
100	1.721851

Consequences of the Effect of Surface Gravity Waves on the Mean Air Flow

Peter A.E.M. Janssen

Department of Oceanography,
Royal Netherlands Meteorological Institute (KNMI),
De Bilt, The Netherlands

Summary

The effect of wind-generated gravity waves on the air flow is discussed using quasi-linear theory of wind-wave generation. In this theory both the effects of the waves and the effect of air turbulence on the mean wind profile is taken into account, but effects of turbulence on the wave-induced air motion are disregarded. In addition, effects of wave breaking on the air-sea momentum transfer are not considered. Nevertheless, this relatively simple model of the momentum transport from air to water is shown to produce realistic results.

The main result of this theory is that the momentum transfer at the air-sea interface is sea-state dependent, explaining the scatter in plots of the experimentally observed drag coefficient as a function of wind speed. As a matter of fact, a very good agreement between observed and modeled stress is found, much better than when the usual Charnock relation for the roughness length is used.

A proper description of the physics of the momentum transfer can therefore only be given by considering a coupled ocean-wave, atmosphere model. Results of the coupling of a simple surface layer model with a third-generation wave model will be discussed. Finally, consequences for coupled climate models of ocean and atmosphere are briefly pointed out.

1. Introduction

For many practical applications, where there is interest in the slow time evolution of random, surface gravity waves, a spectral description is appropriate. The two dimensional spectrum, $F(f,\theta)$, where f is frequency and θ the propagation direction then obeys the energy balance equation

$$\frac{\partial}{\partial t}F + \vec{c}_g \cdot \frac{\partial}{\partial \vec{x}}F = S = S_{in} + S_{NL} + S_{diss}, \tag{1}$$

where $\vec{c}_g$ is the group velocity $\partial\omega/\partial\vec{k}$ (with ω the angular frequency $\omega = \sqrt{gk}$, g acceleration of gravity and k the wave number) and the source terms describe the generation of waves by wind (S_{in}), dissipation due to white-capping (S_{diss}) and energy transfer due to 4-wave interactions (S_{NL}).

Hence, waves extract energy and momentum from the air flow and in general this energy transfer may be sea-state dependent. However, present-day numerical models of the atmosphere assume that this energy transfer from air to ocean only depends on the wind

M. L. Banner, R. H. J. Grimshaw (Eds.)
Breaking Waves
IUTAM Symposium Sydney/Australia 1991

speed resulting in inconsistencies in the energy balance at the air-sea interface when coupling an atmospheric model with a wave model. The purpose of this paper is to overcome this inconsistency and to discuss some consequences of a sea-state dependent energy transfer.

The program of this paper is as follows. In Section 2 we briefly discuss the quasi-linear theory of wind-wave generation and we show (using a parameterised version of quasi-linear theory) that the energy transfer from air to sea is indeed sea-state dependent. In Section 3 we discuss some consequence for wave modeling and we compare the surface stress, as calculated with this theory, with observed stresses. The agreement is good. A summary of conclusions and a discussion of consequences for atmospheric (climate) modeling and storm surge modeling is given in Section 4.

Most of the results in this paper have been published elsewhere. The quasi-linear theory of wind wave generation was discussed in Janssen (1982), results on the sea-state dependent momentum transfer were given in Janssen (1989) and the parameterised version of quasi-linear theory is presented in Janssen (1991). The experimental verification of the sea-state dependent surface stress will be published shortly.

2. Quasi-linear theory of wind-wave generation

A possible mechanism for the generation of water waves by wind is resonant interaction of the gravity waves with a plane-parallel flow (Miles, 1957). Resonance occurs at a critical height z_c if $U(z_c) = c(\omega)$, where U is the air velocity and $c(\omega)$ is the phase velocity of a wave with frequency ω. Only those waves grow for which the curvature of the velocity profile at the critical height is negative. However, Fabrikant (1976) and Janssen (1982) found that while the water waves grow owing to energy transfer from the air flow, the waves in turn try to modify the flow in such a way that the air-sea energy transfer is quenched. Thus, there are two competing mechanisms to control the shape of the wind profile, namely, turbulent diffusion that would give a logarithmic wind profile, and wave diffusion that would result in a linear wind profile. Janssen (1989) discussed in detail the competition of the above mentioned mechanisms and by numerical solution he found that the air-sea momentum transfer is sea-state dependent, in qualitative agreement with observations of Donelan (1982).

In practical application, such as the coupling of a wave prediction model to the surface layer of an atmospheric model, the numerical solution of the momentum balance is too time consuming. Therefore there is need for a parameterisation of the effect of waves on the wind velocity (Janssen, 1991). The basic assumption Janssen made, which was corroborated by his numerical results of 1989, was that even for 'young' wind sea the wind profile has a logarithmic shape, however, with a roughness length that depends on the sea state through the wave induced stress. As already indicated by Miles (1957) the growth rate γ of the waves by wind is then given by

$$\frac{\gamma}{\omega} = \epsilon\beta\{\frac{u_*}{c}\cos(\theta-\phi)\}^2, \tag{2}$$

where ϵ is the air-water density ratio, ϕ is the wind direction and β the so-called Miles' parameter, which depends on the ratio of the friction velocity u_* and phase speed c and on the profile parameter

$$\Omega = \kappa g z_0/u_*^2, \tag{3}$$

where κ is the von Karman constant and z_0 is the roughness length. The functional form of β is given in Janssen (1991), but it is clear that the growth rate γ depends through Ω on the roughness of the air flow, which in its turn depends on the sea state, as we will see in a moment.

From a consideration of the momentum balance of air Janssen (1991) found that the kinematic stress $\tau = u_*^2$ is given as

$$\tau = (\kappa U(L)/ln(L/z_0))^2, \tag{4}$$

where

$$z_0 = \alpha\tau/g\sqrt{1-y},\ y = \tau_w/\tau. \tag{5}$$

Here, $U(L)$ is the wind speed at height L and τ_w is the wave-induced stress

$$\vec{\tau}_w = \rho_w \int df\, d\theta\, \omega\gamma F \vec{l},\ \vec{l} = \vec{k}/k. \tag{6}$$

In practice, the wave stress points in the wind direction as it is mainly determined by the high-frequency waves that respond quickly to changes in the wind direction. Eq. (5) shows that the roughness length is given by a Charnock relation in which the Charnock constant depends on the sea state. Therefore, Ω also depends on the sea state. The constant α was chosen in such a way that for old wind sea the Charnock constant has the value 0.0185 in agreement with the results of Wu (1982) on the surface stress over sea waves.

The input source term S_{in} in the energy balance equation now becomes

$$S_{in} = \gamma F \tag{7}$$

with γ given by Eq. (2). The dissipation source function is based on Hasselmann's theory of white capping and it reads

$$S_{diss} = C_{diss} < \omega > (< k >^2 E)^2[(1-\delta)k/ < k > +\delta(k/ < k >)^2]F, \tag{8}$$

where C_{diss} and δ are constant, E is the wave variance and $< k >$ and $< \omega >$ are mean wave number and mean angular frequency respectively. Using the discrete interaction approximation for the nonlinear interactions and the numerical method of WAMDI (1988), we present in the next Section numerical solutions of the energy balance equation (1). Throughout the rest of this paper we choose the relevant constants as $\alpha = 0.01$, $C_{diss} = 4.5$, and $\delta = 0.5$.

3. Results of coupled wind-wave modeling

In this Section, we discuss two topics. By using a single grid point version of the wave model we show that a) for given wind speed, the surface stress τ and hence the drag coefficient $C_D = (u_*/U_{10})^2$ depends on the sea state and b) that the surface stress produced by the parameterised version of quasi-linear theory is in good agreement with observations.

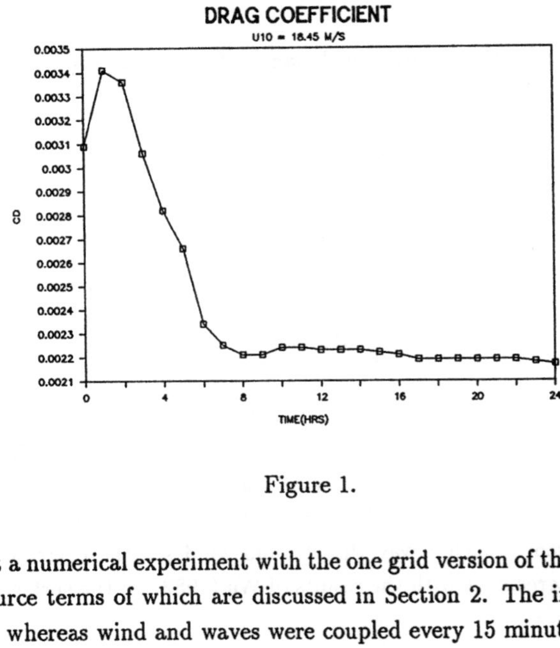

Figure 1.

We carried out a numerical experiment with the one grid version of the coupled wind wave model, the source terms of which are discussed in Section 2. The integration time step was 3 minutes whereas wind and waves were coupled every 15 minutes so that there was sufficient time for the wind to relax to a new equilibrium. The wind speed U_{10} was chosen to be 18.45 m/s, corresponding, in the absence of waves, to a friction velocity of 0.85 m/s. Results of this experiment are discussed in full detail in Janssen (1991). Here, we only show in Fig. 1 the evolution of the drag coefficient as a function of time. Evidently, the drag coefficient varies by about 50% and is larger for small time since gravity waves are then much steeper resulting in a larger wave-induced stress.

In order to check whether this model of the sea-state dependent surface stress produces realistic results we compare in Fig. 2 modeled with observed friction velocity. Here, the modeled friction velocity was obtained from an iterative solution of Eq. (4), where we used observed wave spectra and wind speeds. The observed friction velocity was obtained by independent means through the eddy correlation method. The data were obtained from HEXMAX (Smith et al, 1990). Evidently, the agreement is good, much better than when the Charnock relation for the roughness length is used.

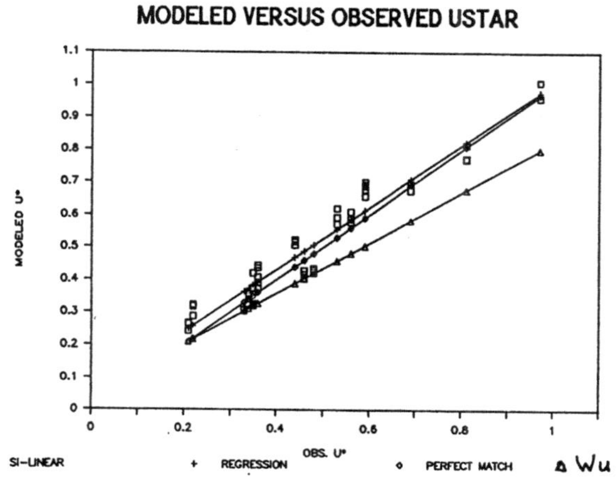

Figure 2.

4. Summary of Conclusions

In this paper we have presented evidence for the effect of gravity waves on the air flow. The momentum flux of air flow over surface gravity waves depends on the sea state because the wave-induced stress τ_w may be a considerable fraction of the total stress. This is especially true for young wind sea when the steepness of the wave is large.

We applied the parameterised version of quasi-linear theory to the problem of the coupling of wind and waves. The same formulation was used in a favourable comparison between observed and modeled surface stress.

The result is that when coupling atmosphere and ocean models we now have a consistent energy balance at the air-sea interface. Allowing for a sea-state dependent drag will most certainly affect the evolution of depressions, while there will also be impact on results from storm-surge models.

5. References

[1] Charnock, J. 1955. Wind stress on a water surface. *Q.J.R. Met. Soc.* **81**, 639-640.

[2] Donelan, M.A. 1982. The dependence of the aerodynamic drag coefficient on wave parameters. Proc. First Int. Conf. on Meteorology and Air-sea Interaction of the coastal Zone, The Hague, *A.M.S.*, 381-387.

[3] Fabrikant, A.L. 1976. Quasilinear theory of wind-wave generation. *Izv., Atmos. Oceanic Phys.* **12**, 524.

[4] Janssen, P.A.E.M. 1982. Quasilinear approximation for the spectrum of wind-generated water waves. *J. Fluid Mech.* **117**, 493-506.

[5] Janssen, P.A.E.M. 1989. Wave-induced stress and the drag of airflow over sea waves. *J. Phys. Oceanogr.* **19**, 745-754.

[6] Janssen, P.A.E.M. 1991. Quasi-linear theory of wind wave generation applied to wave forecasting. Accepted for publication in *J.P.O.*.

[7] Miles, J.W. 1957. On the generation of surface waves by shear flows. *J. Fluid Mech.* **3**, 185.

[8] Smith, S.D., K.B. Katsaros, W.A. Oost and P.G. Mestayer, 1991. Two major Experiments in the Humidity Exchange over the Sea (HEXOS) Program, *BAMS* **71**, 161-172.

[9] The WAMDI-group: S. Hasselmann, K. Hasselmann, E. Bauer, P.A.E.M. Janssen, G.J. Komen, L. Bertotti, P. Lionello, A. Guillaume, V.C. Cardone, J.A. Greenwood, M. Reistad, L. Zambresky and J.A. Ewing, 1988. The WAM model, a third generation ocean wave prediction model. *J. Phys. Oceanogr.* **18**, 1775-1810.

[10] Wu, J. 1982. Wind-stress coefficients over sea surface breeze to hurricane. *J. Geophys. Res.* **87**, 9704-9706.

The Dissipation Subrange of Wind Wave Spectra

Sergei A. Kitaigorodskii

Department of Earth and Planetary Sciences,
Johns Hopkins University,
Baltimore, MD 21218, U.S.A.

Summary

In Kitaigorodskii (1983), it was suggested that due to the breaking of wind waves in deep water, the dissipation of wave energy is restricted to a range of wavenumbers much higher than the wavenumbers typical for the so-called equilibrium range, or much higher than the peak wavenumber k_p and peak frequency ω_p. This prediction, similar to the prediction of the existence of dissipation subrange of Komogoroff's three dimensional turbulence, was never before properly verified for obvious reasons: difficulties related to the measurements and interpretation of the random wave field properties at very high frequencies (or wavenumbers). In this paper an attempt is made to summarize the results of recent field experiments (Leykin and Rosenberg 1989; Tang and Shemdin, 1983; Birch and Ewing, 1986; Hansen et al., 1990; Banner et al., 1989; Banner, 1990; and some others) with the purpose of demonstrating that the rapid spectral fall off needed for determination of the boundaries of dissipation subrange in the wavenumber frequency domain, seems to be an intrinsic property of rather well-developed seas.

It is shown that in the dissipation sub-range, the spectrum most likely, has the form $S(\omega) = \beta g^2 \omega^{-5}$; $\beta \approx 0.0025$, g is gravity and the dissipation of energy is restricted to a range of frequencies $\omega > \omega_g$ much higher than the frequencies of the dominant waves. The empirical characteristics of the transition from the rear high frequency and high wavenumber parts of the wave spectra to the dissipation subrange are summerized, and it is demonstrated that they favour Kitaigorodskii (1983) prediction about the movement of the transitional region towards low frequencies and low wavenumbers as waves grow.

1. Introduction

Let us define the region in 2-D wavenumber space $\mathbf{k} = (k, \theta)$ as the equilibrium range of $F(\mathbf{k})$, $\mathbf{k} = (k\cos\theta, k\sin\theta)$, where $F(\mathbf{k})$ is a statistically averaged characteristic of the wave field, if

$$\frac{DF}{Dt} = \frac{\partial F}{\partial t} + c_g \nabla_{\mathbf{x}} F \equiv O \qquad (1)$$

for (k, θ) belonging to the equilibrium range.

In practice we are met with situations wherein a certain part of the $\mathbf{k}$ domain

$$\left|\frac{DF}{Dt}\right| \ll S(\mathbf{k}), \qquad (2)$$

where $S(\mathbf{k}) = S(\mathbf{k})_{nl} + S(\mathbf{k})_{in} + S(\mathbf{k})_{diss}$, represents the sum of the so-called source terms

M. L. Banner, R. H. J. Grimshaw (Eds.)
Breaking Waves
IUTAM Symposium Sydney/Australia 1991

in the wave transport equation

$$\frac{DF}{Dt} = S(\mathbf{k}). \tag{4}$$

The condition (2) is fulfilled in a fetch growth situation for major parts of rear faces of wind-wave spectra, which means that they are in quasi-equilibrium (can only depend parametrically on t or $\mathbf{x}$). It is also possible that (2) can be applied to the whole energy containing range of wavenumbers and frequencies, and that is why the similarity descriptions of wind-wave development according to Kitaigorodskii's scaling (Kitaigorodskii, 1960; Pierson and Moskowitz, 1964), and parametric description of wave spectra according to JONSWAP (Hasselmann et al., 1976) are both applicable and successful in the prediction of the growth of the wind-wave field with fetch (or duration).

2.

The fact that we must talk about an equilibrium, not only with respect to k, but also to θ, i.e. above a certain region k, θ in the spatial spectrum $\psi_s(\mathbf{k})$ was first pointed out in Kitaigorodskii et al., (1975), and later on was strengthened by Banner (1990) in connection with a description of equilibrium in wave spectra. *One* of the possible forms of equilibrium conditions in a wind-wave field, related to the limitation in the growth of waves imposed only by *gravitational* instability — surface wave breaking (Phillips, 1958), is often called *saturation*, or *saturation spectra*, described as:

$$\psi_s(k, \theta) = Bk^{-4}\varphi(\theta) \tag{5}$$

$$S(\omega) = \beta g^2 \omega^{-5}, \tag{6}$$

where $B = 2\beta$, and $\psi_s(k, \theta)$ is a symmetrical wavenumber spectrum. The range of scales where (5, 6) are valid approximations must be, however, distinguished from what we introduce below as a general definition of the dissipation subrange.

As the boundaries of the saturation region in the k, θ plane are not known, *a priori*, no unambiguous conclusion about the shape of the wavenumber modulus spectra can be drawn even if the function in $\varphi(\theta)$ in (5) satisfying the standard normalization condition $\int_{-\infty}^{\infty} \varphi(\theta) d\theta = 1$ is known, unless one makes an additional assumption not emanating from the similarity theory itself.

3.

On the other hand, contrary to the definition of *saturation* based on pure dimensional arguments (Phillips, 1958), the dissipation subrange can be introduced according to Kitaigorodskii (1983), and Hansen et al. (1990) in the framework of equations (2, 3), i.e. as the range of wavenumbers $k \geqslant \hat{k}_g$ if

$$S(\mathbf{k})_{nl} = O \qquad \text{for} \quad k \leqslant \hat{k}_g, \tag{7}$$

and

$$S(\mathbf{k})_{nl} - S(\mathbf{k})_{diss} = O \qquad \text{for} \quad k \geqslant \hat{k}_g. \tag{8}$$

Here it is assumed that the wind energy input is negligible near the transitional wavenumber $\hat{k}_g$ and that wave breaking is important only at wavenumbers higher than the wavenumber $\hat{k}_g$, where gravitational instability starts to play a cruicial role. The $\hat{k}_g$ value is, supposedly, much higher than the spectral peak wavenumber k_p and can depend also on direction θ. The basic role of weak nonlinear interactions in the whole range of scales is then in redistributing energy from the range of $k_p \leqslant \mathbf{k} \leqslant \hat{k}_g$ to new waves with $k \leqslant k_p$ (Kitaigorodskii, 1983; Zakharov and Zaslavskii, 1982) and to dissipation $k > \hat{k}_g$ in such a way that nonlinear divergence of energy in the range $k_p \leqslant k \leqslant \hat{k}_g$ is either in balance with the wind energy input S_{in} in an equilibrium field (Phillips, 1985), or simply equal to zero, as in the type of cascade regime (Kitaigorodskii, 1983). Here by $\hat{k}_g$, it is worthwhile to understand the average value $\hat{k}_g = \int_\theta k_g(\theta)d\theta$, so that angular distribution of the boundaries of the dissipation subrange can be weighted towards $\hat{k}_g$ and also the transitional frequency ω_g can be related to $\hat{k}_g$ in the usual way through the isotropic dispersion relationship $\omega_g = (g\hat{k}_g)^{1/2}$ if Doppler shifting is not taken into account.

Thus from such a definition of the dissipation subrange, it follows that in the wavenumber (or frequency) domain, the subrange has to occur as a more rapid spectral fall-off compared with one in the equilibrium range of the wave spectra, and that without the latter, there will be no dissipation subrange at all if we follow the above introduced terminology. According to it, $S(\mathbf{k})_{nl} \approx S(\mathbf{k})_{in}$ in the equilibrium range of spectra, based on the so-called balance of source terms (Phillips, 1985), whereas

$$S(\mathbf{k})_{nl} = S(\mathbf{k})_{in} \equiv 0 \tag{8a}$$

is just a partucular type of statistical equilibrium which can be called Kolomogroff's type of statistical equilibrium and is statistically governed by the so-called wave kinetic equation

$$S(\mathbf{k})_{nl} \equiv 0. \tag{9}$$

4.

If the region of energy input from wind in the case of rather well-developed waves corresponds approximately to

$$k_{inp} \simeq \frac{(4 \sim 6)g}{u_a^2} \approx (2 \sim 3)k_p, \tag{10}$$

where k_p is a peak wavenumber, then in 2-D wavenumber space (k, θ) there is a region

$$(9 \sim 10)k_p \geqslant k \geqslant (2 \sim 3)k_p \approx k_{in}, \tag{11}$$

where, in spite of the directionality of the wave field, the nonlinear interactions can still play a major role both in redistributing energy between different directions, as well as giving rise to smaller scale waves. The region (8) characterized by strong directionality and symmetry relative to the direction of propagation of the dominant waves (or mean wind direction) can still be called an equilibrium range of wind-wave spectra (contrary to the terminology of Phillips (1985) and Banner (1990)). The boundaries $k_{diss}(\theta)$ of this equilibrium range inside which dissipation due to the breaking becomes important, may also depend on the polar angle θ, i.e.

$$k_{diss} = k_{diss}(\theta) = k_g(\theta) \tag{12}$$

with effective value $\hat{k}_g$ equal to

$$\hat{k}_g = \int_\theta k_g(\theta)d\theta = \int_\theta k_{diss}(\theta)d\theta. \tag{13}$$

The transitional frequency ω_g, which is supposed to mark the transition from the non-dissipative part of the equilibrium range and the dissipative part, can then be defined as

$$\omega_g = \sqrt{g\hat{k}_g}. \tag{14}$$

The value of ω_g (defined through $\hat{k}_g$) must be close, but not necessarily equal to experimentally derived ω_g values as a transitional frequency or beginning of rapid spectral fall-off on the rear face of the frequency spectra, associated with the asymptotic approach to saturation form (3, 4).

Actually (Glazman, 1990), the spectral fall-off can identify a more rapid decrease of energy with k or ω than in (5, 6). However, by careful analysis of existing data, we (Kitaigorodskii, 1991) were able to present the results of different experiments in a way which allowed us to determine the transition to the dissipation subrange as a function of the stage of wave development.

5.

Some of the results of such data analysis are presented herein (Fig. 1 and Table 1). It follows from the data in this figure, which shows the variation of transitional frequency $\tilde{\omega}_g = \frac{\omega_g U_a}{g}$ with non-dimensional peak frequency $\tilde{\omega}_p = \frac{\omega_p U_a}{g}$, that the assumption about the existence of the dissipation sub-range (Kitaigorodskii, 1983) is well-proven since in most of the cases the transition found above can be definitely associated also with the existence of an energy cascade pattern in wind-wave spactra at the scales larger than the transitional scale $\lambda_g = \frac{2\pi}{k_g}$. This conclusion is a consequence of the observed (see Fig. 1) trend of the movement of the lower (in wavenumber and frequency) boundary of the dissipation subrange towards the lower frequencies with wind-wave growth (or decrease of $\tilde{\omega}_p$), which is in agreement with the prediction (Kitaigorodskii, 1983), that

$$\hat{k}_g = C\frac{g}{\epsilon_0^{2/3}} \tag{13a}$$

and

$$\omega_g = C^{1/2}\frac{g}{\epsilon_0^{1/3}}, \tag{14b}$$

where C is a numerical constant, not necessarily of the order of one (because the dissipative process associated with wave breaking is not well defined either physically or formally). In (13a, 14b) $\epsilon_0 = \int_\theta \epsilon(\theta)d\theta$ represents a constant (independent of k) energy flux from the region of energy input through the non-dissipative part of the spectra towards high wavenumbers which, according to the general expectations as well as the results of numerical calculations, must increase with wave growth (with a decrease in $\tilde{\omega}_p$).

The value of C can be, in principle, related to the values of universal constant B (5) and another universal constant A in the expression

$$\psi_s(k,\theta) = A\epsilon_0^{1/3}g^{-1/2}k^{-7/2}\phi(\theta) \tag{15}$$

according to Kitaigorodskii, (1983); Hansen et al., (1990) in the following way:

$$C = \left(\frac{B}{A}\right)^2. \tag{16}$$

With $B = 2\beta = 0.005$, $A = 0.12$ (Hansen et al., 1990) it leads to $C \approx 1.6 \times 10^{-3}$ in agreement with the first guess in Kitaigorodskii, (1983).

6. References

Banner, M.L., I.S.F. Jones and J.C. Trinder, 1989 Wavenumber spectra of short gravity waves. *J. Fluid Mech.* **198**, 321-344.

Banner, M.L., 1990 Equilibrium spectra of wind-waves. *J. Phys. Oceanography* **20**, 966-984.

Birch, K.G., and T.A. Ewing, 1986 Observations of wind-waves on a reservoir. Report No. 234. *Institute of Oceanographic Sciences, Wormley.*

Glazman, R.E., 1990 Near-nadir radar backscatter from a well-developed sea. *Radio Science* **25**(6), 1211-1219.

Hansen, C., K.B. Katsaros, S.A. Kitaigorodskii, S.E. Larsen, 1990 The dissipation range of wind-wave spectra observed on a lake. *J. Phys. Oceanography* **20**, 1264-1277.

Hasselmann, K., D. Ross, P. Muller and W. Sell, 1976 A parametric wave prediction model. *J. Phys. Oceanog.* **6**, 200-228.

Kitaigorodskii, S.A., 1961 Application of the theory of similarity to the analysis of wind-generated wave motion as a statistic process. *Bull. Acad. Sciences, U.S.S.R., Geophys. Ser.* **1**, 73-80.

Kitaigorodskii, S.A., 1983 On the theory of the equilibrium range in the spectrum of wind-generated gravity waves. *J. Phys. Oceanography* **13**(5), 817-827.

Kitaigorodskii, S.A., 1991 The dissipation subrange in wind-wave spectra (experimental evidence). *J. Phys. Oceanography* (submitted).

Leykin, I.A., and A.D. Rozenberg, 1984 Sea tower measurements of wind-wave spectra in the Caspian Sea. *J. Phys. Oceanography* **14**, 1168-1176.

Phillips, O.M., 1958 The equilibrium range in the spectrum of wind-generated ocean waves. *J. Fluid Mech.* **156**, 426-434.

Phillips, O.M., 1985 Spectral and statistical properties of the equilibrium range in wind-generated gravity waves. *J. Fluid Mech.* **4**, 505-531.

Pierson, W.J. and L. Moskovitz, 1964. A proposed spectral form for fully developed wind-seas based on the similarity theory of S.A. Kitaigorodskii. *J. Geophys. Res.* **69**, 5181-5190.

Tang, S., and O.H. Shemdin, 1983 Measurements of high frequency waves using a wave follower. *Geophys. Res.* **88**, 9832-9840.

Zakharov, V.E. and M.M. Zaslavskii, 1982 The kinetic equation and Kolmogorov's spectra in weak turbulence theory of wind-waves. *Atmospheric and Oceanic Phys.* **18**, 747-753.

TABLE 1

The summary of characteristics of the transition to the dissipation subrange in wind wave spectra

NN	Ratio of transitional frequency ω_g to peak frequency ω_p (average value and range of variability)	Nondimensional transitional frequencies $\frac{\omega_g U_a}{g}$ (average value and range of variability)	Nondimensional peak frequency $\frac{\omega_p U_a}{g}$ (average value and range of variability)	Comments	Source
1	4.8 - 5.25 5.0	6.0 5.76 - 6.3	1.20	Run 108 N IV in the source	Tang & Shemdin, 1983
2	5.57 4.98 - 5.16	5.35	0.96	Run 108 N III in the source	Tang & Shemdin, 1983
3	5.06	4.44 - 4.59	0.89	Run 326 N II in the source	Tang & Shemdin, 1983
4	5.48	2.14	0.39	Run 328 not clearly defined ω_g (in this case) $\omega_g \equiv 6$ rad/sec.	Tang & Shemdin, 1983
5	4.48 - 5.97 5.21	3.0 - 4.0 3.5	0.54	Exp. 3, Td = Tp	Banner et al., 1989
6	3.59 - 4.10	7.0 - 8.0	1.54	Exp 4 Td ≡ Tp in Banner et. al.	Banner et al., 1989
7	1.78	1.28	0.78	See (1) below	SWOP
8	2.85	7.36	2.58		Hansen et al., 1990 al 8 run 246
9	3.90	7.45	1.91		Hansen et al., 1990 run 248
10	3.29	5.73	1.74		Hansen et al., 1990 run 252
11	3.28	4.05	1.23		Hansen et al., 1990 run 254
12	3.2 3.0	4.48 - 6.4 5.4 3.75 - 6.0	1.25 - 2.05 (1.65)		Leykin and Rosenberg (1984)

(1). These numbers are for $k_g = 0.2m(\frac{k_g}{k_p} = 3.2)$; $U_a = 9$m/sec., $\frac{\omega_g}{\omega_p} = 1.78$; $\frac{\omega_g U_a}{g} = 1.28$. If instead of $k_g = 0.2$, accept P.M. fully developed, or $\omega_p = 1.42g/U_a$, then $\omega_g = 3.56$ or 1.57, that means that $k_g = 0.2$m is small. For larger $k_g \frac{\omega_g U_a}{g}$ will be larger than 1.28.

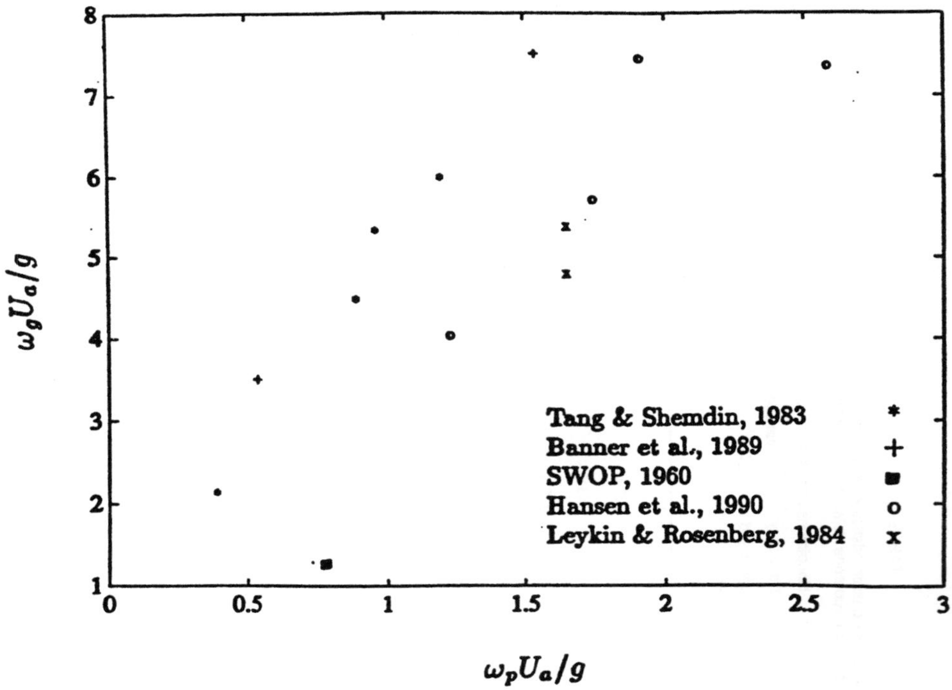

Figure1. Nondimensional transitional frequency $\omega_g U_a/g$ vs. nondimensional peak frequency $\omega_p U_a/g$

A Model for the Response of Wave Directions to Changing Wind Directions for Random Breaking Waves

LUPING LI BIN GONG

First Institute of Oceanography
State Oceanic Administration
Qingdao, P.R. China

Abstract
On the basis of the wave energy balance equation, a model for the response for random breaking waves is derived. The results show that, the dissipations of white-capping tend to increase rather than to slow down the turning rate induced by the wind generation. For fully developed wind waves, the dissipation due to white-capping forces the mean wave directions to turn to the wind directions. This model can be used in the case of random nonbreaking waves, and the models for slowly turning wind directions and in inhomogeneous situations are derived with the same methods.

Introduction

The information of wave directions is very important for engineering. However, it is poorly understood because of the lack of theoretical researches and observations. Especially, the response of wave directions to changing wind directions has not been deeply researched. In our knowledge, reasonable models for this response have not been presented or combined with numerical wave modellings. But in practical situations winds always change their directions. Therefore, many researchers have devoted to studying these problems in recent years and several models are presented (Gunther et al., 1981; Hasselman et al., 1980; Allender et al., 1983; Holthuijsen et al., 1988; and Li Luping et al., 1990). Among these models, the Holthuijsen et al.'s is the first one to be derived theoretically. This approximation model is based on the mean wave direction defined with Pitch-and-Roll Buoy observations and derived using the wave energy balance equation. It is pointed out that the time scale of the response can be estimated from the rate of wave growth.

Li Luping et al. (1990) directly derived a new model for the response based on the wave energy balance equation under the same assumptions as

M. L. Banner, R. H. J. Grimshaw (Eds.)
Breaking Waves
IUTAM Symposium Sydney/Australia 1991

Holthuijsen et al. did. This model can be used for a sudden change in wind directions and for slowly turning wind directions in homogeneous and inhomogeneous situations. The results show that the time scale of the response is not only related to the rate of wave growth, but also to the directional energy distributions and the angles between the mean wave direction and the wind direction. From the models the regulation of the change in the mean wave direction is derived theoretically. Therefore, the change of mean wave directions with time can be numerically calculated. However, these models presented by now only can be used in the case of random nonbreaking waves, that means that the time scale of the response for these models includes no contributions of the dissipation due to white-capping. It is noted that G. Ph. Van Vledder et al. (1988) estimated the effects of wind generation and white-capping and nonlinear wave-wave interactions on the directional response of waves in a turning wind field using a numerical model to compute to a high degree of accuracy nonlinear wave-wave interactions of wind generated waves. It is pointed out that nonlinear wave-wave interactions and dissipations due to white-capping tend to slow down the turning rate induced by the wind generation.

The purpose of this paper is to estimate theoretically the contribution of white-capping to the directional response of waves on the basis of the wave energy balance equation. A model for the response of wave directions to changing wind directions for random breaking waves has been derived. From this study some results are presented for fully developed wind waves. The results show that, under the turning wind directions, the dissipation due to white-capping like wind generation increases the rate of wave direction turning to the wind direction and for fully developed wind waves the dissipations due to white-capping force the mean wave direction to continue turning to the wind direction. In this paper quantitative version of calculating the mean wave directions for fully developed wind waves has been given.

Derivation of a Model for the Response

On the basis of the wave energy balance equation for deep water, we have:

$$\frac{\partial E(f,\theta)}{\partial t} + \nabla \cdot [C_g(f,\theta) \cdot E(f,\theta)] = S(f,\theta) \qquad (1)$$

in which: $E(f,\theta)$ --- two dimensional wave spectrum
$Cg(f,\theta)$ --- energy propagation velocity of frequency f and in direction θ in x-y space
$S(f,\theta)$ --- source function
∇ --- gradient operator in x-y space

For a homogeneous wave field Eq.(1) reduced to:

$$\frac{\partial E(f,\theta)}{\partial t} = S(f,\theta) \tag{2}$$

Here the source function $S(f,\theta)$ represents the sum of all processes of wave generation (wind input) and dissipation (white-capping) and non-linear wave-wave interaction.i.e.,

$$S(f,\theta) = S_{in}(f,\theta) + S_{nl}(f,\theta) + S_{dis}(f,\theta) \tag{3}$$

To arrive at a simple model, the following assumnptions are made here. That is, the shapes of the directional distributions of $E(f,\theta)$ and $S(f,\theta)$ are frequency independent, equal to each other and symmetric around their mean directions. Therefore we have:

$$E(f,\theta) = E(f) \cdot D(\theta) \tag{4}$$

$$S_{in}(f,\theta) = S_{in}(f) \cdot D'(\theta) \tag{5}$$

$$S_{nl}(f,\theta) = S_{nl}(f) \cdot D''(\theta) \tag{6}$$

$$S_{dis}(f,\theta) = S_{dis}(f) \cdot D'''(\theta) \tag{7}$$

From the knowledge available it is reasonable to assume that the two-dimensional spectrum, the dissipation, and nonlinear wave-wave interaction source functions have the same directional disrtributions. Therefore we have:

$$D(\theta) = D''(\theta) = D'''(0) \qquad (8)$$

Eq.(8) means that in the past study it is not correct to assume that the *source function of wave generation* S_{in} *(f, θ), the source function of dissipations due to white-capping* S_{dis} *(f, θ)* and *nonlinear wave-wave interaction* S_{ne} *(f, θ) are of the same directional distribution.* But up to now, all of the models presented use this assumption. In fact, these models equivalently did not consider the influence of the dissipations of white-capping and nonlinear wave-wave interaction on the response of wave directions. According to the above assumptions the directional distributions of two dimensional spectrum can be written:

$$D(\theta) = K \cdot cos^2(0 - \theta_0) \qquad -\frac{\pi}{2} \leqslant 0 - 0_0 \leqslant \frac{\pi}{2} \qquad (9)$$

$$D(\theta) = I(s) \cdot cos^{2s}\left(\frac{0 - \theta_0}{2}\right) \qquad -\pi \leqslant 0 - 0_0 \leqslant \pi \qquad (10)$$

In general, Eq.(10) has a narrow directional distribution because s > 10 is suitable for wind waves. For developing wind waves, we would like to use Eq.(9) for convenience. Therefore the directional distribution of wind generation source functions is

$$D'(\theta) = K \cdot cos^2(\theta - 0_w) \qquad -\frac{\pi}{2} \leqslant 0 - 0_w \leqslant \frac{\pi}{2} \qquad (11)$$

in which θ_w *is the new wind direction.* From Eq. (2) to (11) the wave energy balance equation can be written as:

$$\frac{\partial}{\partial t}[E(f) \cdot Kcos^2(\theta - \theta_0)] = S_{in}(f) \cdot Kcos^2(\theta - 0_w) + S_{nl}(f) \cdot Kcos^2(\theta - \theta_0) + S_{dis}(f) \cdot Kcos^2(0 - \theta_0) \qquad (12)$$

Based on the wave energy balance equation, for a fixed point, we have:

$$\int_0^\infty \frac{\partial E(f)}{\partial t} df = \int_0^\infty [S_{in}(f) + S_{nl}(f) + S_{dis}(f)] df \qquad (13)$$

in which, nonlinear wave-wave interaction has a property of conservation. That is:

$$\int_0^\infty S_{nl}(f) df = 0 \qquad (14)$$

Therefore, Eq. (13) reduces to:

$$\int_0^\infty S_{dis}(f)df = \frac{\partial m_0}{\partial t} - \int_0^\infty S_{in}(f)df \qquad (15)$$

in which $m_0 = \int_0^\infty E(f)\,df$ is the total wave energy. For any intervals with $0<\alpha<\frac{\pi}{2}-|\theta_w-\theta_0|$ $|\theta_w-\theta_0|<\frac{\pi}{2}$ Eq.(12) can be written as:

$$\frac{\partial}{\partial t}[\int_{\theta_w-\alpha}^{\theta_w+\alpha}\int_0^\infty E(f)\cdot K\cos^2(\theta-\theta_0)dfd\theta]' = \int_{\theta_w-\alpha}^{\theta_w+\alpha}\int_0^\infty S_{in}(f)\cdot K\cos^2(\theta-\theta_w)dfd\theta$$

$$+\int_{\theta_w-\alpha}^{\theta_w+\alpha}\int_0^\infty S_{dis}(f)\cdot K\cos^2(\theta-\theta_0)dfd\theta \qquad (16)$$

in which θ_w and α are independent to time t. After some straightforward manipulations, Eq.(16) can be reduced as:

$$\frac{\partial}{\partial t}[m_0(2\alpha+\sin2\alpha\cdot\cos2(\theta_w-\theta_0))] = \int_0^\infty S_{in}(f)[2\alpha+\sin2\alpha]df$$

$$+\int_0^\infty S_{dis}(f)[2\alpha+\sin2\alpha\cdot\cos2(\theta_w-\theta_0))]df \qquad (17)$$

Using Eq.(15), then Eq.(17) reduces to the following model for random breaking waves:

$$\frac{\partial\theta_0}{\partial t} = \frac{1}{m_0}[\frac{\partial m_0}{\partial t} - \int_0^\infty S_{dis}(f)df]\cdot\frac{1}{2\cos(\theta_w-\theta_0)}\cdot\sin(\theta_w-\theta_0) \qquad (18)$$

It should be pointed out that Eq.(18) is directly derived from the wave energy balance equation which includes the influence of the dissipations due to white-capping on the response of wave directions to changing wind directions. It is called a model for the response for random breaking wave in order to make a distinction from the presented models. Furthermore, using the Eq.(19), we can get the time scale of the response for random breaking waves:

$$\frac{\partial\theta_0}{\partial t} = \frac{1}{\tau}\cdot\sin(\theta_w-\theta_0) \qquad (19)$$

$$\tau = \left\{\frac{1}{m_0}[\frac{\partial m_0}{\partial t} - \int_0^\infty S_{dis}(f)df]\right\}^{-1}\cdot 2\cos(\theta_w-\theta_0) \qquad (20)$$

Because the dissipation of white-capping $\int_0^\infty S_{dis}(f)\,df$ is negative, we can come to the following conclusion: the dissipation of white-capping tends to increase rather than to slow down the turning rate induced by the wind generation. Eq.(20) shows that the time scale of the response is not only related to the rate of wave growth, directional energy distribution and the angle between the mean wave direction and the wind direction, but also to the dissipation of white-capping.

Nunerical Results of the Model for Fully Developed wind Waves

All of the models presented by now have shown that the time scale of the response approaches to unlimited for fully developed wind waves. That means that for fully developed wind waves the mean wave direction does not response to the change in wind directions. This conclusion is obviously disagreement with the observations.

For fully developed wind waves, we have:

$$\frac{\partial m_0}{\partial t} = \int_0^\infty S_{in}(f)df + \int_0^\infty S_{dis}(f)df = 0 \tag{21}$$

Eq.(21) is replaced in Eq.(18) we have:

$$\frac{\partial \theta_0}{\partial t} = \frac{1}{m_0}[-\int_0^\infty S_{dis}(f)df] \cdot \frac{1}{2cos(\theta_w - \theta_0)} \cdot sin(\theta_w - \theta_0) \tag{22}$$

Therefore, the time scale of the response can be written as

$$\tau = [-\int_0^\infty S_{dis}(f)df/m_0]^{-1} \cdot 2cos(\theta_w - \theta_0) \tag{23}$$

Eq.(22), (23) show that for fully developed wind waves the dissipation due to white-capping forces the mean wave directions to turn to the wind directions. This shows theoretically that the dissipation due to white-capping plays an important balance role in the response of wave directions.

On the basis of the general form for the dissipation due to small scale white-capping processes (Hasselmann, K., 1974), We have:

$$S_{dis}(\omega.\theta) = -1.6\overline{\omega}(\frac{\omega}{\overline{\omega}})^2 \cdot \hat{\alpha}^2 \cdot E(\omega.\theta) \tag{24}$$

in which $\hat{\alpha} = m_0\overline{\omega}^4/g^2$, $\overline{\omega} = m_0^{-1}\iint E(\omega.\theta)\omega \; d\omega \; d\theta$

For fully developed wind waves, one dimensional spectrum can be chosen PM spectrum which has the following form

$$E_{PM}(\omega) = \alpha g^2 \omega^{-5} exp\left\{-\beta(\frac{g}{U_{19.5}\omega})^4\right\} \tag{25}$$

in which $\alpha = 8.10*10^{-3}$, $\beta = 0.74$, $U_{19.5}$ is the wind speed at 19.5m above the sea level.

After some straightforward manipulations, the dissipations due to white-capping for fully developed wind waves can be calculated:

$$\int_0^\infty S_{dis}(f)df = -1.6\frac{m_1^{\ 7} \cdot m_2}{m_0^{\ 5} \cdot g^4} \tag{26}$$

The total wave energy is:

$$m_0 = 27.36 \times 10^{-4} \cdot g^{-2} U_{19.5}^{\ 4} \tag{27}$$

According to Eq.(23) the time scale of the response of wave directions to changing wind directions for fully developed wind waves can be obtained:

$$\tau = (0.43 \times 10^{-4} g / U_{19.5})^{-1} \cdot 2cos(\theta_w - \theta_0) \tag{28}$$

The maximum of the dimensionless time scale of the response is

$$\tilde{\tau} = 4.65 \times 10^4 \cdot A \tag{29}$$

in which $\tilde{\tau} = g\tau/U_{10}$, $A = U_{19.5}/U_{10} < 1.1$, U_{10} *is the wind speed at 10* meters above the Sea level.

Based on Eq.(22), for fully developed wind waves, the changes of the mean wave directions can be obtained:

$$\theta_0(t) = \theta_w - arcsin[e^{-\frac{t}{\tau}} \cdot sin(\theta_w - \theta_0(0))] \tag{30}$$

Eq.(30) gives the regulation of change of the mean wave directions and shows that under the change in wind direction the wave directions for fully developed wind waves are agreement with the wind direction with the increase in time.

Conclusion

In this paper, a model for the response of wave directions to changing wind directions for random breaking waves has been directly derived based on the wave energy balance equation. The results show that the dissipations due to the white-capping tend to increase rather than to slow down the rate of the response of wave directions to changing wind directions reduced by the wind generations.

For fully developed wind waves, the dissipations due to white-capping force the mean wave directions to turn to the wind directions. The change of the mean wave directions has been obtained and the maximum nondimensional time scale is about $\tilde{T} = 46.5 * 10^3$.

This model can be used in the case of random nonbreaking waves, and slowly turning wind diections and inhomogeneous situations. But, it should be pointed out that in this model the contribution of nonlinear wave-wave interaction to the turning rate of the response isn't included because of the property of conservation of nonlinear wave-wave interactions.

References

1. Gunther, H.; Rosenthal, W.; Dunckel, M. 1981.(eds.) The response of surface gravity waves to changing wind directions. J. Phys. Oceanogr. 11, 718-728.
2. Hasselmann, D. E.; Dunckel, M.; Ewing, J. A.(eds.) 1980. Directional wave spectra observed during JONSWAP 1973. J. Phys. Oceanogr. 10. 1264-1280.
3. Allender, J. H.; Albrecht J.; Hamilton, G. (eds.) 1983: Observations of directional relaxation of wind sea spectra. J. Phys. Oceanogr., Vol. 13, 8, 1519-1525.
4. Holthuijsen, L. H.; Mosselman, E.; Kuik, A.J. (eds.) 1984. A model for the response of wave directions to changes in wind direction. Proc. Symp. on Description and modelling of directional seas. June 18-20, 1984. Technical University, Denmark, Paper No. C-4, 7pp.
5. Holthuijsen, L.H.; Kuik, A.J.; Mosselman, E. (eds.) 1987. The response of wave directions to changing wind directions. J. Phys. Oceanogr. 7(7), 845-853.
6. Holthuijsen, L.H.; Smith, D.R. (eds.) 1988. An evaluation of model estimates of ocean wave directions. Ocean Engng. Vol. 15, No.2, pp 127-137.
7. Li, Luping; Gong, Bin (eds.) 1990. The response of wave directions to changing wind directions. J. Ocean Univ. Qingdao, China, will be published.

8. Li, Luping; Gong, Bin (eds.) 1990. Research on the response of wave directions to slowly turning wind directions. Submitted to J. Ocean Univ. Qingdao, China.

9. Van Vledder, G. Ph.; Holthuijsen, L.H. (eds.) 1988. Waves in turning wind fields. Twenty-1st Coastal Engineering Conference, Vol.1. 602-611

Mechanisms of Water Wave Transformation in Shear Currents

Victor I. Shrira
P.P.Shirshov Institute of Oceanology USSR Academy of Sciences,
Krasikova 23, Moscow 117218, USSR

Summary

The paper presents a brief review of the author's recent results on wave transformation and instability due to a shear current. The review covers both linear and nonlinear mechanisms in the models with vertical or/and horizontal shears.

1. Critical-layer instability

Linear instability of a free surface of a fluid influenced by an air shear flow was discovered by Miles (1957) to explain wind wave generation and hereafter will be referred as Miles' instability. This instability takes place owing to the existence of the so-called critical layers, where real part of surface wave phase velocity for chosen harmonic component matches the flow velocity. Here we shall briefly examine the geophysically important problem of stability of drift oceanic currents with free surface (for details see Shrira (1991)).

Mathematical statement of the problem is as follows. We start with the boundary-value problem consisting of the Raleigh equation and standard boundary conditions at the free surface at infinity or bottom. The exact solution to this boundary-value problem can be obtained in terms of a converging series in certain parameter ε, defined as a ratio of characteristic mean flow vorticity gradient to the product of wave frequency and wave number. ε proved to be small for the whole range of wave scales and shear flow parameters relevant for the ocean, which provides very rapid convergency of the series and allows us within the context of our problem to confine ourselves to consideration of the solution leading order term only. If Re C belongs to flow velocity interval $[U_{min}, U_{max}]$, the eigenvalues C ($C = \omega / k_x$) are *always complex*. The criterion of instability is:

$$U''(z_c) > 0,$$

where

$$\mathrm{Re}\, C = U(z_c)$$

M. L. Banner, R. H. J. Grimshaw (Eds.)
Breaking Waves
IUTAM Symposium Sydney/Australia 1991

is the definition of the critical layer $z=z_c$. Growth rate Γ is given by a simple explicit formula

$$\Gamma = (Im\ C)/C_{(0)} = \pi\ C_0/C_{(0)} \left.\frac{e^{2kz}(\mathbf{U}''\cdot\mathbf{k})}{2k\ |(\mathbf{U}'\cdot\mathbf{k})|}\right|_{z=z_c}$$

where C_0 is the wave phase velocity in the still water and

$$C_{(0)} = U_0 - U_0'/2k + C_0(k)$$

(the subscript (0) denotes values taken at the surface). The instability described by these formulae is a direct analogue of Miles' instability. However, the instability of surface waves upon drift current has its specific features important for water wave dynamics:

i) unstable waves propagate *against* the wind (as their phase velocity is smaller than the wind driven current at the free surface);

ii) Γ does not depend on air/water densities ratio and therefore can noticeably exceed Miles' increments;

iii) the bandwidth of instability is very narrow, the necessary condition

$$C_0 < U_{max},$$

selects narrow band in capillary-gravity range under typical conditions;

iv) Γ depends strongly on the fine structure of the thin subsurface layer;

v) in contrast to Miles' case we have an explicit expression for the growth rate.

However, there is a large gap between our idealized model and reality. Among the factors omitted the most important, we believe, is the effect of long waves upon the wave range under consideration. We expect that taking the effect of long waves into account entails:

i) considerable broadening of the instability band; the necessary condition will become much milder

$$C_0 < U_{max} + U_{wmax},$$

where U_{wmax} is the maximal long wave orbital velocity;

ii) essential increase of growth rate Γ.

If we calculated the effect of long waves and had reliable data on the skin-layer current, we could evaluate the importance of this instability for short wave breaking.

2.Explosive instabilities

The term explosive instability (EI, hereafter) is commonly used for the instabilities, where field amplitude tends to infinity in finite time (within the framework of weakly-nonlinear theory). The simplest way to elucidate the EI phenomenon is to use the 'quasi-mode' concept (Shrira 1989). Indeed, suppose we have a shear flow of uniform density of boundary layer type with vorticity localized near the boundary within the thin layer of scale h. If this current has no inflection points, the corresponding boundary-value problem has only continuous spectrum, but *an arbitrary initial perturbation* of scale exceeding greatly shear scale h *evolves as if belonging to a certain single discrete mode* characterized by the dispersion relation

$$C_x = \frac{(\mathbf{U}_0 \cdot \mathbf{k})}{k} - \beta\, k - i\gamma\, k^2$$

within the temporal interval

$$(k\, h)^{-1} \ll t \ll (k\, h)^{-3}.$$

In other words, the 'quasi-modes' are intermediate asymptotics of the Cauchy problem. This fact allows us to treat nonlinear interaction of surface waves with continuous spectrum perturbations both in wind or in drift current as as standard 3-wave interaction.

Consider, as an example, dynamics of a wave triad constituted by two surface gravity-capillary waves and a quasi-mode in air boundary flow (Romanova, Shrira 1989).

For the explosive instability to develop, the 'negative energy' waves should exist and resonance conditions should be satisfied

$$\omega_1 + \omega_2 + \omega_3 = 0; \qquad \mathbf{k}_1 + \mathbf{k}_2 + \mathbf{k}_3 = 0;$$

where the sign of the frequency ω has been chosen to coincide with the sign of the wave energy. These conditions proved to be easily satisfied. We stress the remarkable feature of these triads: the explosive instability can occur in absence of linear instability.

Wave amplitude evolution with cubic nonlinearity taken into account is governed by the following equations:

$$\mathcal{D}_t A_1 = V A_2^* A_3 + i\varepsilon A_1 (\beta_{11}|A_1|^2 + \beta_{12}|A_1|^2)$$

$$\mathcal{D}_t A_2 = V A_1^* A_3 + i\varepsilon A_2 (\beta_{21}|A_1|^2 + \beta_{22}|A_2|^2)$$

$$\mathcal{D}_t A_3 = V A_1 A_2 \qquad \{ \mathcal{D}_t \equiv \partial_t + (\mathbf{C}_g \cdot \nabla)\}$$

This system predicts that a singularity will appear for all amplitudes in finite time if the cubic terms are neglected, $A_i \sim (t - t_e)$. However, the cubic nonlinearity changes behavior qualitatively: nonlinear frequency shift violates the resonance at large amplitudes and the explosion turns into a sequence of spikes. The heights of the spikes depend on the initial parameters and specific features of the system. For example, in the case of water-air explosive triads the interaction coefficient V is proportional to square root of the air-to-water-densities ratio, i.e. is small. Thus, the cubic terms, being even in the initial moment of comparable order, effectively limit the explosion intensity. For water wave - current interaction there is no *a priori* smallness and one should expect much more pronounced bursts, which can provide a possible mechanism for wave breaking in gravity-capillary range.

3.Wave-packet collapse

Let us consider a weakly nonlinear wave packet propagating through a slowly varying media. Under rather general and well-known assumptions, its evolution is governed by the nonlinear Schrödinger equation (NS), which in the reference frame moving with the group velocity we cast in the form (Malomed, Shrira 1991)

$$iu_t + 2\alpha(t)|u|^2 u + \beta(t)u_{xx} - i\gamma u_{xxx} = 0$$

The collapse can occur when wave-packet passes through the points where *dispersion coefficient* β *changes its sign.* The necessary extrema of group velocity always take place for short internal gravity waves in the ocean and for 'skew' or cross water waves. In the vicinity of the critical point the NS equation is convenient to present in the form

$$iu_t + 2|u|^2 u - tu_{xx} = i\gamma u_{xxx}$$

In this notation t is the distance between the packet and the critical point. We shall focus our attention on the case when β changes sign *from positive to negative*, while α is fixed.In the case of constant α and β ($\alpha\beta > 0$) and zero γ, the NS equation possesses the well-known soliton solutions

$$u = 2\eta\ sech\ (2\eta x)\ exp\ (4i\eta^2 t)$$

In the case of slowly varying media, soliton-like solutions exist and their amplitude is slowly varying. It is easy to calculate adiabatic transformation of soliton parameters (neglecting higher order dispersion term):

$$u \sim t^{-1/2} \text{ and } l_s \sim t^{1/2}.$$

Thus we have described within the *one*-dimensional NS equation the wave-packet collapse or explosion (amplitude tends to infinity in finite time, while the characteristic soliton width l_s contracts down to zero). We note that the wave radiation due to non-adiabatic effects does not affect noticeably this behavior (Malomed, Shrira 1991). Third-order dispersion provides limitation of the amplitude growth, but, typically, at a rather high level.

We expect that the packet collapse can act, where occurs, as a very effective mechanism causing wave breaking.

<u>References</u>

1. V.I.Shrira (1991) Surface waves on shear currents:solution of the boundary value problem (JFM, submitted)
2. V.I.Shrira (1989) On the subsurface waves in the oceanic upper mixed layer. Dokl. AN SSSR, v.308, n3, 732-736 (in Russian)
3. N.N.Romanova, V.I.Shrira (1978) Explosive generation of surface waves by wind. Izvestia USSR Academy of Sci..Atmospheric and Oceanic Physics. v.24, n7, 528-35 (English translation)
4. B.A.Malomed, V.I.Shrira (1991) Soliton caustics. Physica D (to appear)

Nonlinear Acoustic Phenomena in Subsurface Bubble Layers and its Usage for Bubble Measurements

A. M. SUTIN

Institute of Applied Physics USSR Academy of Science, 46 Uljanov st., 603600 Nizhny Novgorod, USSR

Summary

This paper contains a brief review of our recent theoretical and experimental investigations of nonlinear acoustic phenomena in sub-surface bubble layers produced by breaking waves and results concerning linear and nonlinear acoustic methods of bubble measurements.

The review of nonlinear scattering effects.

What do we mean by nonlinear sound scattering? As is well known, a bubble in a liquid is a monopole scatterer with prominent nonlinearity. When a harmonic acoustic wave affects a bubble, the spectrum of the scattered wave contains a second harmonic component. When a biharmonic wave affects a bubble, the spectrum of the scattered wave contains second-harmonic components, and components of sum and difference frequencies. The level of these components can be calculated from the formula given in [1]. We introduced the new parameter-nonlinear cross-section of bubble, which relates the pressure amplitudes of nonlinear components in the scattered waves with the incident waves . The results of calculations for nonlinear cross-sections were given in [2-3].

There are many bubbles in subsurface bubble layer and the nonlinear scattered field from this layer can be obtained by summation of the fields scattered by separate bubbles. The scattered field consists of a coherent and noncoherent part. Non-

M. L. Banner, R. H. J. Grimshaw (Eds.)
Breaking Waves
IUTAM Symposium Sydney/Australia 1991

linear noncoherent scattering is similar to linear noncoherent scattering (acoustic reverberation). The difference of nonlinear reverberation from the linear one is in the scattered frequency and the specific dependence of signal amplitude on distance to scattered volume [2]. The estimates show that the level of scattered field at second harmonic frequency can be quite observable for various types of sonars. The nonlinear reverberation takes place at the difference frequency for parametric array.

The coherent part of the nonlinear scattered field can be calculated in a continual approximation where the bubble-liquid mixture is treated as a continuous medium with some averaged parameters. For the medium with bubbles of different sizes the nonlinear parameter was considered in [4,6]. The value of ε can reach 10^3 and even more while the nonlinear parameter ε of water is equal to 3,5. Such a prominent nonlinearity can, in principle, increase the efficiency of the parametric array in the upper layer of the ocean. Calculations [5,6,8] show that only a two- or three- fold increase of the parametric array field due to bubbles is possible. Such low increase is due to the strong attenuation of primary waves in bubble medium, which decreases the effective length of the parametric array.

An interesting possibility is associated with the interaction of acoustic beams in a bubble layer. In particular, the scattering of a biharmonic wave normal to the layer produces difference-frequency waves propagating in both directions from the layer [9,10]. The cross-interaction of two beams in the layer gives the difference frequency radiation in a direction different from that of the incident waves, thus making it possible to scan the directive pattern of a parametric array [11]. In the case where the frequency of one of the incident waves is twice as large as the frequency of the other, the difference-frequency radiation is phase-conjugate, so that a spherically divergent front turns to a convergent one [12].

Acoustic diagnostics of gas bubbles in the sea. The importance of the problems of bubble spectroscopy in ocean acoustics stems from the role of bubbles as an ocean factor, as well as a possible additional source of sea noise, an indicator of biological objects, etc. Methods of bubble diagnostics are based on the main physical peculiarities, which distinguish bubbles among any other sea particles, namely, their acoustic resonance and strong acoustic non-linearity. These methods can be classified by the measured factors (dam ping,scattering,non-linearity). Linear methods based on the measurement of resonance attenuation and scattering were developed just after World War II and have been used (with some modifications) up to now [13].

Measurements of damping are performed now on a relatively small base (1-2 m) but in a wide frequency interval [14]. An advantage of this method is the independence of the dumping factor on losses in a single bubble, so that it is determined only by the concentration of resonance bubbles. The method permits direct measurement of the distribution function of bubbles in radii. However, this method is effective only for rather high concentrations of bubbles because losses are quite a "rough" parameter. Recently a method based on the linear scattering of sound by bubbles has been used more widely . The first measurements using this method were performed at several single frequencies [15-17] but recently the bubble concentration in a wide range of sizes (from 2 10^{-3} to 10^{-1}cm) has been registered by a parametric array operating in a wide frequency band [18].

Non-linear acoustic methods are based on the features of a bubble as a scatterer with prominent non-linearity. Here the registration of the signal takes place at the second harmonic, as well as at summary or difference frequencies of the incident waves [2,3,19]. The advantages of these methods are associated with a possibility to distinguish the bubbles among other scatterers and especially to register them in the presence of strong reverberation, for example, near reflective boundaries. The sensibility of such methods is sufficient for the registration of single bubbles of measuring 5-50 microns. Comparative analysis of the methods for the second harmonic [2,3] and difference frequency [19] shows that the signal scattered at the second harmonic is several orders of magnitude stronger. This simplifies realization of the method (for example, the receiver may be not so sensitive and the signal/noise level is higher). On the other hand, the difference-frequency method has its own advantages: a relatively easy frequency variation and more precise localization of a bubble due to higher operating frequencies.

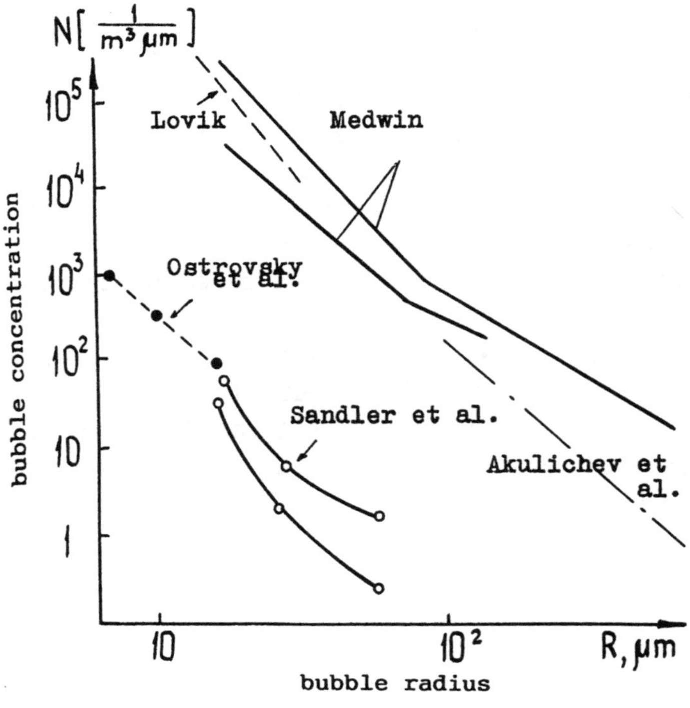

Fig. 1. Ocean measurements of bubble concentration.

Some results of measurements of bubble concentration in the sea by various methods are presented in Fig.1. One can always observe a monotonic growth of bubble concentration with a decrease in their radii. This dependence can be represented by the formula $N \sim R^{-n}$, where the values of n lie within the limits from 2.5 to 4.5 with a most typical mean value 3.5

References

1. Zabolotskaya ,E.; Soluyan ,S.: The radiation of harmonics and combination frequencies by air bubbles. Sov.Phys.Acoustics. 18 (1972) N 3.
2. Ostrovsky ,L.; Sutin ,A.: Nonlinear sound scattering from subsurface bubble layers. Proc. Symp.-Natural physical sources of underwater sound. Cambridge 1990 (to be published)
3. Ostrovsky, L.; Sutin ,A : Nonlinear acoustic methods in diagnostics, in: Ultrasound Diagnostics. Grekhova (ed.), Gorki,(in Russian) 1983.
4. Kobelev, Yu.; Ostrovsky, L,; Sutin A.:Self indused transparency and frequency conversion effects for acoustic waves in water containing gas bubbles. Cavitation and Inhomogeneities in Underwater Acoustics. W.Lauterborn, ed., Berlin-Heidelberg-New York, Springer-Verlag 1980.
5. Naugolnich, K.; Ostrovsky, L.; Sutin, A. . Parametric array, in: "Nonlinear acoustics" . Zverev, Ostrovsky (eds), Gorky 1980 (in Russian).
6. Kobelev, Yu.; Sutin, A. : Difference frequency generation in liquids with different size bubbles. Sov. Phys. Acoustics 26 (1980) N 6.
7. Nazarov, V.; Sutin, A.: Parametric array with bubble layer in far field zone. Sov. Phys. Acoustics 30 (1984) N 6.
8. Lerner, A.; Sutin , A.: Optimization of saturation limited parametric array. Sov. Phys. Acoustics. 34 (1988) N.4.

9. Kustov, L.; Nazarov, V.; Ostrovsky, L.; Sutin, A.; Zamolin, S. :Parametric acoustic radiator with a bubble layer. Acoustics Letters. 6 (1982) 15-17.
10. Donskoi, D.; Zamolin, S.; Kustov, L.; Sutin, A.: Nonlinear backscattering of acoustic waves in a bubble layer. Acoustics Letters, 7 (1984) 134-135.
11. Kustov, L.; Nazarov, V.; Sutin, A.: . Nonlinear sound scattering by a bubble layer. Sov. Phys. Acoustics. 32 (1986) N 6.
12. Kustov, L.; Nazarov, V.; Sutin A.: Phase conjugation of the acoustic wave by a bubble layer. Sov. Phys. Acoustics. 31 (1985) N 6.
13. "The Physics of Sound in the Sea". Techn. Report, N.S.Navy, D.C. 1946.
14. Medwin, H.: In situ acoustic measurements of microbubbles at sea. J.Geophys. Research . 82 (1977) 971-978.
15. Lovik, A.: Acoustic measurements of gas bubble spectrum in water, : Cavitation and Inhomogeneities in Underwater Acoustics", W.Lauterborn, ed., Berlin-Heidelberg-New York, Springer-Verlag 1980.
16. Schippers, Ir.P.:Density of air-bubbles below the sea surface, theory and experiments. Ibid.
17. Thorpe, S.:On the clouds of bubbles formed by breaking wind-waves in deep waters, and their role in air-sea gas transfer. Phys. Trans. R. Soc.Lond. A-304 (1984) 155.
18. Akulichev, V.;Bulanov, V.;Klenin, S.: Acoustic probing of gas bubbles in the sea. Sov. Phys. Acoust. 32 (1986) N 3.
19. Sandler, B.; Selivanovsky, D.; Sokolov, A.: Measurement of gas bubble concentration on the sea surface. Doklady Akademii Nauk SSSR. 260 (1981) 1474-1476 (in Russian).

Unsteady Free Surface Waves

Ib A. Svendsen
University of Delaware
Department of Civil Engineering
Newark, DE 19716, USA

A. K. Otta
Delft Hydraulics Laboratory
De Voorst, Postbus 152, 8300 AD
Emmeloord, The Netherlands

Stephan T. Grilli
Department of Ocean Engineering
The University of Rhode Island
Kingston, RI 02881, USA

Introduction

A Boundary Element Method (BEM) which solves the exact nonlinear equations of potential flow is used to generate irregular waves and follow their development to possible breaking in a "Computational Wave Tank." This is made possible by using the BEM directly in the physical space (rather than in a conformally mapped region). This paper describes some of the problems that need to be solved for such an application and presents results for breaking of solitary waves on plane slopes.

The Boundary Element Method

A detailed description of the basic feature of the BEM method used has been given by Grilli et al., 1989. The method is computational and solves the exact equations of the potential flow of nonlinear water waves. Integration is performed in time and each time step basically consists of two steps.

The first step consists of solving the Laplace equation for the velocity potential $\phi(\mathbf{x},\, t)$ which is formulated as a boundary integral equation

$$\alpha(\mathbf{x},\, t)\phi(\mathbf{x},\, t) = \int_\Gamma \frac{\partial \phi}{\partial n}(\mathbf{x}_o)G(\mathbf{x},\, \mathbf{x}_0) - \phi(x_0)\frac{\partial G(\mathbf{x},\, \mathbf{x}_0)}{\partial n}d\Gamma \qquad (1)$$

where α for a 2D boundary curve is the opening angle between the tangents at the two sides of point $\mathbf{x}$. A free space Greens function is used for $G(\mathbf{x},\, \mathbf{x}_0)$. To solve (1), either ϕ or $\partial\phi/\partial n$ is specified along the boundary of the fluid domain studied and the solution of (1) then provides ϕ_n where ϕ was specified and ϕ where ϕ_n was known.

M. L. Banner, R. H. J. Grimshaw (Eds.)
Breaking Waves
IUTAM Symposium Sydney/Australia 1991

The second step is the integration in time which is governed first of all by the free surface boundary conditions written in the form

$$\frac{D\mathbf{x}}{Dt} = \nabla\phi \tag{2a}$$

$$\frac{D\phi}{Dt} = -gz + \frac{1}{2}|\nabla\phi|^2 - \frac{p_s}{\rho} \tag{2b}$$

where $\mathbf{x}$ is the position vector for points on the free surface.

Other versions using (1) and (2) have been used by, e.g., Longuet-Higgins & Cokelet (1976), Vinje & Brevik (1981) and Dold & Peregrine (1984), which all use mapping or complex variables. The important feature of the present version is that the equations are solved directly in the physical space. This typically implies that, in addition to the free surface and the (arbitrary) bottom boundary, the flow domain studied is enclosed between boundaries stretching from surface to bottom, such as the paddle of a wave maker or a vertical or sloping fixed boundary. Thus corners are created on the boundary where, e.g., the free surface and a wave maker intersects (Fig. 1).

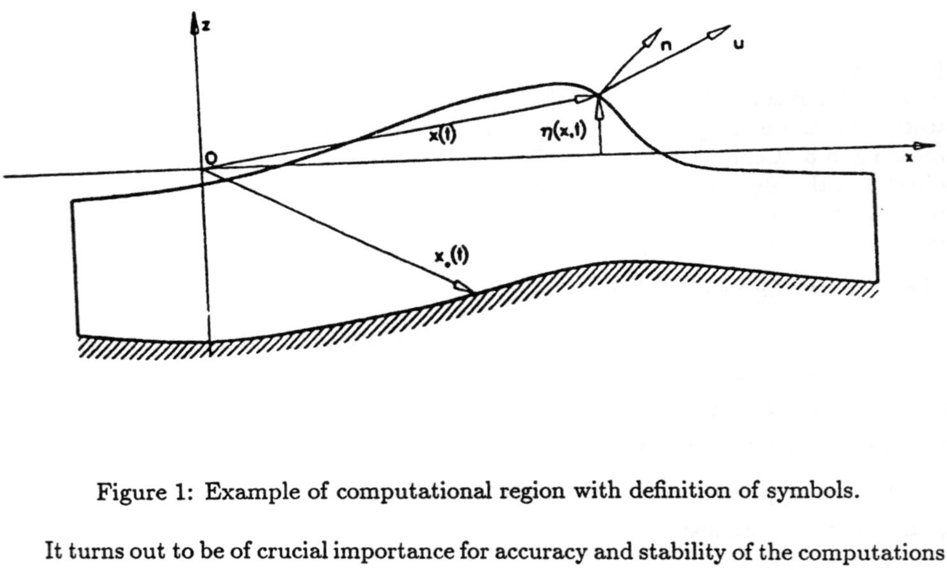

Figure 1: Example of computational region with definition of symbols.

It turns out to be of crucial importance for accuracy and stability of the computations how the equations are approximated numerically near such corners. Equivalent problems occur at other points as well, such as in overturning jets, splashes, etc.

In the following, these accuracy problems are briefly discussed along with solutions that greatly improve the computational performance, and in reality constitute the difference between failure and success of the method. The method is then used to analyze the criteria for breaking of solitary waves on a plane slope and it is shown that the numerical computations yield results substantially different from results obtained by a Boussinesq approximation.

Accuracy Problems in the BEM Method

The critical accuracy problems for solutions in physical space can briefly be divided into three types:

i. Loss of accuracy in the calculations of the integrals of ϕG_n and $G\phi_n$ where the geometry varies significantly between two nodes.

ii. A singularity at the corner between the free surface and a moving body occurring when the angle between the corner tangents of the two surfaces are $\pi/2$.

iii. A high sensitivity to errors in the solution process at the same corners as in ii) when the angle α between the tangents is $\neq \pi/2$.

The accuracy problem described in i) occurs in situations like those shown in Fig. 2. They are all characterized by a significant variation within the interval between each two nodes $(j, j+1)$ on the distance vector $\mathbf{r}$ (originating at collocation point i) and its orientation relative to the normal vector of the boundary. This implies equivalent strong variations of G and G_n in the integrals of (1), whereas the variations of ϕ and ϕ_n are often less important.

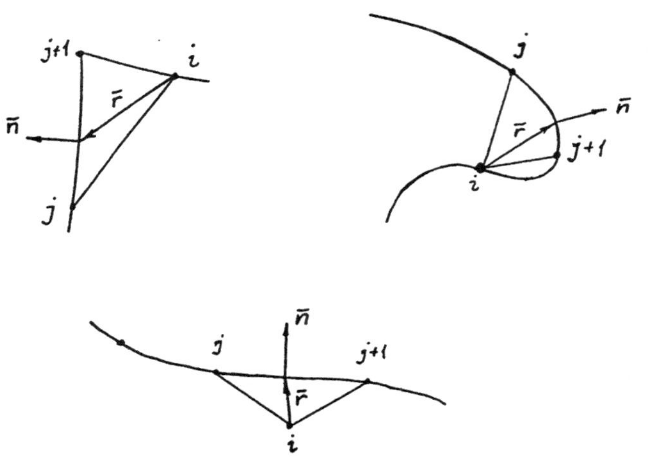

Figure 2: Situations where loss of integration accuracy is significant.

The remedy used against this consists in establishing a high order integration procedure within the intervals in question, including emphasis on a proper approximation of the geometry. This was discussed in detail by Grilli and Svendsen (1990).

The singularity at a 90 degree corner point may occur where a free surface (with ϕ specified for the solution of Laplace's equation) meets a solid body (with ϕ_n specified). In principle, a vertical body surface and a horizontal free surface at rest at $t = 0$ contain all the features of the problem. For our purposes, that situation also occurs at, e.g., the cold start of a wave maker in a wave tank. This was first pointed out by Kravtchenko (1954). Following contributions by Peregrine (1972), Chwang (1982) and others, the cold start of a plate was analyzed in a linear context by Roberts (1987) and later by Joo et al. (1990). Roberts showed that several types of singular behavior are possible, but the problem remains non-singular as long as the acceleration of the solid surface remains bounded. Notice that the conflict leading to the singularity occurs if ϕ_s along the free surface where ϕ is specified differs from ϕ_n on the solid body.

In real physical problems, solid bodies are always associated with a finite mass whence unbounded accelerations would require unbounded force or totally inelastic impacts between two physical bodies (such as a sledge hammer hitting a wall, Greenhow & Lin, 1983). Hence in most physical problems, the acceleration of the body will remain bounded and no singularity occurs.

In numerical computations, however, (for example, a wave starting from rest) finite time steps simulate finite changes in velocity that—over the single time step where the angle remains 90°—will start oscillations that may grow later on in the computation. Therefore in such situations, the motion $\xi(t)$ of the body is specified so that $\ddot{\xi}(t)$ remains bounded (for a cold start of a wave maker $\dot{\xi}(0) = \ddot{\xi}(0) = 0,\ \dddot{\xi} \neq 0$ is used).

The nonlinear case of finite amplitude and non-orthogonal intersection of surface and solid body has not been analyzed analytically in the literature. It is fairly straight forward to see, however, that under such conditions there is no singularity at the corner: If we assume that $\bar{\phi}$ is the specified variation of the velocity at the free surface and $\bar{\phi}_n$ the specified normal derivative on the solid body and $\alpha \neq \pi/2$ then the problem is well posed (Fig. 3). Hence, the Laplace solution will always be able to establish a value ϕ_n of the normal derivative of ϕ at the free surface side of the corner which together with the $\bar{\phi}_s$ (the tangential derivative of the specified $\bar{\phi}$ along the surface), will correspond to the $\bar{\phi}_n$ specified for the solid body. That implies both boundary conditions at the corner point are satisfied and no conflict or singularity occurs.

In the numerical computations, however, small errors occur in the solution of the Laplace equation. Fig. 3 illustrates that situation. When α is close to 90°, the solution is very sensitive to such errors and it turns out that this often induces unwanted oscillations to the solution.

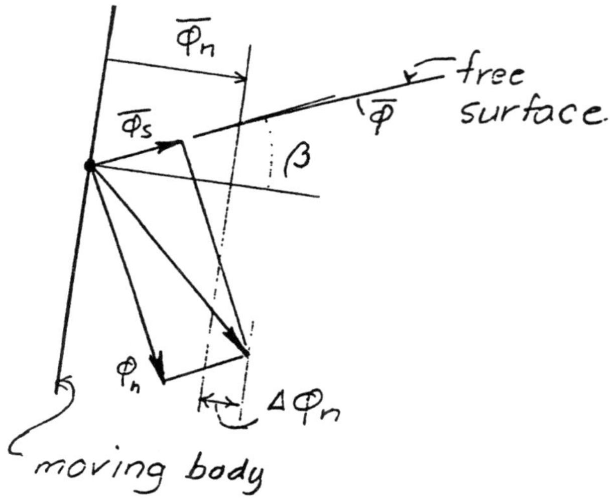

Figure 3: $\underline{\phi}$ and $\underline{\phi_n}$ specified boundary conditions, $\underline{\phi_s}$ is derived from $\underline{\phi}$, ϕ_n determined by solution of Laplace's equation. $\Delta\phi_n$ is discrepancy between specified and computed velocity normal to the body.

The problem is solved by replacing one of the equations in the matrix solution for Laplace (i.e., (1)) by an equation that ensures that the boundary conditions on the two sides of the corner are satisfied by the Laplace solution. In essence, this implies that we slightly modify the specified boundary values $\bar{\phi}$ of ϕ near the corner. The best results are obtained when only $\bar{\phi}$ at the corner point itself is modified.

Breaking Criteria for a Solitary Wave on a Plane Slope

The method has been applied to the run-up and breaking of solitary waves on a plane slope. Using a nonlinear shallow water approximation, this problem was studied analytically by Synolakis (1987). He found that a solitary wave propagating onto a plane slope from a region with constant depth would break in the uprush on the slope if the relative wave height H/h on the constant depth satisfied the relation.

$$\frac{H}{h} > 0.8182(\cot\beta)^{-10/9} \tag{3}$$

where β is the angle of the slope.

The computations using the BEM show that waves much steeper than given by (3) still do not break. As Fig. 4 shows, in general, the waves have to be almost 4 times as steep as indicated by (3) before they break on a given slope. And if the slope is steeper than approximately 12°, even the steepest solitary wave of $H/h \sim 0.775$ which remains

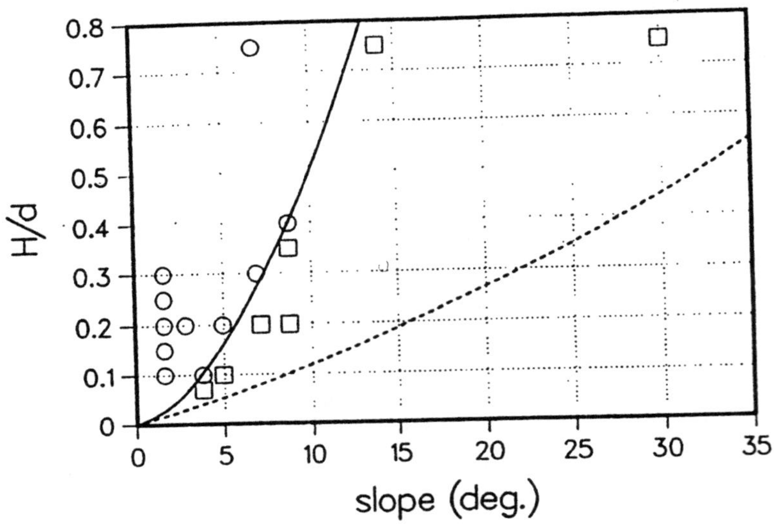

Figure 4: Breaking limit for solitary waves on slopes. ○ breaking waves, □ non-breaking, - - - (3), — (4)

stable on a constant depth (Tanaka et al. 1987) will not break on the uprush. The actual limit shown in Fig. 4 corresponds to breaking occurs when

$$\frac{H}{h} > 8.404 \cot \beta^{-1.62} \tag{4}$$

In order to achieve high accuracy of the results, the solitary waves used in the computations were assumed present at $t = 0$ in the constant depth part of the computational region. Initial values of the surface profile $\bar{\eta}(x, 0)$ and the surface value of the velocity potential $\bar{\phi}(x, 0)$ were determined by the integral method developed by Tanaka (1986). The crest of the wave was initially so far from the toe of the slope that the surface elevation at that point was only 10^{-4} times the wave height.

Fig. 5 shows a wave of height $H/h_0 = 0.75$ on three different slopes. Only the part of the computational domain closest to the shoreline is shown. In part a, the slope angle is 30° and the wave does not break at all but runs up to a total height of $2.5h_0$.

In Fig. 5b the slope angle is 7.12° (1:8) and the wave is breaking almost at the undisturbed shoreline. The height of the wave actually remains almost unchanged through out the process although the water depth decreases from 1 to almost zero in front of the wave at the last profile. At that point the front of the wave essentially forms a vertical wall.

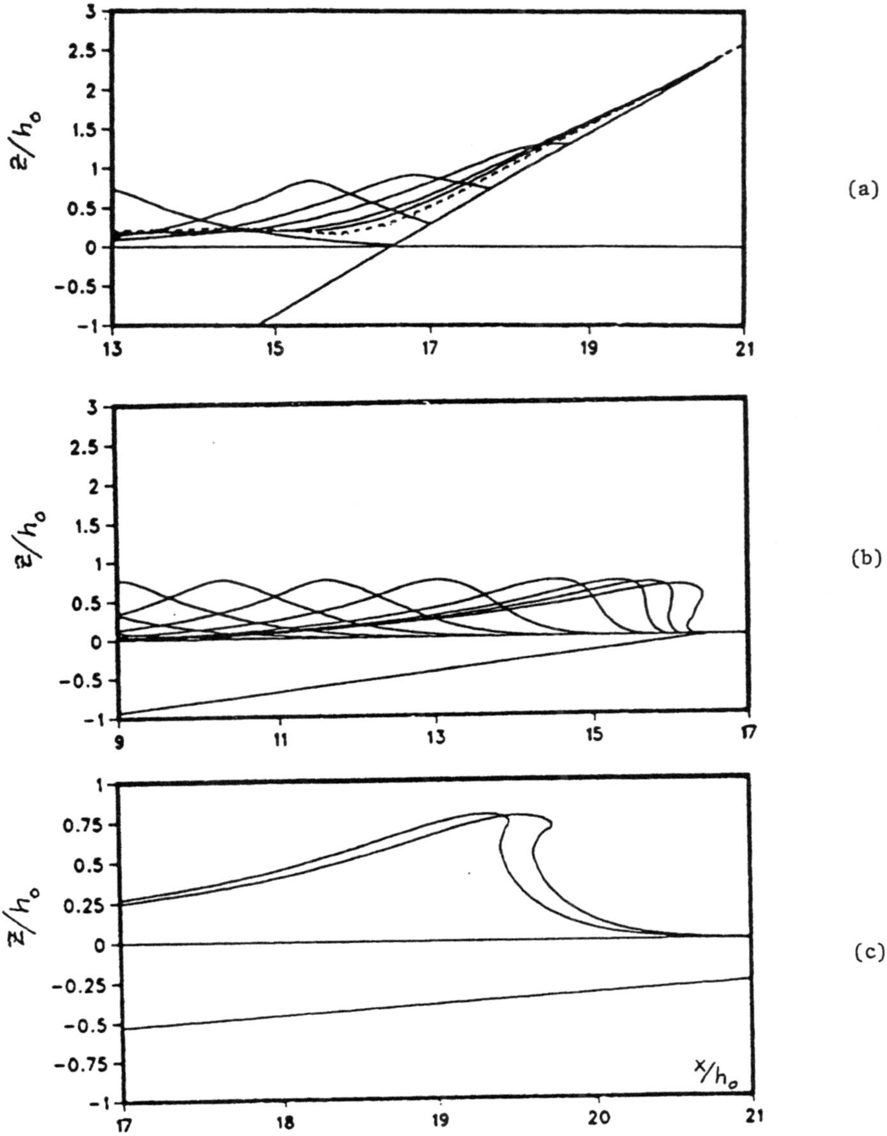

Figure 5: The run-up or breaking of a solitary wave of initial height $H/h_0 = 0.75$. (a) Slope 30° (1:1.732); (b) Slope 7.12° (1:8); (c) Slope 3.81° (1:15). h_0 denotes the water depth in front of the slope.

The last part, Fig. 5c shows a traditional plunging wave breaking occurring on a slope 3.81° (1:15). Again, the wave height has remained virtually unchanged, the maximum value having increased from the initial 0.75 to 0.80 h_0 at breaking. However, due to the decrease in depth, the index H/h at breaking is 2.28.

The three cases in Fig. 5 correspond to waves well to the right of the curve in Fig. 4 (5a), approximately on the curve (5b) and to the left of the curve (5c), respectively.

References

1. Chwang, A.T. (1982). "Nonlinear hydrodynamic pressure on an accelerating plate," *Phys. Fluids*, 26, 383–387.

2. Greenhow, M. and W.-M. Lin (1983). "Nonlinear free surface effects: Experiments and theory," *Report 83-19*, MIT, Dept. of Ocean Eng.

3. Grilli, S.T. and I.A. Svendsen (1990). "Corner problems and global accuracy in the boundary element solution of nonlinear wave flows," *Engineering Analysis with Boundary Elements*, 7, 4, 178–195.

4. Grilli, S.T., J. Skourup and I.A. Svendsen (1989). "An efficient boundary element method for nonlinear water waves," *Engineering Analysis with Boundary Elements*, 6(2), 97–107.

5. Joo, S.W., W.W. Schultz and A.F. Messiter (1990). "An analysis of the initial-value wave-maker problem," *J. Fluid Mech.*, 214, 161–183.

6. Kravtchenko, J. (1954). "Remarques sur le calcul des amplitudes de la houle linéaire engendrée par un batteur," *Proc. 5th Int. Conf. Coast. Eng.*, 50–61.

7. Peregrine, D.H. (1972). "Flow due to a vertical plate moving in a channel," unpublished note.

8. Roberts, A.J. (1987). "Transient free-surface flows generated by a moving vertical plate," *Q. J. Mech. Appl. Math.*, 40, 1, 129–158.

9. Synolakis, C.E. (1987). "The runup of solitary waves," *J. Fluid Mech.*, 185, 523–545.

10. Tanaka, M., J. Dold, M. Lewy and D.H. Peregrine (1987). "Instability and breaking of a solitary wave," *J. Fluid Mech.*, 185, 235–248.

11. Tanaka, M. (1986). "The stability of solitary waves," *Phys. Fluids*, 29, 3, 650.

The Role of Modulational Instability in the Formation of Wave Groups

Mitsuhiro TANAKA

Department of Applied Mathematics, Faculty of Engineering,
Gifu University, 1-1 Yanagido, Gifu 501-11, Japan

Abstract

The effect of modulational instability on the formation of wave groups is investigated for wave fields with various bandwidths of the energy spectrum. By numerical study of the modified K-dV equation and the nonlinear Schrödinger equation derived from it, it is shown that the nonlinearity generally enhances the formation of wave groups when the system is modulationally unstable while suppresses it when the system is stable. It is also shown that this tendency remains noticeable even when the energy spectrum of the wave field becomes broad.

Introduction

It is well known to coastal engineers that a big wave in the ocean rarely comes alone, but it usually comes as a member of a group of successive big waves. In other words, the wave energy is not distributed uniformly along a wavetrain, but it tends to be concentrated from place to place along a wavetrain to form groups of relatively energetic waves. We call this tendency as "grouping" in this paper.

According to the recent numerical study by Yasuda and Ito[1] on the evolution of the field of surface gravity waves with broad energy spectrum, the "groupiness factor" GF which they used as a measure of nonuniformity of energy distribution along a wavetrain behaves in a very much different manner when the water was deep and shallow. From this result, they suggested that the modulational instability was playing an essential role in the formation of wave groups.

However, the concept of modulational instability naturally applies to a wavetrain with a narrow-banded spectrum, and the meaning of 'modulational instability' itself is not quite clear for such wavetrains with broad energy spectra as those studied by Yasuda and Ito. Moreover, the analytical studies on the evolution of random inhomogeneous wave field by Alber[2] (for the NLS equation) and Crawford, Saffman and Yuen[3] (for Zakharov's equation) show that the growth rate of the modulational instability diminishes as the spectral bandwidth increases, and it finally disappears when the bandwidth exceeds some critical value of $O(ak)$. Bearing this result in mind, the conjecture by Yasuda and Ito that the modulational instability can still have a significant effect on the evolution of wave field with such a broad energy spectrum might sound peculiar.

To understand the phenomenon more clearly, we investigate here how strongly the modulational instability can affect the evolution of wave field, especially the formation

M. L. Banner, R. H. J. Grimshaw (Eds.)
Breaking Waves
IUTAM Symposium Sydney/Australia 1991

of wave groups, when the bandwidth of the energy spectrum becomes broad. To this purpose it proves to be beneficial to introduce a system governed by the modified K-dV equation (MKdV for short),

$$\frac{\partial u}{\partial t} - \frac{\partial^3 u}{\partial x^3} = \pm\frac{\partial u^3}{\partial x} \tag{1.1}$$

instead of the complicated water wave equations. Needless to say, the numerical integration of the MKdV is much easier and less time consuming than that of the water wave equations. Another and much more important reason to choose the MKdV is as follows.

The nonlinear Schrödinger equation (NLS for short) for the complex amplitude $A(x,t)$ of a weakly nonlinear, narrow-banded wavetrain with carrier wavenumber k_0 can be written as

$$i\left(\frac{\partial A}{\partial t} + 3k_0^2\frac{\partial A}{\partial x}\right) + p\frac{\partial^2 A}{\partial x^2} = q|A|^2 A, \tag{1.2}$$

and $p = 3k_0$, $q = \mp 3k_0/4$ when the evolution of u is governed by the MKdV. This shows that, in the system governed by the MKdV, any wavetrain is unstable to modulation irrespective of its carrier wavenumber when the sign of the nonlinear term of the MKdV (1.1) is positive, while no wavetrain is unstable when the sign is negative. Thus we can change the stability of the system at our will just by changing the sign of the nonlinear term, which enables us to compare two cases which are identical except for the only difference in the stability to modulation. In the original water wave equation, on the other hand, we have to change the water depth in order to change the stability of a given wavetrain. But this at the same time changes various factors that affect the evolution of the wave field, such as the strength of dispersion relative to the nonlinearity etc., and it would be difficult to single out the effect of modulational instability. In this respect, the system governed by the MKdV seems to be much more suitable for our purpose than the original system of water waves.

Numerical Study of the MKdV Equation

The MKdV (1.1) is integrated numerically by the split-step Fourier method. $u(x,t)$ is assumed to be periodic in x with period L which is taken as 64π, and 512 mesh points are distributed with a constant interval within $0 \le x \le L$. The spatial derivatives are evaluated in the spectral space by using FFT, and the aliasing error which appears in the calculation of the nonlinear term is removed from each product of $u(x,t)$ by doubling the number of Fourier coefficents with zeros for the additional wavenumbers.

We assume that the initial energy spectrum is given by the Gamma distribution

$$E(k) = \alpha k^{\nu-1}e^{-\mu k}, \tag{2.1}$$

and the phase of each Fourier mode is random. As a measure of the bandwidth of the energy spectrum, we use W defined by

$$W \equiv \int_0^\infty (k-\overline{k})^2 E(k)dk \Big/ \overline{k}^2 \int_0^\infty E(k)dk, \qquad (2.2)$$

where

$$\overline{k} \equiv \int_0^\infty kE(k)dk \Big/ \int_0^\infty E(k)dk, \qquad (2.3)$$

By normalizing the initial energy spectrum such that $\overline{k} = 1$, it reduces to the following form

$$E(k) = E(1) \times \exp\{(1-k)/W + (1/W - 1)\ln k\}, \qquad (2.4)$$

which are shown graphically in Fig.1 for various values of W. The spectral peak k_p is at $1 - W$. $E(1)$ is always determined in such a way that the 'significant' steepness $k_p a_{1/3}$ of the initial wave field is equal to 0.3. When $W = 1/3$, $E(2k_p)/E(k_p)$ is about 54%, and it is about 29% when $W = 1/5$, indicating that the initial spectrum is fairly broad for these values of W. It may be worth noting that for the Wallops spectrum with $m = 7 \sim 11$, which is often observed for a swell which has propagated a long distance, W is $0.35 \sim 0.17$.

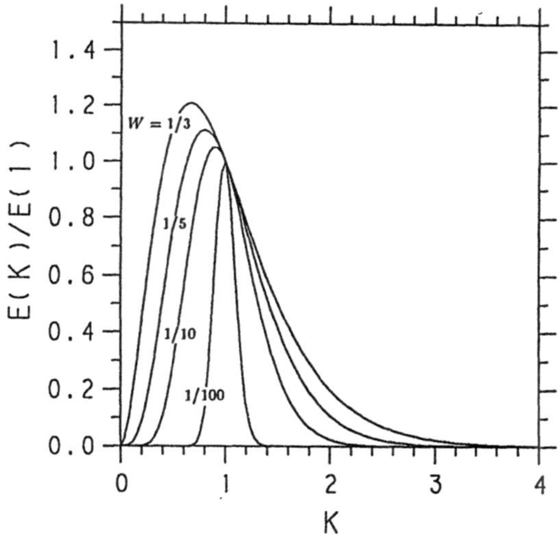

Fig.1 Initial energy spectrum for various values of W_0

We evaluate the degree of wave group formation by the "groupiness factor" GF which is defined from the envelope function $a(x,t)$ by the relation

$$GF(t) = \sqrt{\frac{1}{L}\int_0^L \{a(x,t) - \overline{a}(t)\}^2 dx} \Big/ \overline{a}(t), \qquad (2.5)$$

where $\overline{a}(t)$ is the average of $a(x,t)$ with respect to x.

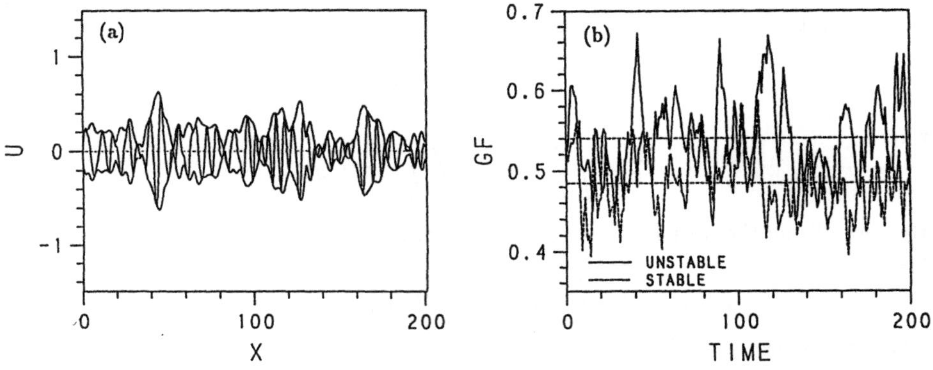

Fig.2(a) Initial wave profile $u(x,0)$ with its envelope function $a(x,0)$; (b) The evolution of GF given by the MKdV. ($W_0 = 0.1$).

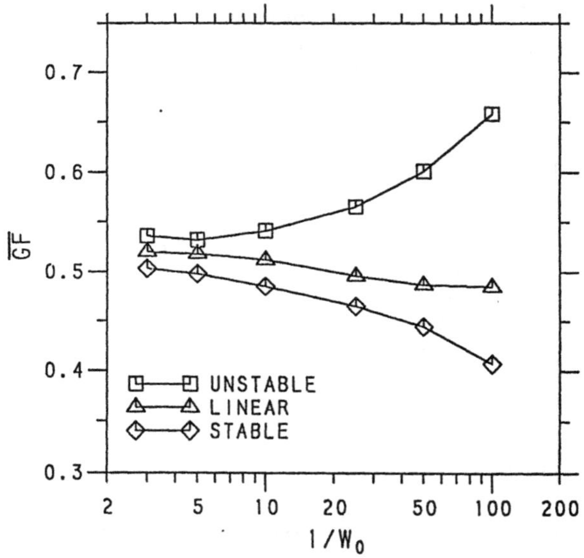

Fig.3 $\overline{GF}$ vs. $1/W_0$ given by the MKdV.

The evolution of GF is shown in Fig.2 when $W_0 = 0.1$. Figure 2a shows the initial profile $u(x,0)$ together with its envelope function $a(x,0)$, while Fig.2b shows the evolution of GF for the unstable and the stable case. Figure 3 shows $\overline{GF}$ as a function of W_0. This figure clearly shows that the nonlinearity has an opposite effect on the formation of wave groups depending on the stability of the system to modulation. When the system is unstable to modulation, nonlinearity enhances the formation of wave groups, while it suppresses when the system is stable. Although this effect seems to be diminishing as the bandwidth increases, it still remains noticeable for the case of the broadest spectrum treated here. Alber[2] and Crawford, Saffman and Yuen[3] show analytically that the modulational instability disappears when the bandwidth of a random wave field exceeds some critical value of $O(ak)$. Therefore it should be quite unexpected that the modulational instability still has some noticeable effect on the evolution of wavetrain with such a broad energy spectrum.

Comparison with NLS Theory

In this section, we study the behaviour of GF not by the MKdV (1.1) itself but by the NLS (1.2) derived from the MKdV. As the bandwidth is assumed to be $O(ka)(\ll 1)$ when the NLS is derived, we expect the NLS to give a correct prediction for GF only when W is very small. The initial condition $A(x,0)$ for the NLS can be obtained from $u(x,0)$ by the standard method based on the Hilbert transform (or numerically by just one FFT and inverse FFT), and the spectral peak k_p of the initial condition (i.e., $k_p = 1 - W_0$) is used as the carrier wavenumber k_0 which appears in the NLS.

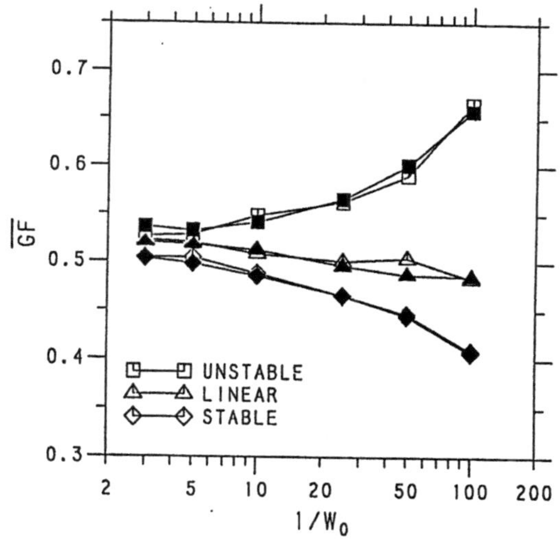

Fig.4 $\overline{GF}$ vs. $1/W_0$ given by the MKdV (solid symbols) and the NLS (open symbols).

Figure 4 shows $\overline{GF}$ given by the NLS together with those given by the MKdV as a function of W_0. At the first sight, it appears to be a very crude approximation to express a wave field with such a broad energy spectrum with $W \geq 1/5$, say, as a modulated wavetrain and try to describe its evolution by the NLS. Nevertheless, the $\overline{GF}$ predicted by the NLS agrees surprisingly well with those given by the MKdV for the whole range of W_0 treated here, i.e., $1/100 \leq W_0 \leq 1/3$.

We have restricted our attention to the system governed by the MKdV so far. However, the fact that the NLS derived from the MKdV also shows the same behaviour of GF strongly suggests that a similar type of dependence of GF on the stability characteristics of the system as welll as the spectral bandwidth would be observed generally in any nonlinear dispersive system which supports the modulational instability. We have confirmed this statement by studying the same problem for another system which is governed by the model equation for waves in moderately deep water (see Tanaka[4])

$$\frac{\partial u}{\partial t} + \frac{3}{2}\sqrt{\frac{g}{h}}u\frac{\partial u}{\partial x} + L(u) = 0, \tag{3.1}$$

where u is the free surface displacement, h the water depth, g the gravity, and L is a linear operator corresponding to the dispersion relation $\omega = \sqrt{gk\tanh(kh)}$. (See Fig.5)

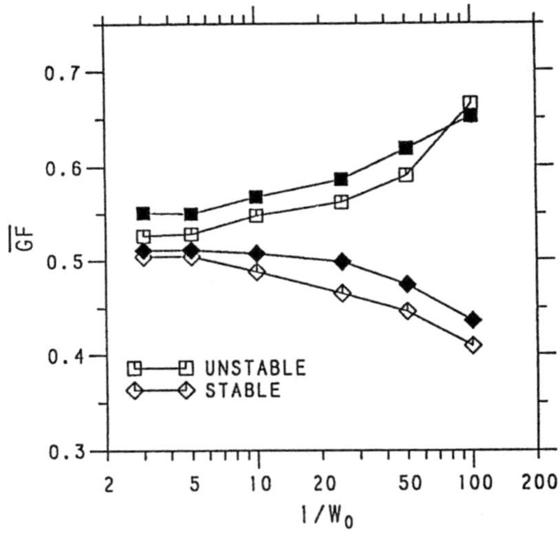

Fig.5 $\overline{GF}$ vs. $1/W_0$ given by the Model equation (solid symbols) and the NLS (open symbols).

References

[1] Yasuda, T. and Ito, K.: Nonlinear effects on the run of unidirectional high waves. Proc. 37th Conf. Coastal Eng. Jpn, 1990, 106–110. (in Japanese)

[2] Alber, I.E.: The effects of randomness on the stability of two-dimensional surface wavetrains. Proc. R. Soc. Lond. A363 (1978), 525–546.

[3] Crawford, D.R., Saffman, P.G. and Yuen, H.C.: Evolution of a random inhomogeneous field of nonlinear deep-water gravity waves. WAVE MOTION 2 (1980), 1–16.

[4] Tanaka, M.: On a model equation for waves on moderately deep water. Res. Rep. Fac. Eng., Gifu Univ. No.40 (1990), 75–84.

Action of Windstress and Breaking on the Evolution of a Wavetrain

KARSTEN TRULSEN[1] AND KRISTIAN B. DYSTHE
University of Tromsø
P.O.Box 953
N-9001 TROMSØ
NORWAY

Summary
A numerical simulation of the one dimensional development of a moderately steep wavetrain is investigated. We use evolution equations that are correct to fourth order in the wave steepness, [1], [2]. To these we add terms simulating the effects of wave breaking and windstress. When only breaking is taken into account, we find the so-called down-shift, frequently observed in wavetank experiments, [3]. When both wind and breaking are taken into account the evolution is quantitatively the same for low winds. With increase of the wind, the modulational instability is delayed, and for larger winds seems to disappear altogether as found experimentally by Bliven et al. [4].

1 Introduction

The so-called modulational instability (M.I.) of Stokes waves has been extensively studied following its prediction by Lighthill [5], and experimental verification by Feir [6]. Today it is well documented experimentally and fairly well understood theoretically.

In connection with the M.I. there are two rather characteristic phenomena that have long remained unexplained. The first is the frequency down-shift first reported by Lake et al. [7]. In their experiment a continous wavetrain of frequency ω is generated at one end of a wavetank. At some distance from the paddle the M.I. appears as the growth of essentially two sidebands $\omega \pm \Delta\omega$. The resulting groups develop a characteristic skew shape. Further down the wavetank an approximate recurrence seems to occur, except the dominant frequency is now shifted from ω to $\omega - \Delta\omega$. This down-shift seems to happen only if the initial steepness, ka_0, of the wavetrain is sufficiently large (Melville [8] gives $ka_0 > 0.16$). It seems that whenever down-shift occurs, there is also some wavebreaking.

A second phenomenon was described by Bliven et al. [4]. In an experiment with wind along a wavechannel they investigated the evolution of a paddle generated wavetrain. They found rather to their surprise that the wind reduced, and (for stronger winds) even suppressed the M.I.

In an earlier paper [3] we studied the down-shift, using evolution equations correct to $\mathcal{O}(\epsilon^4)$ (ϵ is a characteristic wave steepness), to which we added a term simulating the effect of breaking. The numerical simulations [3] did show a down-shift. We believe that the

[1]Current address: Parsons Laboratory, Department of Civil Engineering, Massachusetts Institute of Technology, Cambridge, MA 02139, USA.

M. L. Banner, R. H. J. Grimshaw (Eds.)
Breaking Waves
IUTAM Symposium Sydney/Australia 1991

role of breaking is to selectively damp the upper sideband. This seems possible because: 1) Spatial separation of the different frequencies occur during the development of a M.I. (see Melville [9], and c.f. figure 1, and 2) It so happens that the steepest parts of the modulated wavetrain contain largely frequency components from the upper sideband.

Hara and Mei [10] using a fourth order equation with windaction, a current shear in the water, and viscous damping found a frequency down-shift. They also found that the combined action of wind and shear in the water reduces the initial development of the M.I. The magnitude of this effect seems to be rather sensitive to the type of shear chosen.

In the present paper we modify our previous model equations [3], to include the effect of wave growth induced by wind. This is done by adding a linear growth term with an empirical growth rate factor. Numerical simulation with this model show qualitative agreement with the observations of Bliven et al. [4].

2 The model equations

For the slowly evolving wavetrain, we assume a Stokes-like expansion for the elevation, ζ, and the velocity potential, ϕ:

$$\begin{aligned} \phi &= \bar{\phi} + Ae^{kz+i\theta} + A_2 e^{2(kz+i\theta)} + \cdots + \text{c.c.}, \\ \zeta &= \bar{\zeta} + Be^{i\theta} + B_2 e^{2i\theta} + \cdots + \text{c.c.}, \end{aligned} \tag{1}$$

where $\theta = \mathbf{k}\cdot\mathbf{x} - \omega t$ and $k = |\mathbf{k}|$. The complex coefficients A, B ($\mathcal{O}(\epsilon)$), and A_n, B_n ($\mathcal{O}(\epsilon^n)$) are assumed to be varying slowly, on the scales ϵt and ϵx, where ϵ is some characteristic wave steepness ka_0. Inserting this into the basic equations (the kinematic and dynamical boundary conditions, and Laplace's equation) a hierachy of evolution equations are found. To $\mathcal{O}(\epsilon^3)$ the N.L.S. equation emerges. To $\mathcal{O}(\epsilon^4)$ the modified equations described by Dysthe [1], and Lo and Mei [2] are found.

Transforming to a coordinate system moving with the group velocity, dimensionless variables are introduced as follows [2]:

$$\left.\begin{array}{rclrcl} \epsilon\gamma(2kx - \omega t) & \to & \xi, & \epsilon^2 kx & \to & \eta, \\ A & \to & -\frac{ia_0\omega}{2k}A, & \bar{\phi} & \to & \omega a_0^2\bar{\phi}, \\ B & \to & a_0 B, & \zeta & \to & a_0\zeta \end{array}\right\} \tag{2}$$

All simulations are done with periodic boundary conditions on ξ, and γ is a scale factor to normalize the domain of ξ to $(0, 2\pi)$.

The transformed equations read

$$\frac{\partial A}{\partial \eta} + i\gamma^2\frac{\partial^2 A}{\partial \xi^2} + iA|A|^2 + 8\epsilon\gamma|A|^2\frac{\partial A}{\partial \xi} + 4i\epsilon\gamma A\left.\frac{\partial\bar{\phi}}{\partial \xi}\right|_{z=0} = 0, \tag{3}$$

$$4\frac{\partial^2\bar{\phi}}{\partial\xi^2} + \frac{\partial^2\bar{\phi}}{\partial z^2} = 0 \quad \text{for} \quad -\infty < z \le 0, \tag{4}$$

$$\frac{\partial\bar{\phi}}{\partial z} = \frac{\partial}{\partial\xi}|A|^2 \quad \text{for} \quad z = 0, \qquad \frac{\partial\bar{\phi}}{\partial z} = 0 \quad \text{for} \quad z \to -\infty. \tag{5}$$

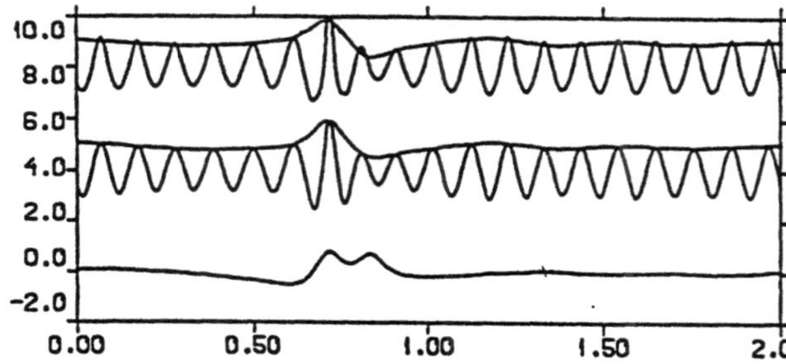

Figure 1: Wavetrain perturbation by the M.I. The upper profile is the surface displacement, ζ, superimposed with $|A|$, both with zero level shifted to 8.0. The middel profile is the first harmonic, $Be^{i\theta} + c.c.$, superimposed with its amplitude, $2|B|$, both with zero level shifted to 4.0. The lowest graph is the wavenumber of the modulation, ψ_ξ, where $B = |B| \exp i\psi$.

The surface elevation in the new coordinates, is

$$\zeta(\xi,\eta) = \epsilon^2\gamma \left.\frac{\partial\bar{\phi}}{\partial\xi}\right|_{z=0} + \left(\frac{1}{2}A - \frac{i\epsilon\gamma}{2}\frac{\partial A}{\partial\xi} - \frac{3\epsilon^2}{16}A|A|^2\right)e^{i\theta} + \left(\frac{\epsilon}{4}A^2 - i\epsilon^2\gamma A\frac{\partial A}{\partial\xi}\right)e^{2i\theta} + \left(\frac{3\epsilon^2}{16}A^3\right)e^{3i\theta} + c.c., \tag{6}$$

where the phase function becomes $\theta = \xi/\epsilon\gamma - \eta/\epsilon^2$. For a homogeneous wavetrain ($\partial/\partial\xi \equiv 0$) (6) reduces to a third order Stokes wave.

We now introduce the effects of wind and breaking in an ad hoc manner by adding two source terms S_1 and S_2 to the right hand side of (3).

For S_1, representing the action of wind, a natural choice is a linear term βA, where the dimensionless coefficient β is related to the wind induced spectral energy growth rate, γ_w, by $\beta = \gamma_w/(\epsilon^2\omega)$. For γ_w we use an empirical estimate given by Plant [11]: $\gamma_w = m\omega(u_*/c)^2$, where the dimensionless constant m is estimated as 0.04 ± 0.02, u_* is the friction velocity, and c the phase velocity. Thus

$$S_1 = \frac{m}{\epsilon^2}\left(\frac{u_*}{c}\right)^2 A. \tag{7}$$

The wave breaking term S_2 we model as follows [3]:

$$S_2 = -\frac{1}{\tau}A\left[\left(\frac{|A|}{A_c}\right)^r - 1\right]\mathrm{H}(|A| - A_c), \tag{8}$$

where H is the unit step function, τ, r and A_c are real positive constants. If $|A|$ exceeds the critical value A_c, the term S_2 ensures a relaxation towards A_c on the timescale τ/r, which should be of the same order of magnitude as the duration of a typical breaking event.

Simulations show that as long as τ/r is much less that a typical timescale for the M.I., the results are not sensitive to the values of τ and r chosen. The important parameter in S_2 is therefore A_c. Since the simulated wave steepness is somewhat larger that $\epsilon|A|$,

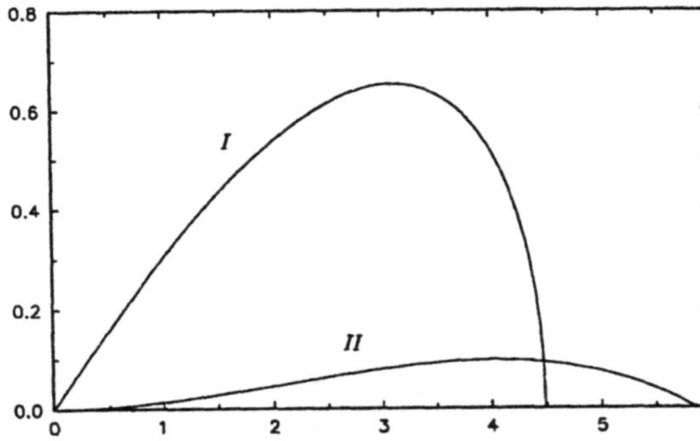

Figure 2: Growth rate of the M.I. as a function of perturbation wavenumber. I: for initial wave steepness 0.23 and $u_* = 0$. II: for wave steepness $kA_c(1+\beta\tau)^{1/r}$ (slightly supercritical).

as can be appreciated in figure 1, the "critical wave steepness" ϵA_c should therefore be smaller that the limiting steepness of $\simeq 0.39$ (see Melville [8]). In our simulations we have used $\epsilon A_c = 0.35$, $r = 2$, and $\tau = 0.125$.

If $S_1 + S_2$ are substituted into the right hand side of (3), and the complex equation is written as two coupled real evolution equations (for amplitude $|A|$ and phase $\arg A$) it is seen that the source terms enter the equation for $\partial|A|/\partial\eta$ only. This means that the influence of wind and breaking terms on the phase is only indirectly through their action on the wave amplitude.

3 Numerical simulation and conclusions

We have chosen $\epsilon = 0.23$, $\gamma = 0.229$, and initial carrier amplitude $A_0 = 1$. A linear stability analysis for a uniform wavetrain with these parameters, and with $u_* = 0$, give a growth rate for the different perturbation wavenumbers shown in figure 2. A quasispectral method is applied [2] with periodic boundary conditions in ξ. The spectral representation of the perturbation have only integral wavenumbers of which only $\pm1, \pm2, \pm3$ and ±4 are unstable, with ±3 being the most unstable. The initial profile [3] is taken as a carrier wave, with all unstable and some stable modes present at the 1% level.

In figure 3a a simulation with $u_* = 0$ is shown. It is identical to the evolution shown in [3], and the down-shift can be clearly seen. For an explanation of the physical mechanism producing this effect, we refer to [3].

For small winds ($u_*/c < 0.3$), nothing much seems to happen. When the wind increases somewhat, as in the figures 3b and 3c, the down-shift effect becomes more dramatic, and the two lower sideband modes -3 and -2 are competing for dominance. It is also seen that the wind growth of the carrier wave is bringing its amplitude near to the critical one (which is a little less than 1.5 on the scale shown).

For still stronger winds ($u_*/c > 0.5$), the evolution changes more drastically as shown in figure 4. The carrier wave now reaches the critical level before the initial growth due to the M.I. has produced appreciable sidebands. Once the critical level is reached the sideband development becomes very slow, thus delaying the development of the more

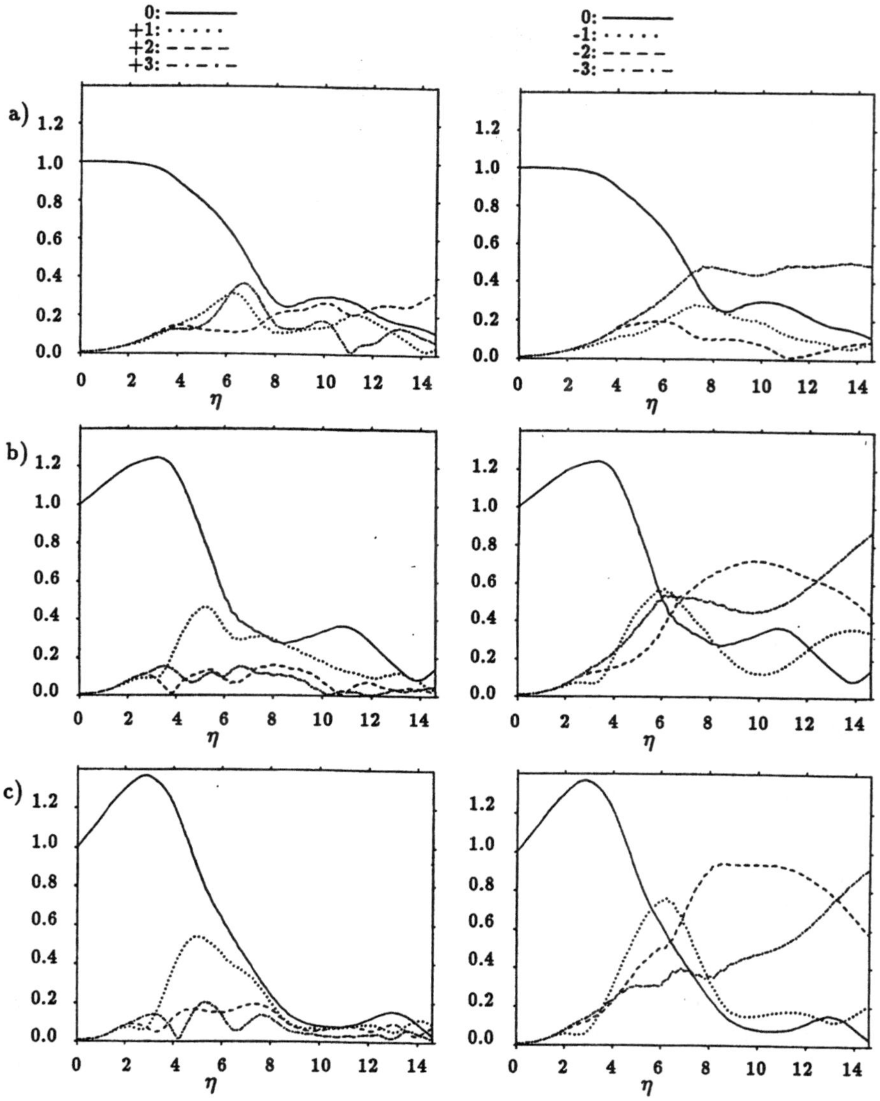

Figure 3: Evolution of wavetrain for different wind velocities. The two columns show spectral development with upper sidebands (left column) and lower sidebands (right column). In a) $u_*/c = 0$, b) $u_*/c = 0.33$ and c) $u_*/c = 0.40$

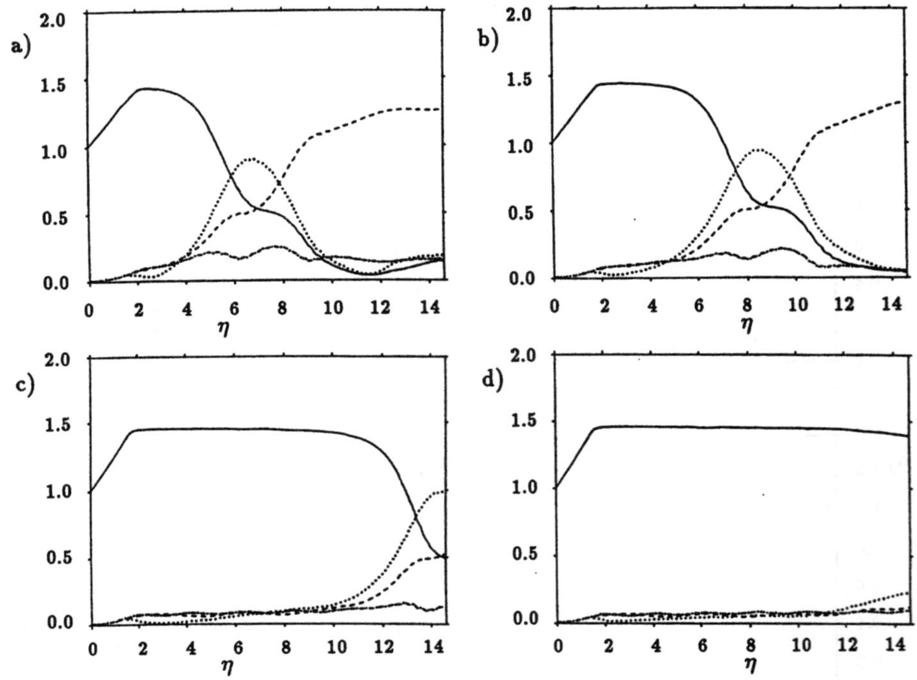

Figure 4: Spectral development of carrier and lower sidebands for different u_*/c a) 0.46 b) 0.48 c) 0.50 d) 0.52

rapid phase of the M.I. To understand how this happens, observe that the evolution equation has a uniform wavetrain solution when $|A| = A_c(1+\beta\tau)^{1/r}$, which is slightly supercritical (giving an "equilibrium" between windgrowth and breaking). Analysing the stability of this solution we find that it is unstable with a growth rate shown in figure 2, which is practically independent of windspeed. Observe that the growth rate is less that approximately 1/8 of that for the initial wavetrain of amplitude A_0 (and no wind). Thus when the carrier has grown to a critical value, the growth rate of the sidebands is drastically reduced. They continue to grow at this slow rate till the sidebands have a sufficient amplitude for the rapid phase of the M.I. to start.

In the experiment by Bliven et al. the initial wave steepness was approximately 0.24, and the M.I. was found to be drastically reduced at $u_*/c = 0.34$, and to disappear altogether at $u_*/c = 0.43$. The length of their wave channel corresponds to $\eta \simeq 11$. In our simulation the appearance of the M.I. is moved beyond the point $\eta \simeq 11$ when $u_*/c \simeq 0.5$. Thus there is a qualitative agreement with [4], but quantitatively we seem to need somewhat too large winds. Hara and Mei [10] found that a shear current in the water would tend to lower the initial growth rate of the M.I. Any effect that would reduce the initial M.I. growth rate, would make the "saturation delay" described above to happen for smaller winds.

Financial support from The Norwegian Research Council for Science and the Humanities (NAVF 412.90/024) and ONR Fluid Mechanics Program (N 000 14–90 J–1163) is acknowledged by K.T.

REFERENCES

[1] DYSTHE, K. B. 1979 Note on a modification to the nonlinear Schrödinger equation for application to deep water waves. *Proc. R. Soc. Lond.* A **369**, 105–114.

[2] LO, E. & MEI, C. C. 1985 A numerical study of water-wave modulation based on a higher-order nonlinear Schrödinger equation. *J. Fluid Mech.* **150**, 395–416.

[3] TRULSEN, K. & DYSTHE, K. B. 1990 Frequency down-shift through self modulation and breaking. In Water Wave Kinematics (A. Tørum & O. T. Gudmestad (eds.)) Kluwer Academic Publishers.

[4] BLIVEN, L. F., HUANG, N. E. & LONG, S. R. 1986 Experimental study of the influence of wind on Benjamin-Feir sideband instability. *J. Fluid Mech.* **162**, 273–260.

[5] LIGHTHILL, M. J. 1965 Contributions to the theory of waves in nonlinear dispersive systems. *J. Inst. Math. Appl.* **1**, 269–306.

[6] FEIR, J. E. 1967 Discussion: some results from wave pulse experiments. *Proc. R. Soc. Lond.* A **299**, 54–58.

[7] LAKE, B. M., YUEN, H. C., RUNGALDIER, H. & FERGUSON, W. E. 1977 Nonlinear deep-water waves: theory and experiment. Part 2: Evolution of a continuous wave train. *J. Fluid Mech.* **83**, 49–74.

[8] MELVILLE, W. K. 1982 The instability and breaking of deep-water waves. *J. Fluid Mech.* **115**, 165–185.

[9] MELVILLE, W. K. 1983 Wave modulation and breakdown. *J. Fluid Mech.* **128**, 489–506.

[10] HARA, T. & MEI, C. C. 1991 Frequency downshift in narrowbanded surface waves under the influence of wind. *J. Fluid Mech.* (to be published)

[11] PLANT, W. J. 1982 A relationship between wind stress and wave slope. *J. Geophys. Res.* **87** (C3), 1961–1967.

A Mechanism for Wave Deformation and Breaking Intermediated by Resonant Side-Bands

Marshall P. Tulin and Jiyue J. Li
Ocean Engineering Laboratory, University of California, Santa Barbara

Introduction

Breaking waves at sea are approximately twice as energetic as the waves of spectral mean height, and their steepness lies mainly in the range $.05 < a_0k < .20$, Holthuijsen & Herbers (1986). It has been repeatedly observed at sea that the energetic breaking waves tend to occur within wave groups and, typically, near and just forward of the group center: Donelan, Longuet-Higgins & Turner (1972); Thorpe & Humphries (1980); Su (1986); Holthuijsen & Herbers (1986). It has also been observed, both at sea and in the laboratory, that breaking is immediately preceded by a deformation of the waveform in which the trough is reduced and the crest moves forward, causing a steepening of the forward face: Bonmarin (1989), Figure 1 and Kjeldsen & Myrhaug (1980). The latter have defined a waveform deformation parameter corresponding to breaking, see Figure 2. It is important that this deformation has a strong asymmetric component.

The laboratory observations of Su & Green (1984) on wave instability and breaking in a long wave tank (100 m) strongly suggest a connection between wave-breaking and side-band instability, see Figure 3, as first noted by those authors. This suggestion is supported by the short tank experiments of Melville (1982), and the recent numerical calculations of Cointe, et al (1991). These results connect with the correlations between breaking and wave groups, since there is growing evidence that wave groups at sea are a consequence of side-band (Benjamin-Feir (1967)) instability. The observed connections are summarized in Figure 4. We are left, however, with questions as to the details of the underlying mechanisms leading to breaking, and with the need for a quantitative theory. The overlap between observations at sea and in the laboratory and in theory, gives hope that the phenomena involved are not crucially dependent on the effect of wind. Therefore, standing with the point of view that waveform deformation is crucial to breaking inception, the present authors have undertaken to provide a quantitative theory of wave deformation in its relation to side-band instability. We utilize a method for the evaluation of energy change within a half-wave control volume moving through the three wave system; the formulation is initially exact, and the validity of the results only depends upon the approximations made in describing the pertinent wave system.

M. L. Banner, R. H. J. Grimshaw (Eds.)
Breaking Waves
IUTAM Symposium Sydney/Australia 1991

Quantitative results are obtained which show that within a three wave system, intra-wave energy transfers intermediated by the resonant side-bands cause a substantial excess of energy to accumulate within the breaking quadrant (front face) of the wave when it reaches the center of the wavegroup. The results, which are consistent with both at sea and laboratory measurements, provide a quantitative understanding of the mechanics of wave deformation leading to breaking.

The Three Wave System: Basic Wave plus Resonant Side Bands

At the onset of Benjamin-Feir instability, the finite basic wave, $\eta_a = a\cos\chi + 1/2\, ka^2 \cos 2\chi$ gives rise to the two nascent side bands $\eta_\varepsilon = \varepsilon^+ \cos\chi^+ + \varepsilon^- \cos\chi^-$, where the phases χ^+, χ^- are shifted in wavenumber by $\pm\delta k/k_0$ and in frequency by $\pm\delta\omega/\omega_0$, and shifted in phase χ, too. Of course, the perturbation solutions corresponding to three waves, include higher order terms in a, ε^+, ε^-, and their products. For the analysis of intra-wave transfer we require, for reasons which will be apparent later, that terms of the form $\cos 3\chi$ (or $3\chi^+$, $3\chi^-$) which appear in the products η^2 or $\eta\phi$ be correct to fourth order in the generalized wave amplitude a_i ($a_i = a, \varepsilon^+, \varepsilon^-$). This must be done taking into account that $\delta\omega/\omega_0 = 1/2\, \delta k/k_0 = (ak)_0$ as may be shown from the dispersion relation which arises for this system; this requires that terms proportional to a_i, ka_ia_j, and $(\delta\omega/\omega_0)ka_ia_j$ be included in the perturbation expansions. The expansions for η and ϕ are given in the appendix, including the nonlinear dispersion relation.

Energy Transfer in the Wave System

If R is the region occupied by liquid and it is bounded by the surface S (which may be in motion with the velocity v_n normal to S), then the sum of kinetic and potential energy of the fluid within R, E, is (John (1949)):

$$\frac{dE}{dt} = \iint_s \rho\Big[\frac{\partial\phi}{\partial t}\frac{\partial\phi}{\partial n} - \Big(\frac{\partial\phi}{\partial t} + \frac{p-p_0}{\rho}\Big) v_n\Big] dS \tag{1}$$

We take R as a column bounded above by the free surface, and the right and left sides by vertical planes; the water here is infinitely deep. When R is taken moving with the basic wave, then (1) becomes

$$\frac{dE}{dt} = \left[\int_{-\infty}^{\eta} \rho\left(\phi_t\,\phi_x + [\tfrac{1}{2}(\phi_x^2 + \phi_y^2) + gy]\, c_p\right) dy\right]_{\chi=kx_1}^{\chi=kx_2} - \int_{kx_1}^{kx_2} \frac{\rho}{k}\Big[\frac{p-p_0}{\rho}\frac{\partial\eta}{\partial t}\Big]\, d\chi \tag{2}$$

where x_1 and x_2 are horizontal coordinates in the moving system.

Upon substitution of the appropriate expressions for η and ϕ, and integrating in depth, the results are expressed as the sum of 6 terms, [1] - [6] consecutively,

$$\frac{dE}{dt} = \Big\{ \beta_{i,j}\, a_ia_j \cos(\chi_i-\chi_j) + \beta^*_{i,j}\,(\delta\omega/\omega_0)\, a_ia_j \cos(\chi_i-\chi_j) + \beta_{i,j,k}\, a_ia_ja_k \cos(\chi_i \pm \chi_j \pm \chi_k)$$

$$+ \beta^*_{i,j,k}\,(\delta\omega/\omega_0)\, a_i a_j a_k \cos(\chi_i \pm \chi_j \pm \chi_k) + \beta_{i,j,k,l}\, a_i a_j a_k a_l \cos(\chi_i \pm \chi_j \pm \chi_k \pm \chi_l)\Big\}_{k_0 x_1}^{k_0 x_2}$$

$$+ \int_{k_0 x_1}^{k_0 x_2} \left[\beta^*_{i,j,k,l}\, a_i a_j a_k a_l \sin(\chi_i \pm \chi_j \pm \chi_k \pm \chi_l)\right] d\chi \; + \cdots \qquad (3)$$

where the β coefficients are known in terms of the perturbation expansions. Inter-wave transfer arises from the limits $x_2 = x_1 + \lambda_0$, which are non-zero for terms [1], [2], [5], and [6]: term [1] produces the wave group envelope corresponding to the sum of the three linear waves; term [2] and [5] produce a second and third order distortion of the linear envelope, term [6] gives the resonant interactions among waves. Intra-wave transfer arises from the limits $x_2 = x_1 + \lambda_0/2$. The transfer between trough and crest (T-C) or back and front (B-F) may be found through appropriate choice of x_1. These transfers can arise only from terms [3] and [4], but calculations show that term [3], which is of lower order than [4], is null(!).

Intra-Wave Energy Transfer

The evaluation of term [4], with $\varepsilon^+ = \varepsilon^-$, and then integration of $\dot{E}$ gives rise to the following expression for ΔE, the excess of energy within a given half wave; it is seen to depend on position within the wave group, $r = \varepsilon/a$, and $(ak)_0$; we have taken the cumulative phase of the side-bands as $\pi/2$, corresponding to their maximum growth rate:

$$\frac{\Delta E}{E_0} = \frac{r\,(ak)_0}{\pi(1+2r^2)^{3/2}} \sum_{j=1}^{3} \beta_j^{c,f} \left[\cos j[\delta k(X-X_0)+\pi\delta\omega/\omega_0] + \cos j[\delta k(X-X_0)-\pi\delta\omega/\omega_0]\right] \qquad (4)$$

where X is the coordinate moving with the group speed. Below, c and f denote crest and front.

$$\beta^c_1 = 2\sqrt{2}(2+3r^2), \quad \beta^c_2 = 15/2\, r, \quad \beta^c_3 = 2\sqrt{2}r^2 \qquad (5)$$

$$\beta^f_1 = \sqrt{2}(1+6\,r^2), \quad \beta^f_2 = 9/2\, r, \quad \beta^f_3 = 2\sqrt{2}r^2 \qquad (6)$$

These equations show that around the center of the wave group, there occurs a loss of energy in the wave trough, simultaneous with a gain in the front, of similar magnitude, leading to the same type of deformation observed by M & K at sea and by Bonmarin in the laboratory. This change of wave energy (T→C), $\Delta E/E_o$, nondimensionalized by the corresponding half wave energy in the original wave (amplitude a_o) is shown as a function of position within the group, for $\varepsilon/a = 1/\sqrt{2}$, and $(ak)_o = 0.1$, in Figure 5. Also shown is the group envelope amplitude A/a_o.

The loss of trough and steepening of the face are preceded by a gain in trough (C→T) at the beginning of the wave group, and this kind of deformation has also been observed by Bonmarin prior to the breaking deformation. According to the theory, the time of passage of the wave from the beginning of the breaking deformation to its maximum is simply $(4\, a_o k_o)^{-1}T$, where T is the wave period. In the case of Bonmarin, Figure 1, $(ak)_o$ is 0.25, and the entire duration of breaking is one wave period, precisely as predicted.

Breaking occurs in the upper-front wave quadrant, and the excess energy in this quadrant arises both from the transfer (T→C) and (B→F) and is their sum. The total maximum excess energy there is shown in Figure 6 over a wide range of variables $(ak)_o$ and ε/a. These maxima occur at the center of the wave group. The severity of waveform deformation increases with the value of $\Delta E/E_o$, but the present theory does not provide a prediction of limiting values corresponding to breaking, $(\Delta E/E_o)^*$. However, experimental measurements of breaking can be used to estimate $(\Delta E/E_o)^*$, in conjunction with the theoretical results presented in Figure 6. Melville (1982) studied wave instability and observed breaking for a wave of $(ak)_o = .23$ and found that breaking began for values of ε/a in the range = $(1/2 \rightarrow 1/\sqrt{2})$. We show this in Figure 6 as a small darkly shaded region, corresponding to values of $(\Delta E/E_o)^*$ of .35 to .40. More experiments are required to confirm whether breaking actually corresponds to a critical value of $(\Delta E/E_o)^*$, exclusive of other factors. Meanwhile, we note that theory predicts a minimum value of $(ak)_o$ for any particular value of $(\Delta E/E_o)^*$ less than .65. In the case of $(\Delta E/E_o)^*$ as determined above, this value lies in the range .06 -.08, or approximately .10 for the local wave steepness.

Summary and Conclusions

1) The breaking of energetic waves, which occurs over a wide range of wave steepness and usually within a wave group, is believed to result from a very rapid deformation of the waveform, one to four wave periods in duration.

2) This breaking deformation is caused by intra-wave energy transfers which are intermediated by the same two resonant side-bands which result from the Benjamin-Feir instability and are responsible for the formation of the wave groups.

3) This three-wave interaction mechanism leads to transfer of energy between the trough and crest (T→C), and back and front (B→F). Over the middle section of the wave group the energy of the crest and of the front are simultaneously increased at the expense of the trough and back, reaching a maximum at the center of the wave group.

4) Theoretical predictions of the maximum energy excess in the breaking quadrant of the waveform are given, reaching a maximum of about 65% at a value of initial wave steepness $(ak)_0$ of 0.17. Based on measurements, it is deduced that excess energy values of 35-40% are sufficient to induce breaking. In this case, all finite waves of steepness greater than approximately .07 will eventually be caused by that mechanism to deform and break.

5) The predictions of this theory are both quantitatively and qualitatively in accord with the observations of Bonmarin (1989) and consistent with at sea observations of wave deformation and wave breaking in groups, as well as laboratory studies of wave instability, deformation, and breaking.

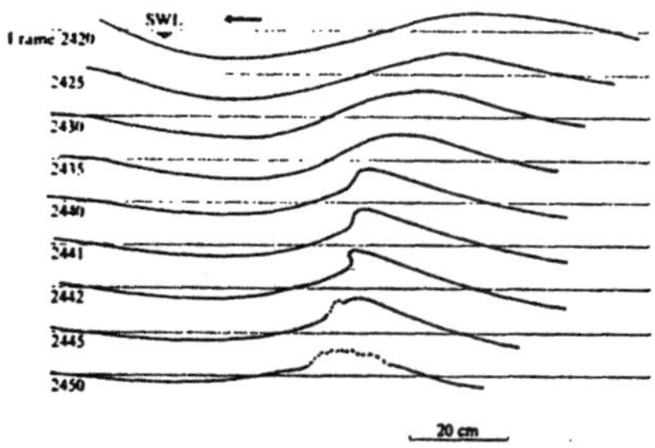

Figure 1. Evolution of a Wave Leading to Breaking (with frame frequency 24 f/s) from Tank Experiments of Bonmarin (1989)

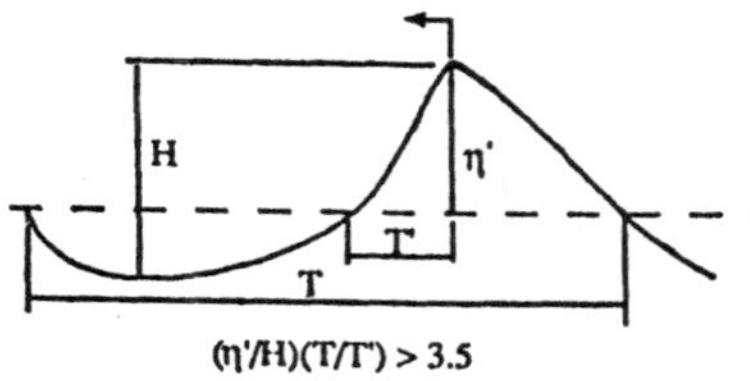

Figure 2. Observations of Myrhaug & Kjeldsen (1980)

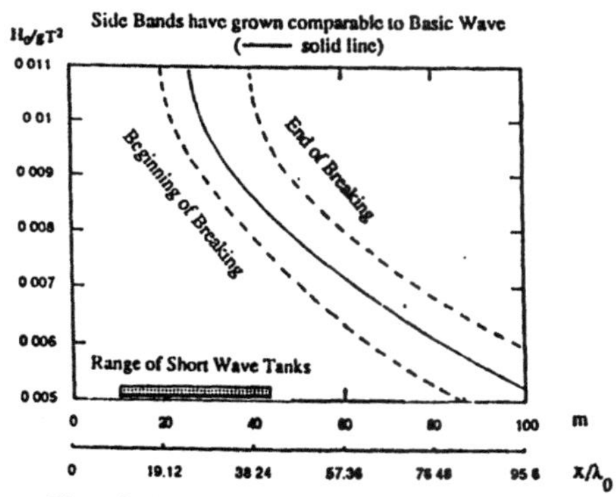

Figure 3. Regions of Breaking vs. Tank Length from Long Tank Experiments of Su & Green (1984)

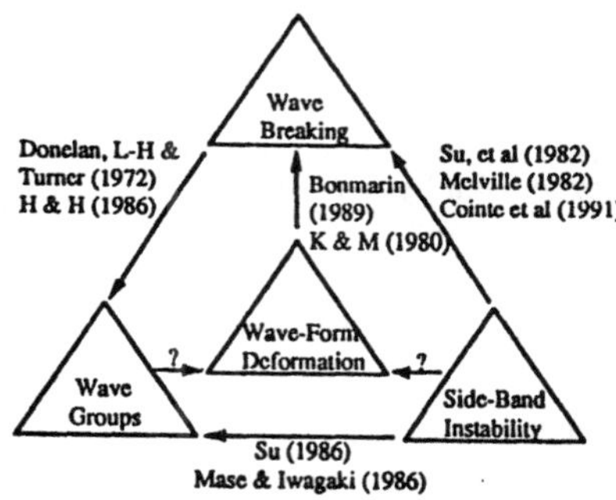

Figure 4. Connections of Wave Breaking to Side-Band Instability, Wave Grouping & Wave-Form Deformation

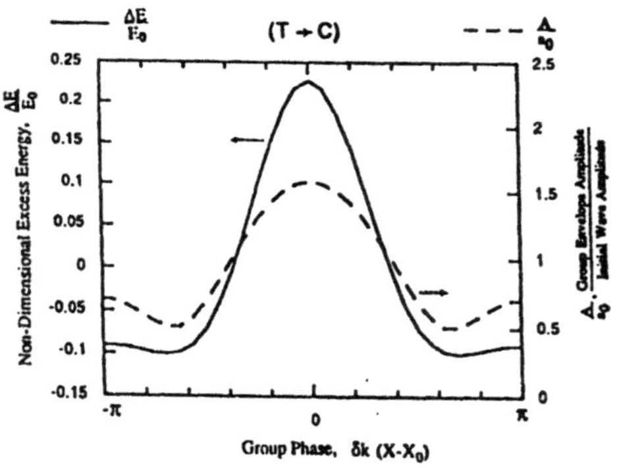

Figure 5. Excess Crest Energy vs. Position in Group ($\varepsilon/a=1/\sqrt{2}$, $(ak)_0=0.1$)

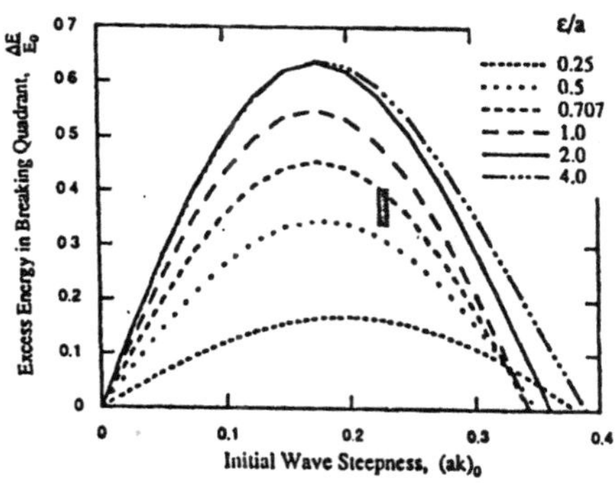

Figure 6. Excess Energy in Breaking Quadrant
[▯ data of Melville(1982), $(ak)_0=0.23$, $(\varepsilon/a)_{br}=1/\sqrt{2}$]

References

Benjamin, T.B.; Feir, J.E. 1967. The Disintegration of wave trains in deep water. Part I. Theory. J. Fluid Mech. 27: 417-430.

Bonmarin, P. 1989. Geometric properties of deep-water breaking waves. J. Fluid Mech. 209: 405-433.

Cointe, R.; Boudet, L. 1991. Nonlinear and breaking waves in bichromatic wave-trains: experiments and numerical simulations. Proc.1st Int.Offsho. Polar Eng.Conf. pp.517-522.

Donelan, M.; Longuet-Higgins, M.S.; Turner, J.S. 1972. Periodicity in whitecaps. Nature. 239: 449-50.

Holthuijsen, L.H.; Herbers, T.H.C. 1986. Statistics of breaking waves observed as whitecaps in the open sea. J. Phys. Oceanogr. 16: 290-297.

John, F. 1949. On the motion of floating bodies. I. Pure & Appl. Math. 2(1): 13-57.

Kjeldsen, S.P.; Myrhaug, D. 1980. Wave-wave interactions, current-wave interactions and resulting extreme waves and breaking waves. Proc. 17th Coast. Eng. Conf. 2277-2303.

Mase, H.; Iwagaki, Y.M. 1986. Wave group property of wind waves from modulational instability. Proc. 20th Coastal Eng. Conf., pp.565-577.

Melville, W.K. 1982. The instability and breaking of deep-water waves. J. Fluid Mech. 115: 165-185.

Su, M.Y. 1986. Large, steep waves, wave grouping and breaking. Proc. 16th Sym. on Naval Hydrodynamics, pp.78-92. (National Academy Press).

Su, M.Y.; Green, A.W. 1984. Wave breaking and nonlinear instability coupling. in The Ocean Surface, pp.31-38.

Thorpe, S.A.; Humphries, P.N. 1980. Bubbles and breaking waves. Nature, 283: 463-465.

Appendix

The solution for wave elevation η,

$$\eta = \eta_1 + \eta_{20} + \eta_{21}$$

with

$$\eta_1 = a\cos\chi + \varepsilon^+\cos\chi^+ + \varepsilon^-\cos\chi^-$$

$$\eta_{20} = 1/2\,ka^2\cos 2\chi + 1/2\,k\varepsilon^{+2}\cos 2\chi^+ + 1/2\,k\varepsilon^{-2}\cos 2\chi^- + ka\,\varepsilon^+\cos(\chi^+ + \chi) + ka\,\varepsilon^-\cos(\chi + \chi^-) + k\varepsilon^+\varepsilon^-\cos(\chi^+ + \chi^-)$$

$$\eta_{21} = (\delta\omega/\omega_0)\,k\varepsilon^{+2}\cos 2\chi^+ - (\delta\omega/\omega_0)\,k\varepsilon^{-2}\cos 2\chi^- + (\delta\omega/\omega_0)\,ka\,\varepsilon^+\cos(\chi^+ + \chi) - (\delta\omega/\omega_0)\,ka\,\varepsilon^-\cos(\chi + \chi^-) - 3/2\,(\delta\omega/\omega_0)\,ka\,\varepsilon^+\cos(\chi^+ - \chi) - 1/2\,(\delta\omega/\omega_0)\,ka\,\varepsilon^-\cos(\chi - \chi^-) - 2\,(\delta\omega/\omega_0)\,k\varepsilon^+\varepsilon^-\cos(\chi^+ - \chi^-)$$

and solution for the potential ϕ,

$$\phi = \phi_{10} + \phi_{11} + \phi_{21}$$

with

$$\phi_{10} = \omega_0/k\,a\sin\chi\,e^{ky} + \omega_0/k\,(1+\delta k/k)\,\varepsilon^+\sin\chi^+\,e^{(k+\delta k)y} + \omega_0/k\,(1-\delta k/k)\,\varepsilon^-\sin\chi^-\,e^{(k-\delta k)y}$$

$$\phi_{11} = (\delta\omega/\omega_0)\,\omega_0/k\,(1+\delta k/k)\,\varepsilon^+\sin\chi^+\,e^{(k+\delta k)y} - (\delta\omega/\omega_0)\,\omega_0/k\,(1-\delta k/k)\,\varepsilon^-\sin\chi^-\,e^{(k-\delta k)y}$$

$$\phi_{21} = -\,\omega_0 a\varepsilon^+\sin(\chi^+ - \chi)\,e^{\delta ky} - \omega_0 a\varepsilon^-\sin(\chi - \chi^-)\,e^{\delta ky} - \omega_0\varepsilon^+\varepsilon^-\sin(\chi^+ - \chi^-)\,e^{2\delta ky}$$

where

$$\chi^+ = \chi + \Delta - \gamma^+, \quad \chi^- = \chi - \Delta - \gamma^-, \quad \chi = kx - \omega t, \quad \Delta = \delta k\,x - \delta\omega\,t.$$

And also the nonlinear dispersion relation becomes,

$$\omega = \omega_0\,[1 + 1/2\,(ka)^2 + (k\varepsilon^+)^2 + (k\varepsilon^-)^2 + (k^2\varepsilon^+\varepsilon^-)\cos(\gamma^+ + \gamma^-)].$$

Vorticity Generation at a Fluid Interface

HARRY H. YEH

University of Washington
Seattle, Washington, 98195
USA

Summary

Based on the generalized circulation theorem, it is shown that there are two possible mechanisms to create rotational fluid motion (i.e. vorticity) at a two-fluid interface: baroclinic torque and viscous-shear torque. When the surface-tension forces are included, an additional torque is possible due to the discontinuity in stress tensor at the interface. Both baroclinic and viscous-shear torques require the presence of fluid-density gradients. Vortical flows which appeared at wave breaking must have been created by those mechanisms.

Introduction

The majority of water-wave problems are described based on the potential-flow theory. The theory is adequate since vortical flows generated at a solid boundary remain in the thin boundary layer due to the oscillatory nature of the flow; the rotational part of the flow plays little role in the irrotational-flow dynamics. Even during the process of wave breaking, flow irrotationality persists . Numerical results based on the boundary-integral-equation method (e.g. Longuet-Higgins & Cokelet, 1977) have evidently demonstrated an over-turning wave-front motion of irrotational flow. Nevertheless, once the over-turning fluid touches down and reconnects to the water in front of the wave, the potential-flow theory breaks down; strong flow circulation appears and the fluid motion becomes vortical and generates turbulence.

A jet entering a quiescent ambient fluid body also causes eddy motions; the initially quiescent (irrotational) ambient fluid becomes vortical. In this case, fluid rotation is created in the boundary layer at the wall of jet nozzle and is then advected into the ambient fluid body. The same is true for a barotropic mixing layer. Vorticity is introduced to the fluid domain from the upstream boundary layers such as that created along a split plate. On the other hand, eddies associated with wave breaking are not caused by vorticity advection from the solid boundary. During wave breaking, fluid rotation must be created at the free surface (i.e. the air-water interface) from the initially irrotational state; the rotational-flow formations are evidently different from the case of a jet or mixing layer. The question is how the vorticity is generated.

First, vorticity creation mechanisms are reviewed in terms of flow circulation. The generalized circulation theorem is then applied to a two-layer fluid system. In a continuous stress field, there are only two possible mechanisms which produce fluid rotation at the interface of two

M. L. Banner, R. H. J. Grimshaw (Eds.)
Breaking Waves
IUTAM Symposium Sydney/Australia 1991

fluids; namely, baroclinic torque and viscous-shear torque. When the surface-tension effect is included, an additional torque is possible due to a discontinuity in stress tensor at the interface.

The air-water interface is a two-fluid interface and is often treated as a 'free' surface in a traditional water-wave problem; i.e. the stresses vanish at the free surface and, consequently, it is a barotropic surface. The concept of 'free' surface is an approximation based on the fact that the air density is much smaller than that of water, hence the air dynamics do not influence the dynamics of water flow. Contrary to this, it is shown that the free-surface condition of vanishing stresses is not a good approximation in terms of the production of fluid rotations. In order to correctly evaluate vortical-flow production at the interface, the surface must be viewed as an air-water two-layered fluid system; i.e. the existence of air cannot be neglected in the analyses.

Generalized Flow-Circulation Theorem

Fluid rotation can be measured in terms of flow circulation Γ, which is defined by

$$\Gamma = \int_c \mathbf{u} \cdot d\mathbf{x} = \int_s \boldsymbol{\omega} \cdot d\mathbf{A} \,, \tag{1}$$

where **u** is the fluid particle velocity, ω is the vorticity (i.e. curl **u**), **x** is the position vector of a closed integration contour c, and **A** is the area vector of the surface s whose boundary is a single closed curve c, see Fig. 1.

Figure 1. A closed integration contour c on a surface S; the surface s is enclosed by c and is defined by the area vector **A**: s ∈ S; **n** is the normal unit vector in the two-dimensional space S pointing outward from the surface s.

The rate of change in circulation Γ following fluid parcels that make up the curve c can be found to be

$$\frac{D\Gamma}{Dt} = \int_c (D\mathbf{u}/Dt) \cdot d\mathbf{x}, \tag{2}$$

where D/Dt is the material derivative. Applying Newton's second law of motion to the fluid parcels along c, (2) becomes

$$\frac{D\Gamma}{Dt} = \int_c \left(\frac{1}{\rho} \operatorname{div} \sigma_{ij} + \mathbf{f}\right) \cdot d\mathbf{x} , \tag{3}$$

where ρ is the fluid density, σ_{ij} is the stress tensor, and **f** is the body force per unit mass. Under the action of conservative body force, the integration of **f** along a closed path vanishes and (3) becomes

$$\frac{D\Gamma}{Dt} = \int_c \left(\frac{1}{\rho} \operatorname{div} \sigma_{ij}\right) \cdot d\mathbf{x} . \tag{4}$$

For Newtonian fluids, the stress tensor σ_{ij} is expressed as

$$\sigma_{ij} = - P\, \delta_{ij} + S_{ij} , \tag{5}$$

where P is the pressure, δ_{ij} is the Kronecker delta, and S_{ij} is the deviatoric stress tensor:

$$S_{ij} = 2\,\mu \left(e_{ij} - \frac{1}{3} (\operatorname{div} \mathbf{u})\, \delta_{ij}\right) , \tag{6}$$

μ is the dynamic viscosity of the fluid, and e_{ij} is the rate-of-strain tensor:

$$e_{ij} = \frac{1}{2} \{\operatorname{grad} \mathbf{u} + (\operatorname{grad} \mathbf{u})^T\} , \tag{7}$$

where the superscript T denotes the transpose.

Assuming the fluid to be incompressible, inviscid, and homogeneous (the fluid density ρ being uniform in the entire fluid domain) , (4) is reduced to the well-known Kelvin theorem:

$$\frac{D\Gamma}{Dt} = -\frac{1}{\rho} \int_c (\operatorname{grad} P) \cdot d\mathbf{x} = -\frac{1}{\rho} \int_c dP = 0 . \tag{8}$$

Equation (8) states that flow circulation Γ around a closed curve c moving with the fluid remains constant. In other words, if the flow field is initially irrotational, flow circulation cannot be created in an inviscid, homogeneous fluid domain. Because the integration surface s is arbitrary in (1), Kelvin's theorem indicates that no vorticity is created in the fluid domain.

If the fluid is homogeneous and viscous, then (4) becomes

$$\frac{D\Gamma}{Dt} = \frac{\mu}{\rho}\int_c \nabla^2 \mathbf{u}\cdot d\mathbf{x} = \frac{\mu}{\rho}\int_s \nabla^2 \omega \cdot d\mathbf{A} = \frac{\mu}{\rho}\int_c \left(\mathrm{grad}_{\parallel}\,\omega_{\perp}\right)\cdot \mathbf{n}\; dc\,, \tag{9}$$

by using the Stokes theorem and the two-dimensional form of the divergence theorem. In (9), ∇^2 is the Laplacian operator, $\omega_{\perp}$ is a component of the vorticity vector which is normal to the surface s, $\mathrm{grad}_{\parallel}$ is the two-dimensional gradient operator taken in the direction on the surface S which contains the integration surface s ($s \in S$, see Fig. 1), **n** is the normal unit vector in the two-dimensional space S pointing outward from the surface domain s. Equation (9) represents the time rate of change in flow circulation due to molecular diffusion of vorticity across the boundary of integration surface s that is enclosed by the line c. Hence the flow circulation is changed due to the flux of vorticity from the surroundings. Since the closed line c is arbitrary, suppose that the line c is expanded to the boundary of the entire (homogeneous) fluid domain. Then, it is evident that, if the flow is initially irrotational everywhere, fluid rotation can only be caused by the molecular diffusion flux from the boundary but cannot be created within the fluid domain.

In the case of non-homogeneous and inviscid fluids, flow circulation can be produced within the fluid domain whenever the fluid is displaced from a state in which the pressure gradient and the density gradient are parallel, and (4) becomes

$$\frac{D\Gamma}{Dt} = -\int_c \left(\frac{1}{\rho}\,\mathrm{grad}\,P\right)\cdot d\mathbf{x} = \int_s \frac{1}{\rho^2}\left(\mathrm{grad}\,\rho \times \mathrm{grad}\,P\right)\cdot d\mathbf{A}\,, \tag{10}$$

where $\times$ denotes the vector (cross) product. This is often called the Bjerknes theorem (see, for example, Lamb, 1932, art. no. 166 a), and represents the time rate of change in circulation by 'baroclinic torque' following fluid parcels that make up the curve c : see, for example, Yeh (1991) for a physical interpretation of baroclinic torque.

Assuming an incompressible Newtonian fluid and a conservative body force, but allowing variations of fluid density and viscosity, (4) becomes

$$\frac{D\Gamma}{Dt} = \int_s \frac{1}{\rho}\Big[\frac{1}{\rho}\left(\mathrm{grad}\,\rho \times \mathrm{grad}\,P\right) + \mathrm{div}\left(\mu\,\mathrm{grad}\,\omega\right) - \frac{1}{\rho}\left(\mathrm{grad}\,\rho \times \mathrm{div}\left(2\mu\, e_{ij}\right)\right) + \mathrm{grad}\,\mu \times \nabla^2\mathbf{u}\Big]\cdot d\mathbf{A}\,. \tag{11}$$

The first term in the integrand represents the production of fluid rotation by baroclinic torque; the second term represents the diffusion of vorticity from the surroundings; the third term represents the creation of fluid rotation due to non-uniform fluid acceleration caused by inhomogeneous fluid media (this mechanism is termed as viscous-shear torque -- see Yeh

(1991) for a physical interpretation); the last term represents the torque produced by a spatial variation of viscous force.

Two-Layer Fluid Interface

Consider a fluid domain which consists of two homogeneous fluids separated by a sharp (discontinuous) interface. As discussed earlier, within a homogeneous fluid domain, there is no mechanism to create fluid rotation; any rotational motion must be created at the boundary and then transferred into the fluid domain via diffusion. This is also true for a two-layer fluid system. The only source for fluid rotation is at the boundaries and the interface of two fluids. The production of fluid rotation at a solid (no-slip) boundary was discussed by Lighthill (1963) and here I focus on a two-layer-fluid interface. Again assuming the body force to be conservative and using the Stokes theorem, (4) can be expressed as

$$\frac{D\Gamma}{Dt} = \int_s \left(-\frac{1}{\rho^2}\left(\text{grad}\,\rho \times \text{div}\,\sigma_{ij}\right) + \frac{1}{\rho}\,\text{curl div}\,\sigma_{ij}\right) \cdot dA \;. \tag{12}$$

Note that the gradient of fluid density is always in the direction normal to the interface of two fluids. Suppose that a contour path c is taken along the interface as shown in Fig. 2.

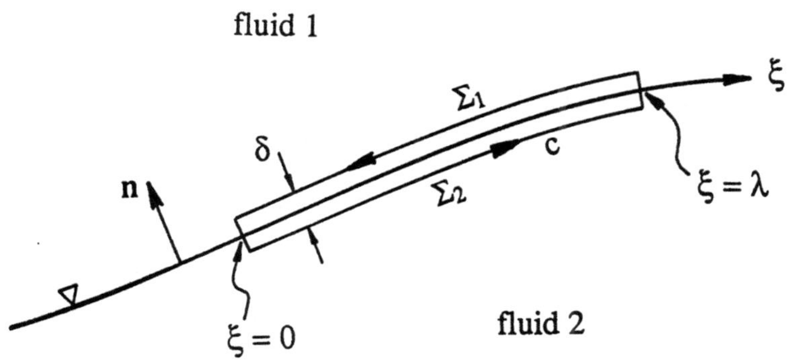

Figure 2. Integration contour c enclosing the interface of fluids 1 and 2; the ξ-coordinate points in the direction along the interface; Σ_1 and Σ_2 are the segments of the contour c, which are parallel to the interface $\xi \in [0,\lambda]$ and are separated by δ; **n** is the normal unit vector pointing outward from fluid 2.

Taking the limit as the thickness $\delta \rightarrow 0$ of the narrow strip area enclosed by the contour, (12) can be reduced to

$$\frac{D\Gamma}{Dt} = \lim_{\delta \to 0} \int_S \left\{ \frac{1}{\delta}\, \Delta\!\left(\frac{1}{\rho}\right) \mathbf{n} \times \operatorname{div} \sigma_{ij} + \frac{1}{\rho} \operatorname{curl} \operatorname{div} \sigma_{ij} \right\} \cdot d\mathbf{A} \tag{13a}$$

$$= \int_0^{\lambda} \left\{ \Delta\!\left(\frac{1}{\rho}\right) \mathbf{n} \times \operatorname{div} \sigma_{ij} \right\} \cdot \mathbf{n_A}\; d\xi\,, \tag{13b}$$

where $\Delta(1/\rho) = (1/\rho_1 - 1/\rho_2)$, $\mathbf{n}$ is the normal unit vector pointing outward from fluid 2 as shown in Fig. 2, and $\mathbf{n_A} = \mathbf{A}/|\mathbf{A}|$ is the unit vector on the interface pointing perpendicular to the integration curve from 0 to λ. The second integrand in (13a) vanishes as $\delta \rightarrow 0$ because the net force acting on a fluid parcel (= div σ_{ij}) is continuous and finite across the interface (this is not true if the surface-tension effect is considered). Equation (13b) indicates that flow circulation can be produced at the interface due to the singularity in grad $(1/\rho)$. For Newtonian fluids, σ_{ij} can be expressed by (5), hence (13b) is expressed as

$$\frac{D\Gamma}{Dt} = \int_0^{\lambda} \left\{ \Delta\!\left(\frac{1}{\rho}\right) \frac{\partial P}{\partial \xi} - \Delta\!\left(\frac{1}{\rho}\right) (\operatorname{div} S_{ij})_{\xi} \right\} d\xi\,, \tag{14}$$

where the subscript ξ in the second integrand denotes the ξ-direction component of (div S_{ij}), the net viscous force acting on a fluid parcel. Equation (14) evidently indicates that there are only two mechanisms to create fluid rotation at the interface of a two-layer fluid system: baroclinic torque and viscous-shear torque, i.e. the first and third terms in (11), respectively.

The surface-tension force becomes important when the curvature of an interface is large. If the value of surface tension is constant, the dynamic boundary condition (i.e. the equality of forces on the surface of each fluid) becomes

$$(\sigma_{1,ij} - \sigma_{2,ij}) \cdot \mathbf{n} = -\gamma\,(R_1^{-1} + R_2^{-1})\,\mathbf{n}\,. \tag{15}$$

where $\sigma_{1,ij}$ is the stress tensor evaluated in fluid 1 at the interface, $\sigma_{2,ij}$ is the stress tensor evaluated in fluid 2 at the interface, R_1 and R_2 are the principal radii of curvature of the interface, γ is the surface tension (assumed to be constant) between the two fluid media, and $\mathbf{n}$ is the unit normal vector pointing outward from fluid 2 as shown in Fig. 2. Then, (14) is no longer valid since div σ_{ij} is discontinuous across the interface. Note that as long as γ is constant, the surface-tension force only affects the normal component of stresses but not the tangential components. Considering a two-dimensional flow domain for simplicity and assuming the normal component of viscous stresses to be negligible (e.g. in the case of a thin boundary-layer flow), the surface-tension effect (15) can be reduced to

$$P_1 - P_2 = \gamma \frac{1}{R}$$

$$= \gamma \left[\frac{\partial_{xx}\eta}{\left(1 + (\partial_x \eta)^2\right)^{3/2}} \right] , \tag{16}$$

where R is the radius of curvature, η is the elevation of interface which is a function of the horizontal position x and time t, $\eta = \eta\ (x,t)$, and ∂ denotes partial differentiation with respect to the letter subscript. Instead of the surface-integral form of (12), the contour integration in the form of (4) is used to evaluate the time rate of circulation production. Taking the closed contour as shown in Fig. 2 with diminishing thickness, i.e. $\delta \to 0$, the contribution to the closed contour integral is the integrals along the lines, Σ_1 and Σ_2, parallel to the interface. Then,

$$\frac{D\Gamma}{Dt} = \int_c \left(\frac{1}{\rho} \operatorname{div} \sigma_{ij}\right) \cdot dx$$

$$= \int_{\Sigma_1} \frac{1}{\rho_1} \left\{ \frac{\partial P_1}{\partial \xi} - (\operatorname{div} S_{ij})_\xi \right\} d\xi - \int_{\Sigma_2} \frac{1}{\rho_2} \left\{ \frac{\partial P_2}{\partial \xi} - (\operatorname{div} S_{ij})_\xi \right\} d\xi . \tag{17}$$

Using (16),

$$\frac{\partial}{\partial \xi}(P_1 - P_2) = \frac{\gamma \left\{ \partial_{xxx}\eta \left(1 + (\partial_x\eta)^2\right) - 3\, \partial_x\eta \left(\partial_{xx}\eta\right)^2 \right\}}{\left(1 + (\partial_x\eta)^2\right)^3} . \tag{18}$$

Substituting (18) into (17) yields,

$$\frac{D\Gamma}{Dt} = \int_0^\lambda \left[\left(\Delta \frac{1}{\rho}\right) \frac{\partial P_2}{\partial \xi} - \left(\Delta \frac{1}{\rho}\right) (\operatorname{div} S_{ij})_\xi + \frac{\gamma \left\{ \partial_{xxx}\eta \left(1 + (\partial_x\eta)^2\right) - 3\, \partial_x\eta \left(\partial_{xx}\eta\right)^2 \right\}}{\rho_1 \left(1 + (\partial_x\eta)^2\right)^3} \right] d\xi , \tag{19a}$$

or

$$= \int_0^\lambda \left[\left(\Delta \frac{1}{\rho}\right) \frac{\partial P_1}{\partial \xi} - \left(\Delta \frac{1}{\rho}\right) (\operatorname{div} S_{ij})_\xi + \frac{\gamma \left\{ \partial_{xxx}\eta \left(1 + (\partial_x\eta)^2\right) - 3\, \partial_x\eta \left(\partial_{xx}\eta\right)^2 \right\}}{\rho_2 \left(1 + (\partial_x\eta)^2\right)^3} \right] d\xi . \tag{19b}$$

Note that (19) reduces to (14) by neglecting the surface-tension effect.

Air-Water Interface

At the air-water interface (considering fluid 1 to be air and fluid 2 to be water in Fig. 2), the fluid density of air ($\approx$ 0.0012 gm/cm^3) is much smaller than that of water ($\approx$ 1.0 gm/cm^3): $\rho_1 << \rho_2$. Based on (14), even though the values of $\partial P/\partial \xi$ and (div $S_{ij})_\xi$ are infinitesimal (as often ignored at the 'free' surface boundary condition in a traditional water-wave problem), the production rate of fluid rotation is not necessarily infinitesimal due to the large value of $\Delta(1/\rho)$. In fact, Yeh (1991) showed using the order-of-magnitude analysis that the production rate of circulation is finite at the air-water interface of a steadily advancing, fully developed bore. He further showed that the dominant generation mechanism of fluid rotation at the front face of a bore is the baroclinic torque caused by the dynamic pressure variation along the interface; the magnitude of viscous-shear torque is insignificant.

The surface-tension effect on circulation production is included in (19). It is emphasized that the surface-tension force acts in the plane of interface; therefore, it cannot form a torque and is not capable of directly creating fluid rotation. However, the (constant) surface-tension can create a variation in the normal component of fluid stresses; the variation in stresses is what produces imbalanced forces (torque) and triggers fluid rotation, i.e. the second integrand in (13a) no longer vanishes. According to (19b), the surface-tension effect can be neglected if the pressure field in the air is known because $\Delta(1/\rho) \approx 1/\rho_1 >> 1/\rho_2$; i.e. the magnitudes of first and second terms in (19b) dominates over the last term. On the other hand, if the circulation production rate is computed with the pressure field in water, the term representing surface-tension effect is not negligible and must be included in (19a). In order to compute the pressure variation along the air-water interface of a bore, Yeh (1991) applied the Bernoulli theorem to the water domain neglecting the surface-tension effect in his order-of-magnitude analysis. Since the evaluation of surface tension effect is difficult, it is not clear whether this deficiency causes significant errors in the results.

The assumption of 'free' surface is not only the absence of stresses at the interface but also the negligence of the fluid (air) itself at the other side of the interface, i.e. hypothetically considered to be a vacuum. Such a state is of course physically impossible. It is emphasized that based on the fundamental analysis of the flow-circulation theorem, there are only two mechanisms to create fluid rotation at a two-fluid interface: baroclinic torque and viscous shear torque. The only possible modification to these mechanisms is due to surface-tension force. The vortical flows which emerge from wave breaking must be created by the mechanisms discussed herein: I can find no other explanation.

This work was completed during my visit to Cornell University. I thank Dr. Philip Liu for his hospitality.

References

Lamb, H. 1932 Hydrodynamics. Cambridge Univ. Press: Cambridge.
Lighthill, M.J. 1963 Introduction. Boundary layer theory. In Laminar Boundary Layers (Ed: L. Rosenhead). Clarendon Press: Oxford.
Longuet-Higgins, M.S. & Cokelet, E.D. 1976 The deformation of steep surface waves on water I. A numerical method of computation. *Proc. R. Soc. Lond. A* **350**, 1-26.
Yeh, H.H. 1991 Vorticity-generation mechanisms in bores. *Proc. R. Soc. Lond. A* **432**, 215-231.

Numerical Experiments on the Evolution of Fetch Limited Waves

I.R. YOUNG and M.L. BANNER

Dept. of Civil & Maritime Engineering,
University College, ADFA,
Canberra, 2600, Australia
and
School of Mathematics,
University of NSW,
Sydney, 2033, Australia

Summary

A numerical wave model with a full solution to the nonlinear spectral transfer terms is used to determine whether the atmospheric input and dissipation source terms proposed in previous wave model studies are consistent with observed fetch limited data. It is shown that existing representations are at variance with observed data and that their spectral form is biased by assumptions made to reduce the computational time. A more appropriate representation is developed with an input source term proportional to the square of the inverse wave age and a dissipation source term proportional to the square of the wave number. The results also raise questions as to the applicability of presently used directional spreading models.

1. Introduction

The source terms currently used in third generation wave prediction models such as the WAM Model (WAMDI group, 1988) have been developed from a combination of theoretical considerations, field data and numerical experiments using fetch limited models with full solutions to the nonlinear terms. These numerical experiments have determined the dissipation term required to balance the other source terms and produce growth rates consistent with measurements. Little effort has, however, been devoted to determining whether the selected dissipation term also reproduces the observed detail structure of the spectrum. With the recent availability of more detailed directional spectral data (Donelan et al., 1985; Banner, 1990), a more critical assessment of presently used source terms is now possible. This paper will present the results of a detailed set of numerical experiments to determine whether existing source terms produce fetch limited spectral development in agreement with the observed data. Alternative representations which may be more suitable will also be explored.

2. Numerical Model

For fetch limited conditions the two-dimensional energy balance equation can be simplified to the one-dimensional form (Komen et al., 1984)

M. L. Banner, R. H. J. Grimshaw (Eds.)
Breaking Waves
IUTAM Symposium Sydney/Australia 1991

$$C_g \frac{\partial E}{\partial x} \cos\theta = S_{in} + S_{nl} + S_{ds} = S_{tot} \tag{1}$$

where C_g is the group velocity and $E = E(f,\theta)$ is the directional wave spectrum which is a function of frequency and direction. The source terms on the right of the equation represent the processes of atmospheric input from the wind, nonlinear interactions within the spectrum and dissipation, respectively. Representations for S_{in} and S_{ds} will be presented below. For the present application, a full solution to the nonlinear term, S_{nl} using the method of Resio and Perrie (1991) has been used. A similar model was used by Komen et al. (1984) to determine an appropriate representation for S_{ds}. Full solutions to S_{nl} are computationally very expensive and to reduce this expense, Komen et al. (1984) performed explicit calculations only for frequencies, $f < 2.5 f_p$, where f_p is the frequency of the spectral peak. For frequencies above this limit a diagnostic tail proportional to f^{-5} was imposed. As we are interested in the detailed behavior of the high frequency spectral region, such a tail has not been imposed in this work. Instead, explicit calculations have been performed to the high frequency spectral limit of the computational grid ($\approx 1Hz$). Due to the nature of the nonlinear interactions, it is important that possible interactions with spectral components at frequencies beyond this cut-off also be considered. In order to account for such interactions, the energy at these non-grid points was estimated by extrapolating the spectrum beyond the high frequency limit with an f^{-4} tail. A number of sensitivity tests were conducted to determine whether this tail had significant influence on the computations. Provided the computational grid is extended to relatively high frequencies, as in the present calculations, the tail has negligible impact.

Equation (1) was integrated using a first order explicit finite difference scheme. The space step Δx was chosen dynamically so as to maintain both accuracy and numerical stability (Komen et al., 1984). The size of Δx was controlled by the the high frequency regions of the spectrum where the waves react rapidly to the source terms. Since explicit calculations are carried out to quite high frequencies, Δx is necessarily small. Typical values of Δx are between 1 and 2 metres.

3. Existing Source Term Representations

In order to assess the adequacy of existing source terms, the forms for S_{in} and S_{ds} proposed by Komen et al. (1984) were used.

$$S_{in} = 3 \times 10^{-4} \omega \left[\frac{28 u_*}{C} \cos\theta - 1 \right] E(f,\theta) \tag{2}$$

$$S_{ds} = 3.33 \times 10^{-5} \overline{\omega} \left(\frac{\omega}{\overline{\omega}} \right)^n \left(\frac{\hat{\alpha}}{\hat{\alpha}_{PM}} \right)^m E(f,\theta) \tag{3}$$

where $\omega = 2\pi f$ is the radian measure of frequency, u_* is the friction velocity, C is the wave phase speed, $\overline{\omega}$ is the mean spectral frequency, $\hat{\alpha}$ is a spectral steepness parameter and $\hat{\alpha}_{PM}$ is the value of $\hat{\alpha}$ at the Pierson-Moskowitz limit. These final three parameters are defined in Komen et al. (1984). Komen et al. (1984) set both n and m in Equation (3) at a value of 2.

Figure (1) shows a plot of the total energy ($\int E(f,\theta) df d\theta$) as a function of the fetch x for a case were the wind velocity $U_{10} = 10ms^{-1}$. The result is also compared with the JONSWAP (Hasselmann et al., 1973) growth relationship and the Pierson-Moskowitz asymptotic limit (Pierson and Moskowitz, 1964).

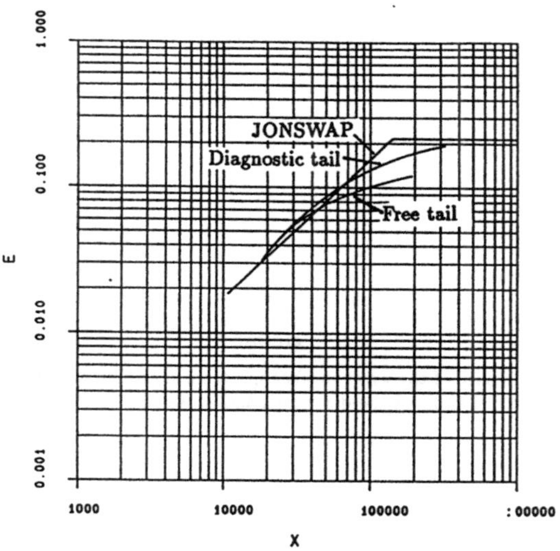

Figure (1) Fetch limited growth curves for a wind speed $U_{10} = 10ms^{-1}$ for models with an unconstrained tail and an f^{-5} diagnostic tail for $f > 2.5f_p$. The JONSWAP and PM relationships are also shown.

It is quite clear that the growth rate is significantly lower than the JONSWAP relationship and that the energy will asymptote to a value lower than the Pierson-Moskowitz limit. Figure (2) shows the wind direction slice of the wave number spectrum normalized by the factor k^4. Based on observational data, Banner (1990) has argued that the high wave number region of the wind direction slice is proportional to k^{-4}. The model yields a high wave number region approximately proportional to k^{-4} ($k^{-3.7}$), but the constant of proportionality is more than a factor of two higher than the Banner (1990) result. Although detailed spectra were not published by Komen et al. (1984), the present calculations do not appear to be consistent with their results. To determine whether the Komen et al. (1984) results were biased by the inclusion of the high frequency diagnostic tail, a second simulation was performed with an f^{-5} tail imposed for $f > 2.5f_p$. The growth rate and wind direction slice of the wave number spectrum are shown in Figures (1) and (2), respectively. The influence of the diagnostic tail is clear, with a significantly enhanced growth rate and a wind direction slice with a significantly reduced proportionality constant at high wave numbers.

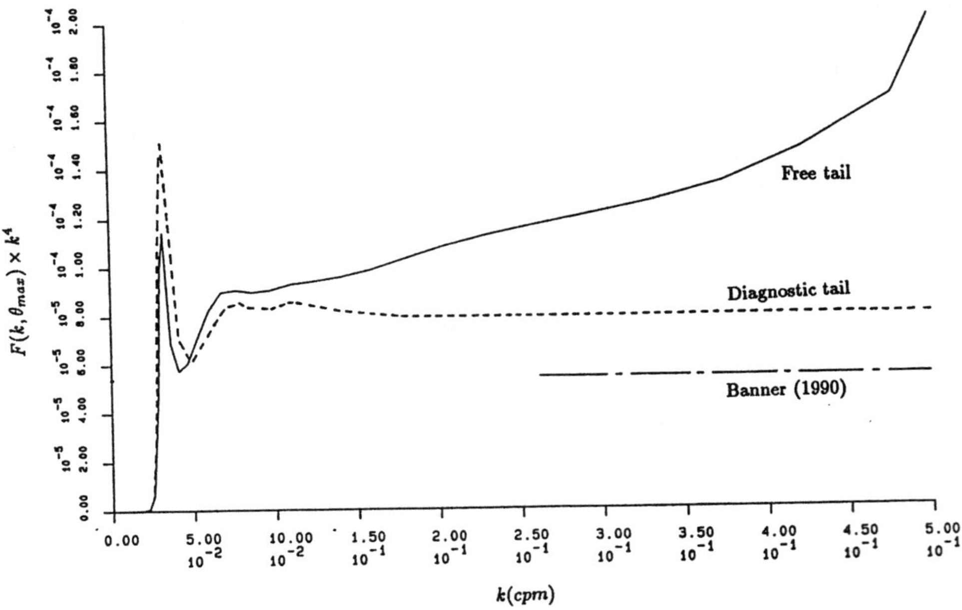

Figure (2) The wind direction slice of the wave number spectrum normalized with the factor k^4. Results for the model with an unconstrained tail and the diagnostic tail as well as the Banner (1990) result are shown.

The reason for this effect is shown in Figure (3) where the one-dimensional frequency spectra and each of the source terms is shown for the two spectra of Figure (2). Although these two spectra have the same total energy and similar values of f_p, the shape of the spectra are quite different. The spectrum with the diagnostic tail has significantly less energy at high frequencies and is also much more peaked. The nonlinear source term, S_{nl} is very sensitive to the peakedness of the spectrum and is consequently much larger for the peaked spectrum. It is the increased magnitude of S_{nl} which is largely responsible for the increased growth rate observed for the run with the diagnostic tail.

As a result, it can be concluded that the Komen et al. (1984) result is heavily biased by the inclusion of the diagnostic tail. Models which use the source terms represented by Equations (2) and (3) can only yield reasonable results if an artificial diagnostic tail is also included.

4. Alternative Representations for the Source Terms

As a result of the poor performance of existing source terms, an effort was made to find more satisfactory representations. Plant (1982) has indicated that at high frequencies, S_{in} is proportional to $(u_*/C)^2$ rather than $(u_*/C - 1)$ as represented by Equation (2).

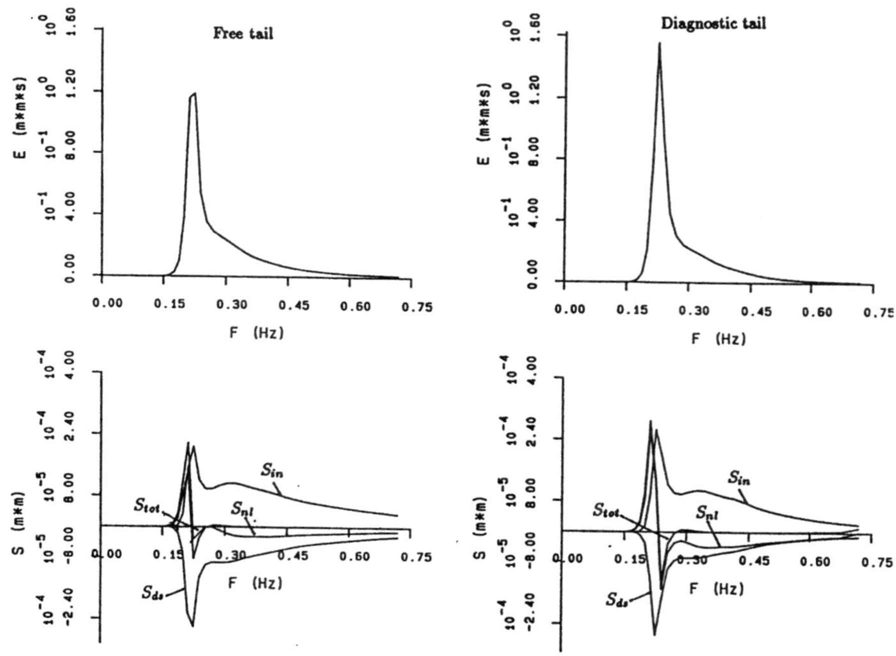

Figure (3) The one-dimensional frequency spectrum as well as each of the source terms. Results are shown for both the free and diagnostic tails.

Yan (1987) has developed a polynomial form for S_{in} which approximates Equation (2) at low values of u_*/C and smoothly transitions to a $(u_*/C)^2$ dependence at high values of u_*/C:

$$S_{in} = \left[0.04\left(\frac{u_*}{C}\right)^2 \cos\theta + 0.00544\left(\frac{u_*}{C}\right)\cos\theta + 0.000055\cos\theta - 0.000311\right]\omega E(f,\theta) \tag{4}$$

Equations (2) and (3) yielded spectra with too much energy at high wave numbers. One means of reducing this energy would be to adopt an S_{ds} representation with a higher power dependence on ω than 2.

Runs were performed with the power n in Equation (3) set at 3 and 4 and with S_{in} represented by Equation (4). The growth rates are shown in Figure (4) and the wind direction slices of the wave number spectra in Figure (5). As expected, increasing the value of n (which increases the dissipation at high frequencies relative to lower frequencies) decreases the energy in the high wave number region of the spectrum. As a consequence, S_{nl} is increased and the growth rate increases.

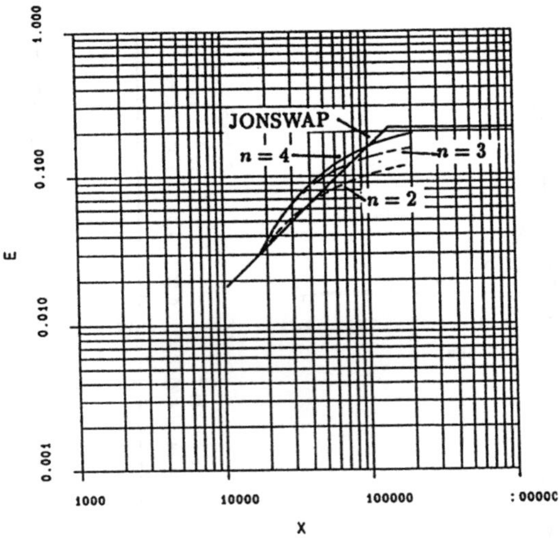

Figure (4) Fetch limited growth curves for a wind speed $U_{10} = 10ms^{-1}$ and dissipation terms proportional to $(\omega/\overline{\omega})^n$. The JONSWAP and PM relationships are also shown.

A value of $n = 4$ produces a growth rate consistent with the JONSWAP (Hasselmann et al., 1973) observations. In addition, the high wave number region of the wind direction slice is approximately proportional to k^{-4} with a proportionality constant consistent with Banner (1990). Hence, this representation for S_{ds} in conjunction with S_{in} represented by Equation (4) yields results in far better agreement with observations than presently used forms.

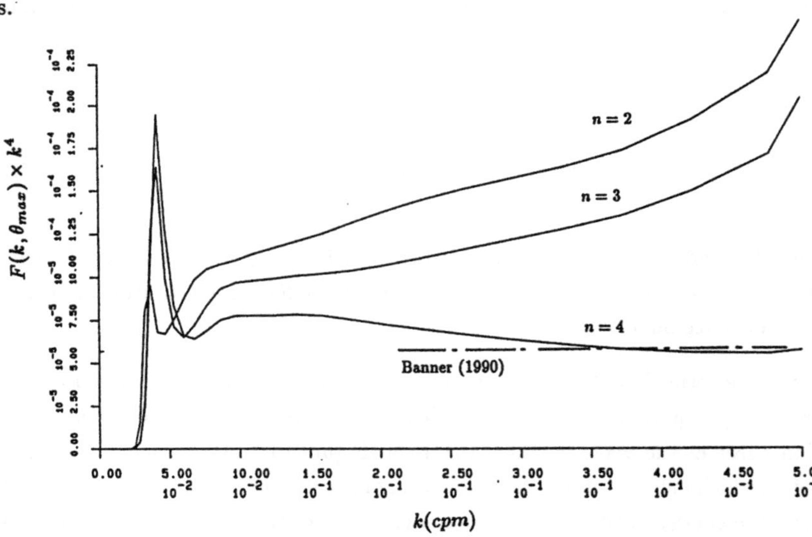

Figure (5) The wind direction slice of the wave number spectrum normalized with the factor k^4. Results are shown for dissipation terms proportional to $(\omega/\overline{\omega})^n$. The Banner (1990) result is also shown.

The present model can also be used to provide information on the direction spread of the spectrum as represented by the directional spreading function $D(\theta;k)$

$$F(k,\theta) = F(k,\theta_{max})D(\theta;k) \tag{5}$$

where $F(k,\theta)$ is the directional wave number spectrum. Figure (6) shows $D(\theta;k)$ at wave numbers equal to k_p, $2k_p$, $3k_p$ and $4k_p$. The spreading function is uni-modal at the spectral peak but with increasing wave number it becomes bi-modal with more energy occuring in off-wind directions than in the wind direction. Such a result is quite different to the traditional uni-modal distributions proposed by Mitsuyasu et al. (1980), Hasselmann et al. (1980) and Donelan et al. (1985). It is, however, questionable whether conventional directional spectral measurement techniques have sufficient directional resolving power to yield such a distribution. The directional spectra presented by Holthuijsen (1983) which were obtained using stereo photography do, however, reveal a similar bi-modal distribution.

The bi-modal distribution is an extremely robust feature of the directional spectra produced by the model. Numerous source term combinations have been tested and, although other features of the spectrum can be altered, the bi-modal directional distribution

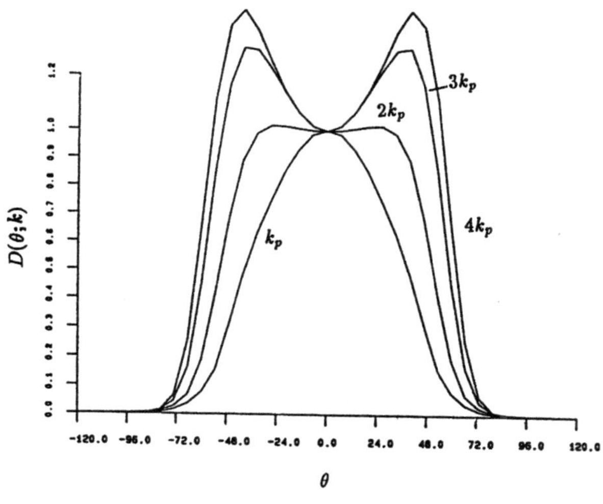

Figure (6) The directional spreading function at different wave numbers.

always remains. The side lobes are caused by S_{nl}. Young and van Vledder (1991) have shown that interactions between wave number components at the spectral peak and approximately 30° to the mean spectral direction are particularly strong. Indeed, source term combinations which produce more peaked spectra and hence large values of S_{nl}, also have larger side lobes in $D(\theta; k)$.

Due to the robust nature of the side lobe structure, it is reasonable to assume they are not simply an artifact of the adopted source terms. Consequently, this may reveal a shortcoming of existing directional spreading models.

5. Conclusions

As a result of the numerical experiments performed with the model it can be concluded that the Komen et al. (1984) results are heavily biased by the diagnostic tail used for the high frequency portion of their spectra. Consequently, the dissipation term derived from their study and used in present day wave prediction models is not appropriate.

A source term combination consistent with observations involves an atmospheric input term proportional to $(u_*/C)^2$ at large values of u_*/C (Plant, 1982) and a dissipation term proportional to ω^4 (ie. k^2 scaling).

The model also yields directional spreading functions which are uni-modal at the spectral peak frequency, but become bi-modal at higher frequencies. This distribution appears very robust and hence challenges proposed spreading models.

6. References

1. Banner, M.L., 1990, "Equilibrium spectra of wind waves", J. Phys. Oceangr., 20, 966-984.
2. Donelan, M.A., J. Hamilton and W.H. Hui, 1985, "Directional spectra of wind-generated waves", Phil. Trans. Roy. Soc. London, A315, 509-562
3. Hasselmann, D.E., M. Dunckel and J.A. Ewing, 1980, "Directional wave spectra observed during JONSWAP 1973", Jnl. Phys. Oceanogr., 10, 8, 1264-1280.
4. Hasselmann, K., et al., 1973, "Measurements of wind-wave growth and swell decay during the Joint North Sea Wave Project (JONSWAP)", Dtsch. Hydrogr. Z., A8(12).
5. Holthuijsen, L.H., 1983, "Observations of the directional distribution of ocean-wave energy in fetch-limited conditions", J. Phys. Oceanogr., 13, 2, 191-207.
6. Komen, G.J., S. Hasselmann and K. Hasselmann, 1984, "On the existence of a fully developed wind-sea spectrum", J. Phys. Oceanogr., 14, 8, 1271-1285.
7. Mitsuyasu, H., et al., 1980, "Observations of the power spectrum of waves using a cloverleaf buoy", J. Phys. Oceanogr., 10, 286-296.
8. Pierson, W.J. and L. Moskowitz, 1964, "A proposed spectral form for fully developed wind seas based on the similarity theory of S.A. Kitaigorodskii", J. Geophys. Res., 69(24), 5181-5190.
9. Plant, W.J., 1982, "A relationship between wind stress and wave slope", J. Geophys. Res., 87, 1961-1967.

10. Resio, D. and W. Perrie, 1991, "A numerical study of nonlinear energy fluxes due to wave-wave interactions. Part 1. Methodology and basic results", J. Fluid Mech., 223, 603-629.

11. WAMDI Group, 1988, "The WAM Model - A third generation ocean wave prediction model", J. Phys. Oceanogr., 18, 1775-1810.

12. Yan, L., 1987, "An improved wind input source term for third generation ocean wave modelling", KNMI, Report WR-NR87-8, 10pp.

13. Young, I.R. and G. Ph. van Vledder, 1991, "A Review of the central role of nonlinear interactions in wind-wave evolution", in preparation.

Theoretical Study of Breaking Wave Spectrum and its Application

Hua Feng *Yuan Yeli*
(First Institute of Oceanography, SOA, Qingdao 266003)

ABSTRACT

Based on breaking wave statistical model, we deduced the relation of original and breaking wave spectra. The anlysis to this expression shows that wave breaking produces the spectral energy loss, which increases with frequency, and peak frequency shifting to lower frequency. Considering that mean frequency between extrema is a logical time interval in a wave event, we obtained a new theoretical dissipation source function which has been successfully used in LAGFD third generation numerical wave model. From the view–point that the spectral form in the equilibrium range is induced by the balance of wind input, nonlinear transfer and dissipation by wave breaking, we derived a new equilibrium spectrum, which shows that the power of frequency of the spectrum changes with frequency and the main part of it is − 4.5.

---

1. Introduction

Wave breaking studies, like the general wave studies, can basically be put into two categories. One is its dynamical study, which involves the strong nonlinear process of wave and its breaking machenism. Many study results of them can be used as the basis of establishing breaking wave statistical model. The other is statistics study based on breaking wave statistical model. Obviously, because of the complicacy of wave breaking, the combination of two studies will provide a useful method.

As an important breaking criterion, Longuet–Higgins (1963) demonstrated that when the progressive Stokes wave reaches the limit state, its vertical acceleration should be $g/2$. Considering that wave breaking, in fact, can only occur around the extreme of the surface elevation ($\zeta'' < -g/2$), Yuan et al. (1988)proposed a new breaking wave statistical model

$$\zeta_b = \zeta\{-[\frac{g/2}{\zeta''}H(\zeta) + \frac{\zeta''}{g/2}H(-\zeta)]H(-\zeta'' - g/2) + H(\zeta'' + g/2) \qquad (1.1)$$

in which H(•) is Heaviside unit step function. It shows that, when breaking occurs on wave crest ($\zeta'' < -g/2$), in the case of $\zeta > 0$, the breaking makes surface elevation approaching the average sea level ($\zeta = 0$)and decreasing in proportion of $(g/2)/|\zeta''|$; in the case of $\zeta < 0$, the breaking makes surface elevation straying from the average sea level and the absolute value of surface elevation increasing in proportion of $|\zeta''|/(g/2)$. On the basis of it, Yuan et al. (1988) derived some important statistical mean quantities about air–sea exchange and Tung et al. (1989) deduced the probability function of surface elevation. Both results agree better with observation data.

As is known to all, an important subject of sea–wave statistical study is spectral form of wind wave in the equilibrium range. Phillips (1958), by means of similarity analysis (gravity being the main dominating parameter), derived a spectral form in the equilibrium, as follows

M. L. Banner, R. H. J. Grimshaw (Eds.)
Breaking Waves
IUTAM Symposium Sydney/Australia 1991

$$F(\omega) = \alpha g^2 \omega^{-5} \tag{1.2}$$

Proceeding from extreme theorem of energy loss by wave breaking, Yuan also derived this spectral form in his 1986 work. It is obvious that the spectrum in the equilibrium is not dominated only by gravity or wave breaking, as supposed earliner by Phillips and Yuan.

Just as indicated by Phillips (1985),it should result from the balance state of synthetical effects including wind input, nonlinear transfer and dissipation by wave breaking. From this view–point,Phillips (1985) derived new results different from his 1958 work,

$$E(\vec{K}) = \beta(\frac{u_*}{c})k^{-4}\cos^p\theta \qquad and \qquad F(\omega) = \gamma u_* g\omega^{-4} \tag{1.3}$$

which are able to fit in with laboratory and field observation (Toba 1973, Kawai et al. 1977, Donelan et al. 1984). However, in his deduction, Phillips (1985), in fact, supposed that dissipation by wave breaking is in accordance with nonlinear transfer in form in the equilibrium range.

2. Statistical Model

In view of consideration in section one, a suitable breaking wave satatistical model should be (1.1)

A nonlinear original waves can be considered stationary in the range of order O(§) (§ is wave slope). Marking $\zeta_i \equiv \zeta(t_i), \zeta''_i \equiv \zeta''(t_i), i=1,2$, according to the theorem of probability theory, the united stationary probability density function can be written as

$$p(\zeta_1,\zeta_2,\zeta_1',\zeta_2') = p(\zeta_1,\zeta_2|_{\zeta_1',\zeta_2'})p(\zeta_1',\zeta_2') \quad . \tag{2.1}$$

The above conditional probability function can be written as

$$p(\zeta_1,\zeta_2|_{\zeta_1',\zeta_2'}) = p(\zeta_1 - a_1\zeta_1' - b_1\zeta_2',\ \zeta_2 - a_2\zeta_1' - b_2\zeta_2') \tag{2.2}$$

where a_1, b_1, a_2 and b_2 can be determined under the condition that the linear combinations $\zeta_1 - a_1\zeta_1'' - b_1\zeta_2''$ and $\zeta_2 - a_2\zeta_1'' - b_2\zeta_2''$ all are irrelevant to both ζ_1'' and ζ_2''. It is easy to be proved that $a_1 = b_2 \equiv a$, $a_2 = b_1 \equiv b$.

3. Breaking Wave Spectrum

① Breaking Wave Spectrum

The spectrum can be derived through Fourier transform on the correlation function $R_{\zeta b}(\tau) = E\{\zeta_b(t_1)\zeta_b(\zeta_2)\}$ of $\zeta_b(t)$, in which $\tau = t_2 - t_1$. From(2.1), we can get

$$R_{\zeta b}(\tau) = E\{\zeta_{b1}\zeta_{b2}\}$$

$$= E\{\zeta_1\zeta_2[-(\frac{g/2}{\zeta_1'}H(\zeta_1) + \frac{\zeta_1'}{g/2}H(-\zeta_1))H(-\zeta_1' - g/2) + H(\zeta_1' + g/2)]$$

$$\cdot [-(\frac{g/2}{\zeta_2'}H(\zeta_2)+\frac{\zeta_2'}{g/2}H(-\zeta_2))H(-\zeta_2'-g/2)+H(\zeta_2'+g/2)]\}$$

$$\approx R_{11}(\tau)M^2(\delta)-2\frac{\mu_2}{\mu_4}R_{22}(\tau)M(\delta)N(\delta)+\frac{\mu_2^2}{\mu_4^2}R_{44}(\tau)M^2(\delta) \tag{3.1}$$

in which $\delta = g/2\sigma_4$. So the breaking wave spectrum is

$$F_b(\omega)=[M^2-2\frac{\mu_2}{\mu_4}\omega^2MN+\frac{\mu_2^2}{\mu_4^2}\omega^4N^2]F(\omega)$$

$$=[M-(\frac{\omega}{\omega_b})^2N]^2F(\omega)=\Psi(\omega)F(\omega) \tag{3.2a}$$

$$\Psi(\omega)=[M-(\frac{\omega}{\omega_b})^2N]^2 \tag{3.2b}$$

in which $\omega_b = (\mu_4/\mu_2)^{1/2}$ is mean frequency between extrema, $F(\omega)$ is original wave spectrum.

$\Psi(\omega)$ in (3.2) can be defined as a breaking filter function, which affects the relation between spectra in front and behind of breaking. It exhibits the fact that the rate of energy loss increases with frequency.

② Peak Frequency Shifting

In order to discuss the effect of breaking on peak frequency shifting, we take the partial derivative of $F_b(\omega)$ with respect to ω

$$\frac{\partial F_b(\omega)}{\partial\omega}=\Psi(\omega)\frac{\partial F(\omega)}{\partial\omega}+\frac{\Psi(\omega)}{\partial\omega}F(\omega) \tag{3.3}$$

Supposing that ω_m is the peak frequency of original waves, we have

$$\frac{\partial F_b(\omega)}{\partial\omega}|_{\omega=\omega_m}=\frac{\partial\Psi(\omega)}{\partial\omega}|_{\omega=\omega_m}F(\omega)<0 \tag{3.4}$$

Because the derivative of spectrum on frequency is positive in the right side of peak frequency or negative in the left side, (3.4) shows that the peak frequency $\omega_{b\ \omega}$ of $F_b(\omega)$ is smaller than that ω_m of $F(\omega)$. It means that breaking makes the spectral peak frequency shifting to lower frequency.

③ Dissipation Souce Function.

Omitting the term of $\exp\{-\frac{\delta^2}{2}\}/\delta^2$ and the higher order from (3.2) we have

$$F_b(\omega)=\psi(\omega)F(\omega) \quad , \quad \psi(\omega)=(1-(\frac{\omega}{\omega_b})^2\varphi)^2 \tag{3.5}$$

in which

$$\varphi(\delta)\equiv\frac{1}{\sqrt{2\pi}\,\delta}exp\{-\frac{\delta^2}{2}\}/\delta^2 \tag{3.6}$$

Considering that the mean period between extrema $T_b=2\pi/(\mu_4/\mu_2)^{1/2}$ is a logical time interval in a wave event, so we obtain the dissipation sounce function $S_{ds}(\omega)$ as follows:

$$S_{ds}(\omega) = -\frac{F(\omega) - F_b(\omega)}{T_b} = -(\gamma_1 \omega^2 - \gamma_2 \omega^4)F(\omega) \tag{3.7}$$

where

$$\gamma_1 = \frac{2\varphi}{T_b \omega_b^2} = \frac{\sqrt{2}}{\pi^{3/2} g} \mu_2^{1/2} exp\{-\frac{g^2}{8\mu_4}\} \tag{3.8a}$$

$$\gamma_2 = \varphi^2 / T_b \omega_b^4 = \frac{\mu_2^{3/2}}{\pi^2 g^2 \mu_4^{1/2}} exp\{-\frac{g^2}{4\mu_4}\} \tag{3.8b}$$

At present, the well–adopted dissipation sousce function is derived by means of the method of data fitting in computer according to π theorem of dimensional analysis Komen et al. (1984). However, the dissipation source function we have derived is expressed theoritically and deduces the energy loss in the higher wave slope, which has been successfully used in LAGFD numerical wave model of the third generation (Yuan et al. 1990).

4. Spectral Form in the Equilibrium Range

JONSWAP (Hasselmann, 1973) experiments make it clear that, in the case of having energy input from wind to wave, wave components around spectral peak frequency (or wave number) increase along with time until wave is fully developed. However, the spectral increment in the range of higher frequency is far slower, and the distribution in phase space is gentler. This range is so called equilibrium range. The balance equation of action spectral density there can be simplified as

$$S_{in} + S_{ds} + S_{nl} = 0 \tag{4.1}$$

It is the balance of these three terms that contral the spectral form in the equilibrium range.

The generally–adopted expression of wind input $S_{in}$is given by Snyder (1980) according to field data. In this paper, we adopt Plant's (1982) result

$$S_{in}(\vec{K}) = m(cos\theta)^{2p} \omega(\frac{u_*}{c})^2 N(\vec{K}) = m(cos\theta)^{2p} gk^{-4}(\frac{u_*}{c})^2 B(\vec{K}) \tag{4.2}$$

which gives a good fitting with measured data in the range of higher frequency (or wavenumber). $B(\vec{K})$ is defined as the degree of saturation,

$$B(\vec{K}) = k^4 E(\vec{K}) = g^{-1/2} k^{9/2} N(\vec{K}) \tag{4.3}$$

According to (3.7), the corresponding dissipation source function can be written as

$$S_{ds}(\vec{K}) = -\frac{N(\vec{K}) - N_b(\vec{K})}{T_b(\theta)} = -(\gamma_1 \omega^2 - \gamma_2 \omega^4) cos^{2p}\theta N(\vec{K})$$

$$= -(\gamma_1 \omega - \gamma_2 \omega^3) cos^{2p}\theta g k^{-4} B(\vec{K}) \tag{4.4}$$

The complicated theoretical expression of $S_{nl}$was represented by Hasselmann (1962, 1963). Kitaigorodskii (1983) and Phillips (1985) gave an approximate expression of nonlirear transfer in the equilibrium range

$$S_{nl} \approx g^{-1/2} k^{19/2} N^3(\vec{K}) = \eta g^{-1/2} k^{19/2} N^3(\vec{K}) = \eta g k^{-4} B^3(\vec{K}) \tag{4.5}$$

η is a positive constant. (4.5) shows that the nonlinear wave–wave interaction produces the net gain of action spectral density in the equilibrium range. It is in accordance with Hasselmann's (1962) discussion. Substituting (4.2), (4.4) and (4.5) into (4.1), we have

$$\eta g k^{-4} B^3(\vec{K}) + m\cos^{2p}\theta g k^{-4}\left(\frac{u_*}{c}\right)^2 B(\vec{K}) - (\gamma_1\omega - \gamma_2\omega^3)\cos^{2p}\theta g k^{-4} B(\vec{K}) = 0 \qquad (4.6)$$

The degree of saturation can be obtained from (4.6)

$$B(\vec{K}) = \frac{1}{\eta^{1/2}}\sqrt{\gamma_1\omega - \gamma_2\omega^3 - m(u_*/c)^2}\,\cos^p\theta$$

$$= \left(\frac{m}{\eta g}\right)^{1/2} u_*\sqrt{k_*^{1/2}[1 - \frac{1}{2}(k/k_b)\varphi] - k^{1/2}}\,k^{1/4}\cos^p\theta \qquad (4.7)$$

in which $k_* = \omega_*^2/g$, $\omega_* = \frac{\gamma_1 g^2}{m u_*^2}$. From (4.3) and (4.7), we gain a wave number spectrum in the equilibrium range of the form

$$E(\vec{K}) = k^{-4}B(\vec{K}) = \left(\frac{m}{\eta g}\right)^{1/2} u_*\sqrt{k_*^{1/2}[1 - \frac{1}{2}(k/k_b)\varphi] - k^{/2}}\,k^{15/4}\cos^p\theta \qquad (4.8)$$

and the frequency spectrum

$$F(\omega) = 2\int_{-\pi/2}^{\pi/2} kE(\vec{k})\left(\frac{dk}{d\omega}\right)d\theta\Big|_{k=\omega^2/g} = \alpha u_* g\sqrt{\omega_*[1 - \frac{1}{2}(\omega/\omega_b)^2\varphi] - \omega}\,\omega^{-9/2} \qquad (4.9)$$

in which

$$\alpha = 4(m/\eta)^{1/2}\, I(p), \quad I(p) = \int_{-\pi/2}^{\pi/2}\cos^p\theta d\theta = B\left[\frac{1}{2}, \frac{1}{2}(p+1)\right],$$

B(m,n) is beta function. (4.9) illustrates that the power of frequency of the equilibrium changes with frequency. The value of the spectum depends on not only α and u_*, but also $\omega_b$and δ.

Because the algebraic expression in the radical sign in (4.9) is a slowly decreasing function, the main part of the equilibrium spectrum is

$$F(\omega) = \alpha u_* g\sqrt{\omega_*}\,\omega^{-9/2} \qquad (4.10)$$

REFERENCES

Hasselmann, K., On the non–linear energy transfer in gravity wave spectrum, Part 1. J. Fluid Mech. 12, 1962, p481–500.

Hasselmann, K. On the non–linear energy transfer in gravity wave spectrum, Parts 2 and 3. J. Fluid Mech. 15, 1963, p273–281; p385–398.

Hasselmann. K. et al. Measurements of wind wave growth and swell decay during the Joint North Sea Wave Project (JONSWAP). Herausgegeben vom Deutsch–Hydrograph. Inst., Reihe A. 1973, no.12.

Kawai, S., K. Okuda, & Y. Yoba, Field data support of three–seconds powerlaw and $gu_*\sigma^{-4}$spectral form for growing wind waves. J. Oceanogr. Soc. Japan 33, 1977, p137–50.

Kitaigorodskii, S. A., On the theory of the equilibrium range in the Spectrum of wind–generated gravity waves, J. Phys., Oceanogr., Vol. 13, 1983, p816–827.

Komen. G. J., Hasselmann, S. & Hasselmann, K, On the existence of a fullydeveloped wind–sea spectrum. J. Phys. Oceanogr. 14, 1984, p1270–1285.

Longuet–Higgins, M. S., The generation of capillasy waves by steep gravity waves, J.Fluid Mech., 16, 1963, p138–159.

Phillips, O. M., The equilibrium rang in the spectrum of wind genergated waves, J. Fluid Mech. 4, 1958, p426–434.

Phillips, O. M. Spectral and statistical properties of the equilibrium range in wind–generated gravity waves, J. Fluid Mech., Vol. 156, 1985, p505–531.

Plant. W. J., A relationship between wind stress and wave slope, J. Geophys. Res., 87, 1982, p1961–1967.

Snyder, R. L., Dobson, F.W., Elliott. J.A. & Long. R.B., Array measurements of atmospheric pressure fluctuations above surface gravity waves. J. Fluid Mech 1981, 102.1–59.

Toba. Y., Local balance in the air–sea boundary processes. III. On the spectrum of wind waves. J. Oceanogr. Soc. Japan, 29. 1973, p209–220.

Tung, C. C., N. E. Huang, Y. Yuan and S. R. Long, Probability function of breaking– limited surface elevation, J. Geoph. Res., Vol. 94, 1989, p967–972.

Yuan Yeli, C. C. Tung and N. E. Huang, Statistical characteristics of breaking wave, in the Wave Dynamicas and Radio Probing of Ocean Surface, edited by O. M. Phillips and K.L. Hsselmann, Plenum, New York, 1986, p265–272.

Yuan Yeli, Hua Feng, Pan Zengdi, N. E. Huang, C. C. Tung., Statistics of Breaking Waves and its Application to Upper Ocean Dynamics. Science in China, Series B, Vol. 33, No. 1, 1990, p 98–110.

Yuan Yeli, Hua Feng, Pan Zengdi and Sun Letao, Dissipation Source Function and an Improvement to LAGFD–WAM Model.

Shoaling and Impacting on Structures

Instability of Surface Water-Wave Solitons Propagating Over an Uneven Bottom

E.S.Benilov

School of Mathematics, University of New South Wales, P.O. Box 1, Kensington, NSW 2033, Australia

Summary

A surface-wave soliton, propagating in a shallow basin with periodic or random bottom irregularities, is considered. It is demonstrated, that if the wave amplitude is sufficiently big, the soliton is unstable due to the resonance between the transverse perturbations, propagating along the crest of the wave, and the "collisions" of the soliton with the bottom irregularities.

1. Introduction and statement of the problem

Long weakly-nonlinear waves, propagating in a basin with an uneven bottom, can be described by the modified Kadomtsev – Petviashvili equation:

$$(2u_x + 3H^{-3/2}uu_\vartheta + {}^1\!/\!_2H^{1/2}u_{\vartheta\vartheta\vartheta} + {}^1\!/\!_2H^{-1}H_x u)_\vartheta + H^{1/2}u_{yy} = 0 \qquad (1)$$

Here u is the elevation of the water surface; H is the depth of the basin; x and y are the horizontal spatial coordinates and $\vartheta = \int H^{-1/2}(x,y)\,dx - t$, where t is the time, is the phase variable in the dominant direction of wave propagation (all quantities are made non-dimensional via the acceleration due to gravity g and the average depth of the basin H). This equation is valid if the the wave field consists of waves propagating approximately in the same direction (close to x-axis), what also enables us to assume (without any loss of generality) that H does not depend on y. In other words, we can consider solitons locally in the vicinity of some straight line $y = y_0$ (cf. [1]).

If $H \equiv$ const, Eq. (1) describes plane solitary wave (soliton) propagating "obliquely" at a phase speed (v_1,v_2):

$$U = 2\,H^{3/2}V \cosh^{-2}[({}^3\!/\!_2H^{-1/2}V)^{1/2}\,\tilde{\vartheta}], \qquad (2)$$

M. L. Banner, R. H. J. Grimshaw (Eds.)
Breaking Waves
IUTAM Symposium Sydney/Australia 1991

where $V = 2v_1 - H^{3/2}v_2^{\,2}$, $\tilde{\vartheta} = \vartheta - v_1 x - v_2 y$. The soliton (2) is *stable in respect to transverse perturbations* (cf. [2]).

Consider now a periodic bottom topography and a soliton, propagating along the x-axis with its crest being parallel to the y-axis ($v_2 = 0$). Introducing the frequency Ω of the "collisions" of the soliton with the periodic bottom irregularities, one can see that *the transversal perturbation with the frequency equal to* $^1/_2\Omega$ *is 'tuned' to a parametric resonance with the bottom irregularities.* The amplitude of this perturbation grows, what can can eventually destabilize the soliton itself. Obviously, this qualitative consideration can be applied to the case of *random* topography as well, where instead of a single unstable frequency, there is a wide spectrum of unstable perturbations, corresponding to continuous spectrum of the random bottom irregularities.

The present paper is devoted to the investigation of the instability of shallow-water solitons over random bottom topography. Since the wave length of the transverse perturbations is much greater than the width of the soliton, the latter is likely to be unstable in respect to the *smooth* irregularities:

$$H = H(\varepsilon x), \tag{3}$$

where $0 < \varepsilon \ll 1$ is the small parameter.

2. Basic equations

Having changed the variables: $\tilde{x} = \varepsilon x$, $\tilde{y} = \varepsilon y$, $\tilde{\vartheta} = \vartheta - \varepsilon^{-1} \cdot S(\varepsilon x, \varepsilon y)$, where S has the physical meaning of the wave phase; we seek the solution to (1), (3) in the form of a series: $u = u^{(0)} + \varepsilon u^{(0)} + \dots$. The zeroth order of the perturbation theory yields: $u^{(0)} = U$, where U is determined by (2) with

$$v = \frac{\partial S}{\partial x} - {}^1/_2 H^{1/2}\left(\frac{\partial S}{\partial y}\right)^2 \tag{4}$$

(here and hereafter tilds over x and y are omitted). Solution (2), (4) describes a surface-wave soliton with the phase S, the speed V and the amplitude $A = 2\,H^{3/2}V$, depending on x and y. The first order of the perturbation theory determines $u^{(1)}$ and relates S to V:

$$H^{13/4}\,\frac{\partial}{\partial y}\left(\frac{\partial S}{\partial y}\,V\right) = \frac{\partial}{\partial x}(H^{15/4}V) \tag{5}$$

(for more details cf. [1]). Eqs. (4) and (5) form a closed system describing the evolution of the soliton over smooth bottom topography. They have an exact solution depending on x only:

$$V_0(x) = v\, H^{-5/2}, \qquad S_0(x) = v \int_{x_0}^{x} H^{-5/2}(x')\, dx'; \tag{6}$$

corresponding to the law of inverse proportionality of the soliton amplitude to the depth of the channel:

$$A(x) = 2H^{3/2}V \sim H^{-1}(x)$$

(cf. [3,4]).

Our goal is the investigation of the stability of the solution (6).

3. Instability of the soliton

For small harmonic perturbations of the solution (6):

$$S = S_0(x) + \tilde{S}(x)\, e^{-iky},$$

where k is the wave number of the transverse perturbations; the linearized system (4), (5) yields:

$$\Psi'' + [H^{-2}\omega^2 - {}^5\!/\!_4 H^{-1}H'' - {}^5\!/\!_{16} H^{-2}H'^2]\Psi = 0, \tag{7}$$

where

$$\omega = ({}^2\!/\!_3 V)^{1/2}|k|, \tag{8}$$

and $\Psi = H^{-5/2}\,\tilde{S}$ and the prime $'$ denotes the differentiation in respect to x. Ordinary differential equation with random coefficients (7) can be easily analyzed if the depth variations are *small*:

$$H = 1 + h(x), \qquad |h| \ll 1$$

(cf. [5,1]). In this case the solution oscillates with the frequency ω, *while its amplitude grows exponentially with the increment*

$$\gamma_+ = {}^9\!/\!_8\, \omega^2\, N(2\omega), \tag{9}$$

where N(ω) is the Fourier transform of the correlation function of h(x):

$$N(\omega) = \int_{\infty}^{-\infty} \langle h(x)h(x + x')\rangle\, exp(i\omega x')\, dx',$$

where $\langle\cdot\rangle$ denotes the average over the ensemble of realizations. Expression (9) indicates the parametric character of the instability – indeed, the increment of the harmonic wave with the frequency ω depends on the spectral density of the irregularities with the doubled frequency $\Omega = 2\omega$.

4. Damping of the unstable perturbations

Since the above perturbation theory was based on the smallness of $\varepsilon = d/\lambda$ (d is the width of the soliton and λ is the wave length of the transverse perturbations), the expressions (8) and (9) are, in fact, the first and the second terms of the expansion of the "true" frequency of transverse waves in powers of ε:

$$\omega_{true} = \omega + \varepsilon\, i\gamma_{+} + \varepsilon^2\, i\gamma_{-} + \ldots \tag{10}$$

Note, however, that γ_{+} depends on the amplitude of the irregularities, and if the latter is sufficiently small ($|h| < \varepsilon$), the third term in the series (10) becomes important. This term describes the energy loss due to the radiation of the perturbations from the soliton and the corresponding damping of the transverse waves. Evidently, γ_{-} does not depend on the irregularities and could be calculated with the help of the "classical" Kadomtsev – Petviashvili equation (cf. [2]):

$$\gamma_{-} = -\left({}^{27}/_{2v}\right)^{-1/2} k^2. \tag{11}$$

The (stabilizing) effect of damping "competes" with the influence of the bottom topography – comparing (9) and (11), one can obtain the following criterion of the stability:

$$max[N(\omega)] \leq \left({}^{32}/_{243}\right)^{1/2} v^{-3/2}. \tag{12}$$

This is the main result of the present paper.

In order to illustrate the usage of (12), we consider the model expression for the correlation function of the irregularities

$$\langle h(x)h(x + x')\rangle = \sigma^2 exp(-x'^2/R^2),$$

where $\sigma^2 = \langle h^2\rangle$ and R is the correlation radius of the irregularities. For this case, the criterion (12) yields:

$$\sigma^2 R\ v^{3/2} \leq {}^4/_9({}^2/_{3\pi})^{1/2} \simeq 0.2\ .$$

In terms of dimensional variables (marked with overbars), this inequality can be rewritten as

$$\bar{A} \leq \bar{H}^3(\bar{\sigma}^2\ \bar{R})^{-2/3},$$

where $\bar{A}$ is the amplitude of the soliton and $\bar{H}$ is the mean depth of the basin.

Thus, bottom topography effects basically the stability of the large-amplitude solitons.

References

1. Benilov, E.S. Stability of plane nonlinear waves in smoothly inhomogeneous random media. Sov. Phys. – JETP **94**, No.11 (1988) 2235-2241.
2. Zakharov, V.E. Instability and nonlinear oscillations of solitons. JETP Lett. **22** (1975) 172-173.
3. Grimshaw, R. The solitary wave in water of variable depth. J. Fluid Mech. **42** (1970) 639-656.
4. Ostrovskiy, L.A.; Pelinovskiy, E.N. Wave transformation on the surface of a fluid of variable depth. Akad. Nauk SSSR, Izv. Atmos. Ocean. Phys. **6** (1970) 552-555.
5. Arnold, L.; Papanicolaou, G.; Wihstutz, V. Asymptotic analysis of the Lyapunov exponent and rotational number of the random oscillator and applications. SIAM J. Appl. Math. **46** (1986) 427-450.

Violent Motion as Near Breaking Waves Meet a Vertical Wall

M.J. COOKER and D.H. PEREGRINE,
Department of Mathematics,
University of Bristol, Bristol BS8 1TW, England.

Abstract

Very steep unsteady water waves are modelled by an accurate numerical computer program. Recent work is discussed with particular emphasis on waves that are nearly breaking as they approach a vertical wall. Extremely violent water motion is calculated in full detail, with water accelerations exceeding 1000 g, and transient pressures exceeding 30 times the hydrostatic pressure.

Introduction

Inviscid, incompressible, irrotational fluid flow has long been considered to be a useful approximation to unbroken water waves, in the absence of wind. It is only in recent years, commencing with the study by Longuet-Higgins and Cokelet (1976), that it has become possible to model, with detailed computations, the early stage of wave breaking in which waves overturn and form a plunging jet. Detailed examples of such flows are given in New, McIver and Peregrine (1985). A substantially faster, more accurate and more stable computational method, described in Dold and Peregrine (1986), has made it easier to study these motions. Two examples, in particular, have increased confidence in the program. These are described in Tanaka, Dold, Lewy and Peregrine (1987) and Cooker, Dold, Vidal and Peregrine (1990).

When modelling waves breaking against a wall we expected to find initial water *velocities* suitable for further studies of impact. However, an unexpected type of water motion is encountered when the conditions required for the most intense impact pressures, as reported by experimenters, are modelled. Accounts of experiments indicate that the highest pressures occur when the face of the breaking wave is parallel to

M. L. Banner, R. H. J. Grimshaw (Eds.)
Breaking Waves
IUTAM Symposium Sydney/Australia 1991

the wall just before impact. In attempts to reproduce this case in computations, we find that there is *no direct impact* of the water surface in these cases. Violent acceleration of the water line up the wall causes the surface to "flip-through" to an upward splash. The pressures which drive this flip-through also act on the wall and are found to be similar to the highest pressures reported from experiments. A preliminary report of this work is in Cooker and Peregrine, 1990a.

On coasts and in laboratory experiments waves that break on walls and structures are usually forced to break by shoaling water depth; see the sketch in figure 1. Although there is no difficulty in computing with a

Figure 1 Sketch of a typical wave impact situation

range of different bed geometries, we considered it best to make the waves break using a simpler approach. In view of our computational results and further theoretical work, which is briefly reported in Cooker and Peregrine (1990b), we are confident that the mechanisms of wave impact depend primarily on the immediate neighbourhood of the impact region and not on large-scale geometrical features or vorticity. This is also borne out by the similarities between Chan and Melville's (1988) experimental results on the impact of deep-water waves and the many "shallow water" observations and experiments.

Computations of wave impact

The boundary geometry is a flat horizontal bed and a vertical wall. In the computations the wall is simulated by colliding two identical waves which have a line of symmetry at the wall. The initial data is a long smooth wave which is a gradual increase in water depth from h to $h + \Delta h$. Initially the wave is long enough for finite-amplitude shallow-water theory to be applicable. This theory is used to given an *initial* velocity distribution such that the wave only propagates toward the wall. The unimpeded motion of one such wave which shows steepening to breaking in the absence of a wall is shown in figure 2. The wave propagates into still water with no wall present and has a height $\Delta h = 1.5h$. The same initial wave is used for the later illustrations *with* a wall.

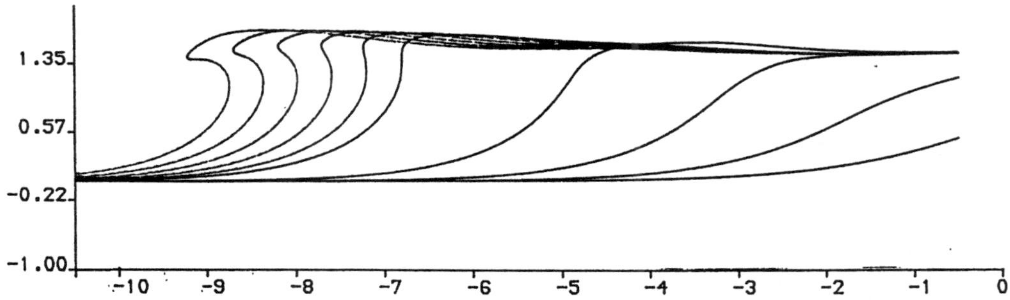

Figure 2 The evolution towards breaking of a long wave of increasing water depth. In this example $\Delta h/h = 1.5$. $t = 0,1,2,3,4,5\ \sqrt{h/g}$ $y = 0$ is the still water level.

With such initial conditions and a still water surface in front of the wave, the wall can be placed at any chosen point. In our computations we place the wall at a chosen place, $-x_0$ in figure 2. The range $6h \leq x_0 \leq 9h$ is the most interesting for this wave.

To identify our initial wave with those in experiments, note that the still water in front of the wave corresponds to the wave trough. Unless a wave has a particularly sharp crest, we suspect that the important parameters governing impact are depth of water at the trough, the maximum height of the wave and its ***shape*** just before impact. Our set of initial conditions allows an exploration of such a parameter space. With such a large parameter space it is difficult to be confident that one is examining typical examples. However, the following example $\Delta h/h = 1.5$ with $x_0 = 7.5h$ is interesting and not atypical. Other computations for larger and smaller values of $\Delta h/h$, and a number with other waves, support our statements below. From consideration of figure 2 one can estimate that $x_0 = 7.5h$ could give the case where an almost flat impact with no air pocket might occur. Figures 3(a,b) shows the motion of the free surface. As may be seen, the wave face does not become vertical. The time from maximum wave steepness of 81.3° at $t = 4.1(h/g)^{\frac{1}{2}}$ to the formation of the upward jet, at say $t = 4.18(h/g)^{\frac{1}{2}}$ is $0.08(h/g)^{\frac{1}{2}}$, which is only 8 milliseconds for a 15 cm wave on 10 cm of water. Spreading out the motion by using the short time intervals in figure 3(b) is deceptive, the motion

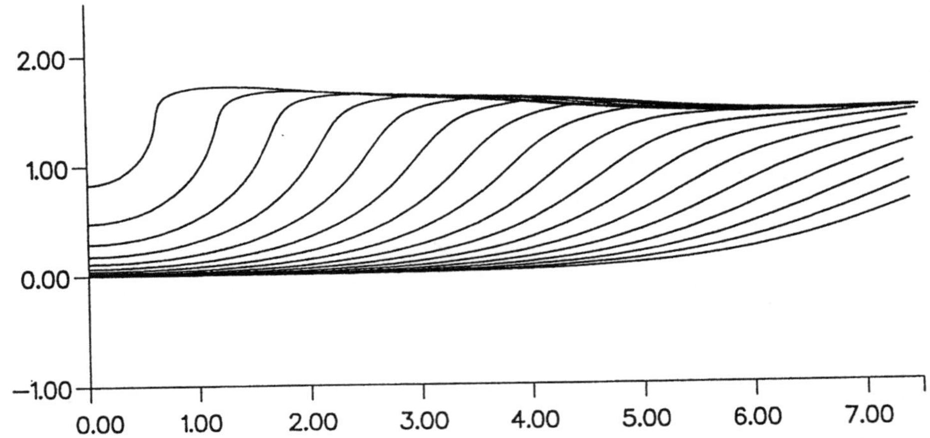

Figure 3(a) Surface profiles at $t = 0,(0.25),4.0(h/g)^{\frac{1}{2}}$ for a wave of initial height $\Delta h/h = 1.5$ and $x_0 = 7.5h$, travelling right to left, against a vertical wall at $x = 0$.

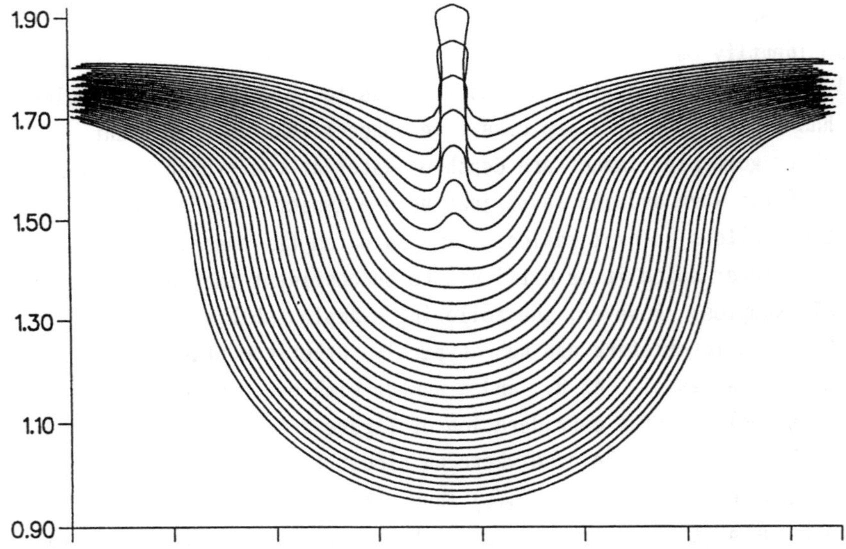

Figure 3(b) As figure 3(a). Detail of motion during flip-through. Surface profiles at $t = 4.0,(0.01),4.21\ (h/g)^{\frac{1}{2}}$. The full symmetrical motion calculated is shown here.

from a steep sided depression at the wall to a vertically moving jet is rapid and the most descriptive words we have found are "flip-through". That is the surface flips through between the wall and the horizontally moving near-vertical surface.

The most violent motion is at the point of jet creation. Maximum acceleration of the surface water is 1850g at $t = 4.185(h/g)^{\frac{1}{2}}$ and is associated with a pressure peak at this time of 26 ρgh. This is illustrated in figure 4. It can be seen that the peak pressure region is very small: about 0.1 h in height. It lasts only about $0.01(h/g)^{\frac{1}{2}}$.

As was noted, this wave with $x_0 = 7.5h$ does not have a vertical face. The same wave with $x_0 = 8.0h$ does have a vertical face in the final flow towards the wall, but as the trough rises up the wall, towards the wave face, the result is a convergence of the free surface to a tiny region of the wall. This case, $x_0 = 8.0h$, gives the most violent water motion we have computed, an energetic jet is formed which shoots up the wall. Among the values of x_0, $x_0 = 8h$ gives the smallest convergent trough and the latest appearance of the vertical jet. The maximum computed rate of change of pressure, and maximum water acceleration, are 3000 $\rho g^{\frac{1}{2}}h^{\frac{3}{2}}$ and *ten thousand g*, respectively. These values are still increasing at the last time computed.

Discussion of results

It was a surprise to find that such violent motions could occur without direct impact of the water surface against the rigid wall. Among reports of experiments there is only one which may confirm this type of water motion: Chan and Melville (1988) show a deep-water wave striking a vertical plate. Surface profiles are shown in their figure 6(a). In addition they describe the water motion as follows:

> "It is observed that wave impact occurred through the focussing of the incident wave front onto the wall; that is through a convergence of the wave crest and the surface intersection point at the wall."

Chan and Melville used photography at 1100 frames per second.

Another experimental report, by Arami and Hattori (1989), shows instances where there is single sharp peak of pressure, and the associated video frames indicate, at first sight, that a vertical wave face hits the wall. However, the corresponding theoretical time scale of a flip-through

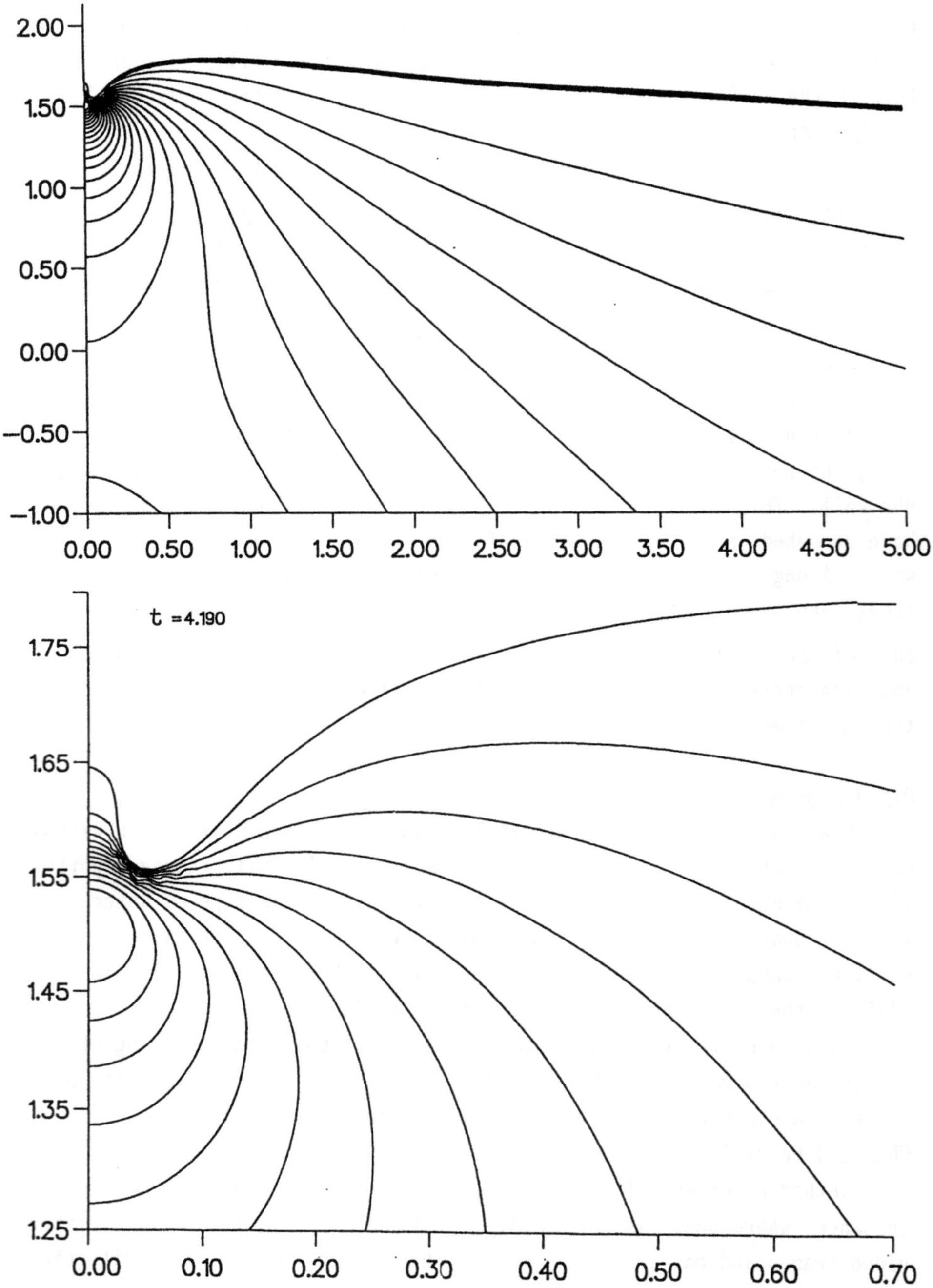

Figure 4 As for figure 3(a),(b). Pressure distribution at $t = 4.19(h/g)^{\frac{1}{2}}$ the instant of maximum pressure. The contour interval is $2\rho gh$. The irregularities are related to limitations in the contouring process

is less than the time between their video frames. Preliminary attempts to make a comparable computation are described by Passoni, Cooker, and Peregrine (1990). An interesting aspect of our results is that there are also relatively high pressures on the *bed* near the wall. We know of no measurements in this area, but consider that the effect of such a short high pressure pulse on bed material is of interest. See Cooker and Peregrine (1991).

Financial support from S.E.R.C. grants GR/F/28298 and GR/G/21032 is gratefully acknowledged.

References

1 Arami, A. & Hattori (1989) Experimental study on shock wave pressures. Internal Report Civil Engng., Chuo Univ., 37-63.

2 Chan, E.S., Melville, W.K. (1988) Deep water plunging wave impact pressures on a plane vertical wall. Proc.Roy.Soc., A 417, 95-131.

3 Cooker, M. & Peregrine, D.H. (1990a) Computations of violent motion due to waves breaking against a wall. Proc.22nd Internat.Conf. on Coastal Engng., Delft.

4 Cooker, M.J. & Peregrine, D.H. (1990b) A model for breaking wave impact pressures. Proc.22nd Internat.Conf. on Coastal Engng., Delft.

5 Cooker, M.J. & Peregrine, D.H. (1991) Wave impact pressures and their effect on bodies lying on the bed. MAST G6 Proj,.1 Workshop, Hannover.

6 Cooker, M. & Peregrine, D.H., Vidal, C., Dold, J.W. (1990) The interaction between a solitary wave and a submerged semi-circular cylinder. J.Fluid.Mech. 125, 1-22.

7 Dold, J.W. & Peregrine, D.H. (1986a) An efficient boundary-integral method for steep unsteady water waves. In "Numerical Methods for Fluid Dynamics II" (Eds. K.W. Morton & M.J. Baines). 671-679 Oxford U.P.

8 Longuet-Higgins, M.S. & Cokelet, E.D. (1976) The deformation of steep surface waves on water. I A numerical method of computation. Proc.Roy.Soc.Lond. A 350, 1-26.

9 New, A.L., McIver, P. & Peregrine, D.H. (1985) Computations of overturning waves. J.Fluid Mech. 150, 233-251.

10 Passoni, G., Cooker, M.J. & Peregrine, D.H. (1990) Comparisons between theory and experiment of water-wave impact on a vertical wall. Internal Report, Mathematics, Univ.of Bristol, AM-90-19, 21pp.

11 Tanaka, M., Dold, J.W., Lewy, M. & Peregrine, D.H. (1987) Instability and breaking of a solitary wave. J.Fluid Mech. 185, 235-248.

Shear Stress Distribution in the Surf Zone

Rolf Deigaard

Institute of Hydrodynamics and
Hydraulic Engineering
Technical University of Denmark

Summary

The vertical distribution of shear stress in the surf zone is described. The significance of the vertical convection of horizontal momentum is discussed. It is explained how this mechanism contributes to the shear stress in case of energy dissipation near the surface, near the bed and in case of no energy dissipation.

Introduction

In a surf zone the wave height decreases towards the shore due to the strong energy dissipation. The gradient in the wave height gives a gradient in the radiation stress, which is largely balanced by a slope of the mean water surface, the wave set-up. It was recognized by Dyhr Nielsen and Sørensen [1] that the radiation stress gradient is not evenly distributed over the vertical, and a balance cannot be obtained without introducing shear stresses. These shear stresses was explained to be the main driving force for a mean circulation current in the vertical plane.

The shear stress distribution

The mean stress can be determined by an Eulerian analysis, considering the momentum balance for the control surface shown in Fig. 1. This technique was applied by Dally and Dean [2]. The effect of surface rollers was included by Svendsen [3], and Deigaard and Fredsøe [4] included the contribution from the organized convection of momentum through the bottom of the control surface.

In the present formulation the wave motion is assumed to be described by linear shallow water wave theory. The wave celerity is then $c = \sqrt{gD}$, where g is the acceleration of gravity and D is

M. L. Banner, R. H. J. Grimshaw (Eds.)
Breaking Waves
IUTAM Symposium Sydney/Australia 1991

the mean water depth. The pressure is hydrostatic and the horizontal and vertical orbital velocities u and w are respectively constant over the depth and varying linearly with the distance from the bed

$$w = \frac{z}{D}\frac{\partial \eta}{\partial t} \tag{1}$$

z is the vertical coordinate measured from the bed and η is the instantaneous water surface elevation, measured relative to the man water surface. η can be written

$$\eta = \frac{H}{2}\cos(kx - \omega t) \tag{2}$$

where H is the wave height, k is the wave number and ω is the angular frequency. The wave height varies due to the energy dissipation. Within a short distance a linear approximation can be applied: $H = H_0 + xH_x'$.

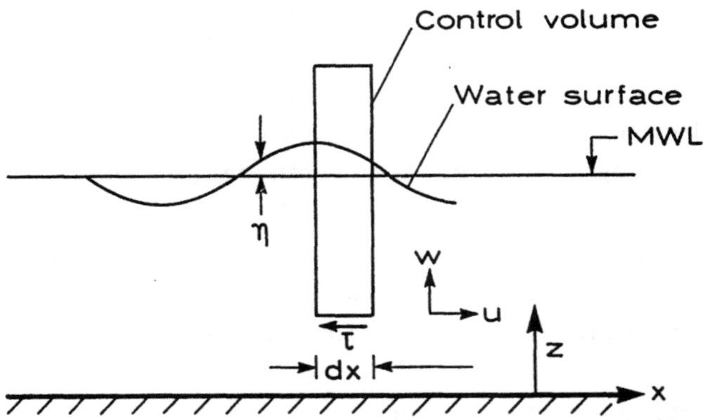

Fig. 1. The control surface for the momentum equation.

When the wave conditions are not uniform (or stationary), the water surface elevation and the horizontal orbital velocity are not in phase. A small phase difference exists, because water must be transferred from the wave tops to the wave troughs. The horizontal velocity can be found from the continuity equation

$$u = \frac{Hc}{2D}\left(\cos(kx-\omega t) - \frac{H_x'}{kH}\sin(kx-\omega t)\right) \tag{3}$$

where it has been assumed that the dissipation is weak : $-H_x'/kH \ll 1$.

In addition to the wave motion surface rollers are present. These have been described by Svendsen [3] as lumps of water that travel with each wave front. The flow velocity is thus equal to c in the roller. The cross sectional area of each roller is A. In addition to the wave motion the mean water surface has a slope S. The magnitude of S is not given apriori, but it is assumed to be a second order term, proportional to the gradient of the wave height squared.

The momentum equation is written by use of integrals over the volume and the area of the control surface

$$\int_V \rho\frac{\partial\vec{u}}{\partial t}dV = -\int_A \rho\vec{u}(\vec{u}\cdot d\vec{A}) - \int_A pd\vec{A} + \int_V \rho\vec{g}dV + \int_A d\vec{T} \tag{4}$$

where $\vec{T}$ is the force from the shear stress on a unit surface element. The momentum equation is averaged over a wave period, and projection on the horizontal is made. Only the lowest order terms are maintained. The averaging makes the left hand side zero due to periodicity, and the gravity term is zero as $\vec{g}$ is vertical. The different elements can then be summarized as follows (an overbar signifies time averaging and $\vec{e}$ is the unit vector in the x-direction).

<u>a. The momentum flux</u>

$$\overline{\vec{e}\cdot\int_A \rho\vec{u}(\vec{u}\cdot d\vec{A})} = \rho\left\{\frac{d}{dx}(\overline{u^2}(D-z) + \frac{d}{dx}\left(\frac{cA}{T}\right) - \overline{uw}\right\}dx \tag{5}$$

The first term is due to the gradient in the horizontal orbital motion, the second term is Svendsen's roller contribution, the third is convection of momentum through the bottom of the control surface. This latter becomes non-zero, and of the same order as the others, because the small phase shift is included in the horizontal orbital motion, cf. eq. (3). Inserting the expressions for the orbital motion, eqs. (1) and (3) gives the following momentum flux

$$\overline{\vec{e}\cdot\int_A \rho\vec{u}(\vec{u}\cdot d\vec{A})} = \rho\frac{d}{dx}\left(\frac{cA}{T}\right)+\rho g H H_x'\left(\frac{1}{4}-\frac{1}{8}\frac{z}{D}\right) \tag{6}$$

b. The pressure force

The pressure force is found from the hydrostatic pressure distribution. It has been demonstrated by Svendsen [3] that the pressure contribution from the rollers is negligible. The pressure force contains therefore two terms, one (constant over the vertical) due to the wave height gradient and a linearly varying term due to the set-up

$$\overline{\vec{e}\cdot\int_A p d\vec{A}} \approx \rho g\left(\eta\frac{\partial\eta}{\partial x}+S(D-z)\right)dx = \rho g\left(\frac{HH_x'}{8}+S(D-z)\right)dx \tag{7}$$

c. The shear stress

The force due to shear stresses is only containing a contribution from the shear stress acting on the bottom of the control surface

$$\overline{\vec{e}\cdot\int_A d\vec{T}} = -\overline{\tau_{xz}}dx \tag{8}$$

Equations (6) to (8) gives an expression for the time averaged shear stress distribution

$$\begin{aligned}\overline{\tau_{xz}} &= -S\rho g D\frac{D-z}{D}-\frac{\rho g}{8}\frac{dH^2}{dx}\left(1+\frac{D-z}{D}\right)-\frac{\rho}{T}\frac{d(cA)}{dx} = \\ &-S\rho g D\frac{D-z}{D}-\frac{1}{c}\frac{dE_{fw}}{dx}\left(1+\frac{D-z}{D}\right)-\frac{\rho}{T}\frac{d(cA)}{dx}\end{aligned} \tag{9}$$

where E_{fw} is the energy flux associated with the wave motion. The shear stress distribution is illustrated in Fig. 2 for the special case where the wave set-up S balances the radiation stress gradient exactly, giving a mean bed shear stress of zero. It is seen from Eq. (9) that it is not possible for any magnitude of the set-up to obtain an equilibrium without shear stresses, and these shear stresses determine the velocity distribution of the mean current cf. Dally and Dean [2] and Svendsen [3]. The near surface shear stress τ_s (at the wave trough level) is independent of the set-up

$$\overline{\tau}_s = -\frac{1}{c}\frac{dE_{fw}}{dx} - \frac{\rho}{T}\frac{dcA}{dx} = \frac{Diss}{c} - \frac{\rho}{T}\frac{d(cA)}{dx} \tag{10}$$

where Diss is the rate of loss of wave energy per unit bed area.

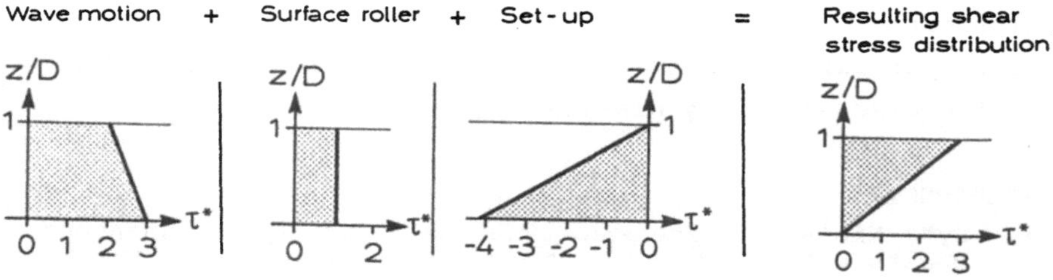

Fig. 2. Example of the shear stress distribution.

The role of the vertical momentum convection

The vertical convection of horizontal momentum, the term $\rho\overline{uw}$ is important for the distribution of the mean shear stress. In the present case with energy dissipation near the water surface it contributes with one half of the first term on the r.h.s. of eq. (10).

In case of no energy dissipation, or energy dissipation in a near bed wave boundary layer, the $\overline{\rho uw}$ term changes sign and cancels out the pressure gradient term near the surface to give a $\overline{\tau}_s$ of zero (Deigaard and Fredsøe [4]). In case of wave shoaling this is obtained simply by using a wave description with the correct phase shift between η and u_x in the momentum equation, and gives the result that the shoaling process will not in itself produce shear stresses. In case of dissipation in a wave boundary layer the vertical motion due to the displacement in the non-uniform wave boundary layer (Longuet-Higgins [5]) must be included in the flow field. It is found that $\overline{\tau}_{xz}$ can be zero over the entire water column outside the wave boundary layer for a wave set up that balances one third of the radiation stress gradient (the rest is balanced by the $\overline{\rho uw}$ term). Inside the wave boundary layer the mean shear stress will increase to the value determined by the streaming at the bed, Longuet-Higgins [5].

Conclusion

The mean shear stresses can be determined by Eulerian analysis, considering the momentum balance for a control surface. The convection of horizontal momentum is important for obtaining a correct shear stress distribution. The location of any energy dissipation is important for the shear stress distribution. The mean near surface shear stress will only be different from zero in the case of energy dissipation near the surface.

References

1. Dyhr-Nielsen, Sørensen, T.: Sand transport phenomena on coasts with bars. Proc. 12th Coastal Eng. Conf. (1970) 855-866.

2. Dally, W.R., Dean, R.G.: Suspended sediment transport and beach evolution. ASCE J. Waterw. Harbors Coastal Ocean Eng. Div., 110(1). (1984) 15-33.

3. Svendsen, I.A.: Mass flux and undertow in a surf zone. Coastal Eng. 8(4) (1984) 347-366.

4. Deigaard, R., Fredsøe, J.: Shear stress distribution in dissipative water waves. Coastal Eng. 13(4) (1984) 357-378.

5. Longuet-Higgins, M.S.: Mass transport in water waves. Philos. Trans. R. Soc. London, Ser. A., 245 (1953) 535-581.

Random Wave Forces on a Vertical Cylinder in the Free Surface Zone at High Reynolds Numbers

Gert Klopman and Jan K. Kostense
Delft Hydraulics, P.O. Box 152, Emmeloord, The Netherlands

Abstract

Large-scale random-wave experiments ($H_s = 1.3$ m, $f_p = 0.17$ Hz, $d = 5.0$ m) on a vertical circular cylinder ($D = 0.5$ m), having a roughness of about 60 μm, were performed to study the wave loading in the free surface zone at high Reynolds numbers (up to $7 \cdot 10^5$). Some first results of the data analysis are presented, showing characteristic time histories of velocities, pressures, and in-line forces. Also the energy density spectra and exceedance curves for both the maximum and the minimum inline-force are compared at two levels, $z = 0$ m and $z = -2.3$ m.

1 Introduction

The wave forces near the free surface give a relative large contribution to the total wave forces and overturning moments of offshore structures. However, our knowledge about these forces is limited, and also few experimental data sets are available. Field measurements have been reported by Dean, Dalrymple and Hudspeth (1981). Results of laboratory experiments are described by Isaacson & Baldwin (1990), Tørum (1989), Kjeldsen, Tørum & Dean (1986).

Chaplin et al. (1991) performed tests in a large-scale wave channel under 'freak' wave conditions, resulting in overturning plunging breakers at the location of the cylinder. This paper describes some first results of the data analysis of an experiment in random waves, using their experimental set-up.

2 Experimental Arrangement

The experiment was carried out in the large-scale wave channel of Delft Hydraulics, having a length of 230 m and a width of 5.0 m. During the tests the water depth was 5.0 m. Wave forces on a vertical cylinder of $D = 0.5$ m diameter were measured by a circumferential ring of 24 pressure transducers, located near the still water level (SWL), and two force sleeves of 0.25 m height, located below trough level. The cylinder was placed in a 2 m deep pit in the channel floor. By moving the cylinder vertically, wave pressures and forces could be measured at several elevations. The cylinder surface was covered with emery cloth, resulting in a small roughness height of approximately 60 μm. Flow kinematics were recorded by three perforated ball velocity meters, at the elevations of the pressure transducer ring and the two force sleeves, and a one-component laser flow meter.

M. L. Banner, R. H. J. Grimshaw (Eds.)
Breaking Waves
IUTAM Symposium Sydney/Australia 1991

The cylinder was subjected to random waves with a narrow-banded spectrum, having a significant wave height of 1.3 m and a spectral-peak frequency of $f_p = 1/T_p = 0.17$ Hz. The resulting Reynolds numbers $Re = (u \cdot D)/\nu$ and Keulegan-Carpenter numbers $K = (u \cdot T_p)/D$ at the mean water level, based on the maximum velocity are $Re_{max} = 7 \cdot 10^5$ and $K_{max} = 18$. Based on the standard deviation of the velocity, they become $Re_{std} = 3 \cdot 10^5$ and $K_{std} = 8$.

Tests were performed with the cylinder at two different elevations: the first elevation with the pressure transducer ring located at SWL, the second one with the transducers at SWL +0.5 m. All signals were sampled at a rate of 25 Hz.

3 Results and Discussion

Here we will only present results of the test with the pressure transducer ring at SWL, with a duration of 53 minutes. Before further analysis, all pressure transducer signals were corrected for a low-frequency drift of low-amplitude.

Figure 1 and 2 show time histories of the pressure transducers respectively at the front and the rear of the cylinder. The inline force, derived by integration of the pressure signal around the cylinder circumference, is given in figure 3. This figure shows that the maximum and minimum forces are of comparable magnitude, contrary to the horizontal velocity signal (figure 4), as obtained from the perforated ball at SWL. Based on the Morison equation the ratio of the maximum drag force to the maximum inertia force is expected to be about 0.5 for the shown time interval of 100 s. This will not yet result in an increase of the maximum force due to drag, due to the phase lag between drag and inertia.

The spectral density of the capacitance-type wave gauge, located 5 m in front of the cylinder, is shown on figure 5. Figure 6 compares the inline force spectra at SWL -2.3 m (force sleeve) and at SWL 0.0 m (integrated pressures). Characteristic features of the force spectrum at SWL are the high energy densities at high frequencies and the high peak at 2 f_p.

The exceedance probability curves of above mentioned signals are plotted in such a way that a Rayleigh distribution results in a straight line (figures 7–9). The exceedance probability curves of the maximum and minimum wave force shows that even for the highest waves (up to $K_{max} = 18$) the effect of drag on the force is not yet apparent.

Acknowledgement

The large-scale wave-flume experiment, initiated and conducted by Prof. J.R. Chaplin, Dr. C.A. Greated and staff of City University, Edinburgh University and Delft Hydraulics, was supported by the U.K. Department of Energy as part of the Marine Technology Directorate's Fluid Loading Programme.

References

Chaplin, J.R., Greated, C.A., Flintham, T.P. & Skyner, D.J. (1991) *Breaking wave forces on a vertical cylinder.* U.K. Dept. of Energy Report OTH90.324.

Dean, R.G., Dalrymple, R.A. & Hudspeth, R.T. (1981) Force coefficients from Wave Project I and II data including free surface effects. *Soc. Petroleum Engineers J.* **21**(6), pp. 779-786.

Isaacson, M. & Baldwin, J. (1990) Measured and predicted random wave forces near the free surface. *Appl. Ocean Res.* **12**(4), pp. 188-199.

Kjeldsen, S.P., Tørum, A. & Dean, R.G. (1986) Wave forces on vertical piles caused by 2- and 3-dimensional breaking waves. *Proc.* 20^t*h Int. Conf. Coastal Engng.*, Taipei, pp. 1929-1942.

Klopman, G. & Kostense, J.K. (1989) The loading on a vertical cylinder in random waves at high Reynolds numbers. In *Proc. of NATO Adv. Res. Workshop on Water Wave Kinematics, Molde, Norway, May 1989*, (ed. A. Tørum & O.T. Gudmestad), NATO ASI Series E178, Kluwer Acad. Publ., Dordrecht, The Netherlands, pp. 679-699.

Tørum, A. (1989) Wave forces on pile in surface zone. *J. Waterway, Port, Coastal and Ocean Engng.* **115**, pp. 547-565.

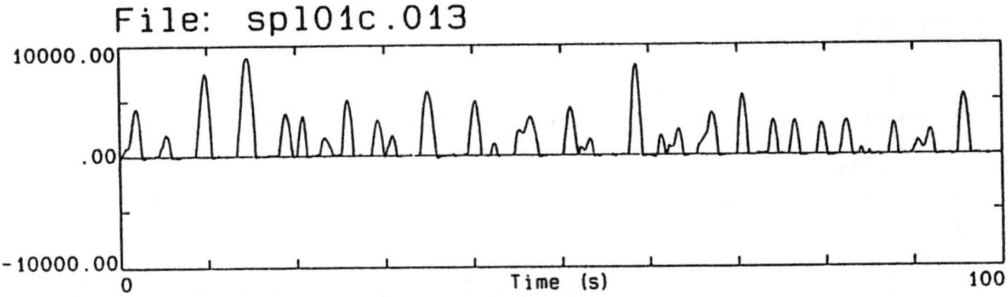

Figure 1: Time history of the pressure at $\alpha = 0°$ (front)

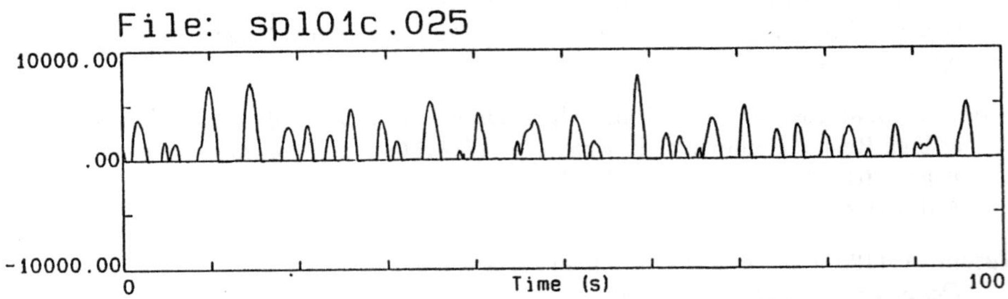

Figure 2: Time history of the pressure at $\alpha = 180°$ (rear)

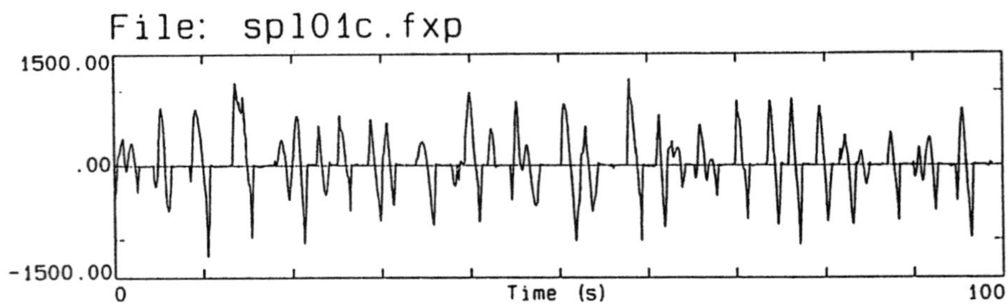

Figure 3: Time history of the inline force at SWL

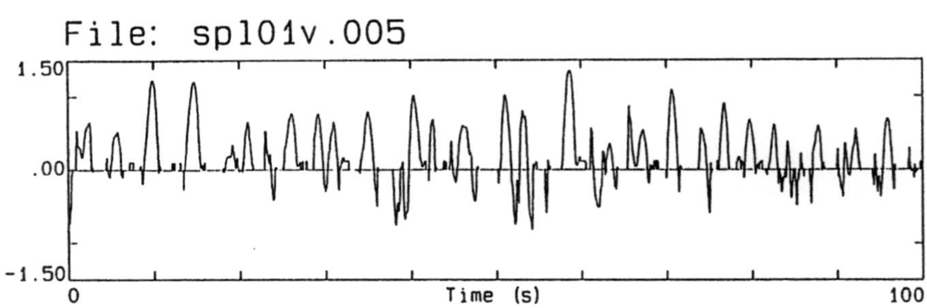

Figure 4: Time history of the horizontal velocity at SWL

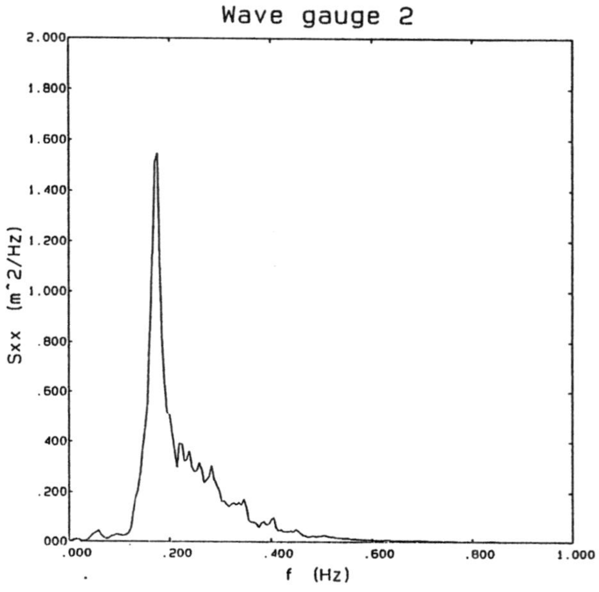

Figure 5: Spectral density of the free surface elevation

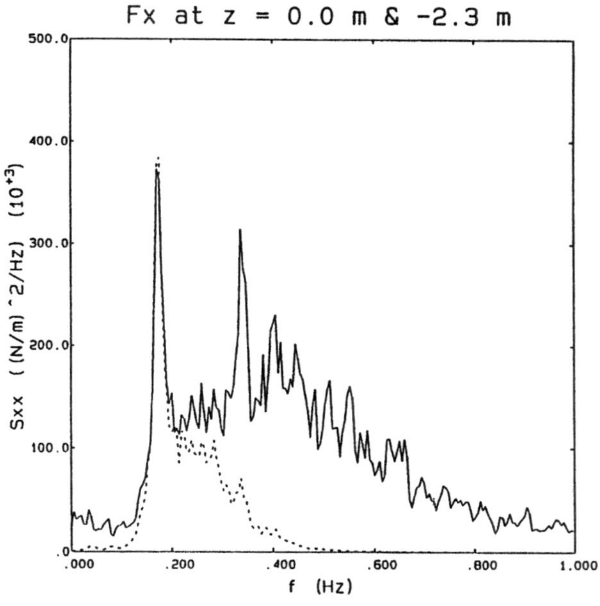

Figure 6: Spectral density of the inline force; — SWL +0.0 m; - - - SWL -2.3 m

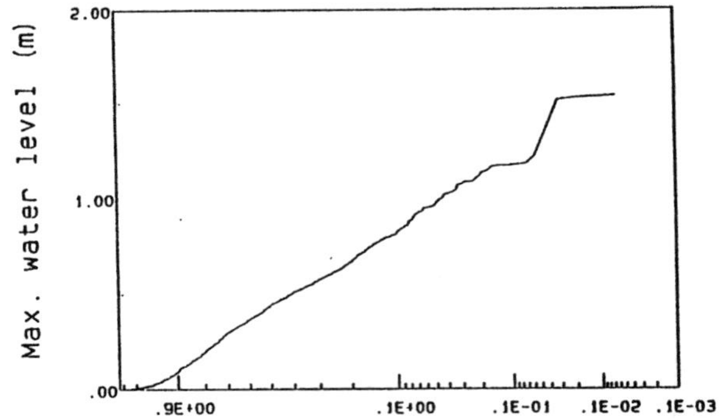

Figure 7: Exceedance probability curve of maximum water elevation

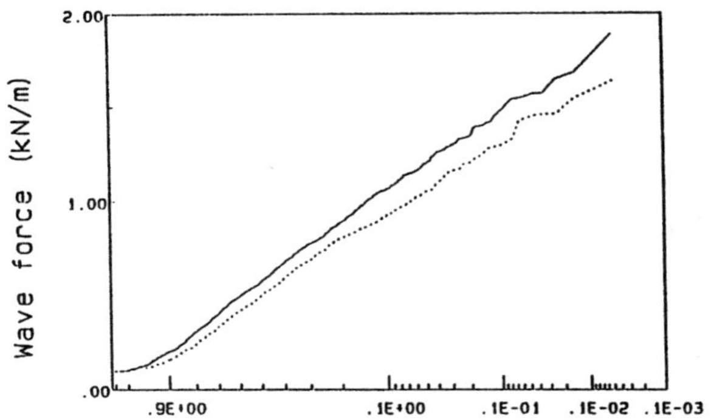

Figure 8: Exceedance probability curve of maximum and minimum inline force at SWL +0.0 m; — maximum forces; - - - minimum forces

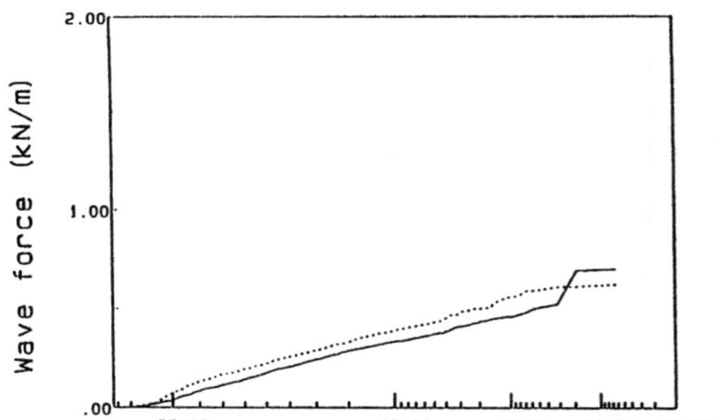

Figure 9: Exceedance probability curve of maximum and minimum inline force at SWL -2.3 m; — maximum forces; - - - minimum forces

Prediction of Deterministic and Random Force on Structures by Plunging Breaking Waves

J.C.Li M.Lin

Institute of Mechanics, CAS, Beijing, China 100080

ABSTRACT

Evolution and transformation of plunging breaking waves are numerically simulated directly on the physical plane, accounting for the effects of surface tension. With velocity and acceleration fields given we have calculated the induced force by breakers and made comparison with those by steep periodic waves. Based on long crest ,frozen flow field and total loss of normal momentum hypotheses, a momentum theorem is suggested for the first time to derive less experiential formulae to evaluate impact forces on structures above MSL. In the end, we have studied and analyzed the impact pressure by breakers of different amplitude in the ocean with variety of wave spectra statistically.

INTRODUCTION

As is well known, breaking waves have aroused more and more attention by scientists owing to their evergrowing importance in the area of ocean engineering design, sediment transport and ripple formation, coastal erosion, wind wave prediction, air-sea interaction and remote sensing. Although they are commonly observed natural phenomenon when you go to beach or travel by ship, some aspects of breaking waves dynamics, among which are the induced and impact force by breaking waves , seem to remain unsolved. The main difficulties arise not so much from the governing equation as from the unknown, time dependent position of the free surface and the relatively complicated boundary conditions imposed there. Of course, observations on severe sea isn't an easy thing either. Since the accidents of structure collapsing or of ship capsizing due to enormous force by breakers are very common rather than exceptions, the problem we are facing at present has been proved of vital significance.

1. FLOW FIELD OF BREAKING AND NONLINEAR PERIODIC WAVES

It was merely until 1976 when Longuet-Higgins succeeded in describing the evolution of breakers for the first time the overturning process of plunging breaking waves had been poorly understood. However, he did not provide any information about velocity and acceleration in the flow field. Recently, we have simulated the evolution of plunging breaking waves directly on the physical plan instead of mapping one with the advantages that the numerical instability Longuet-Higgins happened to meet

M. L. Banner, R. H. J. Grimshaw (Eds.)
Breaking Waves
IUTAM Symposium Sydney/Australia 1991

in his research don't emerge . We not only follow how the wave shape is changing but also calculate both velocity and acceleration in the whole field (see Fig.1) according to the following formulae with velocity potential and stream function given:

$$w(z,t) = \frac{1}{2\pi i} \oint \frac{b(z_0,t)}{(z-z_0)^2} dz_0 \tag{1}$$

$$a_x - i a_y = \frac{\partial w}{\partial t} + w^* \frac{\partial w}{\partial z} \tag{2}$$

where b(z,t) indicates complex potential, w(z,t) complex velocity and a(z,t) complex acceleration. We have investigated in details the breaking of sinusoidal waves with amplitude ak being 0.42 and 0.44 and found that the maximum velocities and accelerations turn out 1.31 c and 1.84 g for the former and 1.45 c and 3.3 g for the latter. The particle velocity seems always larger than phase celerity and the acceleration in the front part is larger than g, while that at the rear part always smaller than g/3, which is in good accord with Peregrine's conclusion. The purpose to study the flow field of nonlinear periodic waves is twofold. On the one hand, we are going to see what is the difference between breaking waves and nonlinear periodic ones. On the other hand, there are some reports on the measured velocity in laboratories such as:

"Recent measurement of wave kinematics show among other things that the absolute magnitude of the horizontal velocity under the wave crest is less than that under the wave trough at the same depth."(refer to [5]).

"Recent measurements of wave kinematics indicate that the horizontal wave velocity is smaller at the crest and higher (more negative) in the

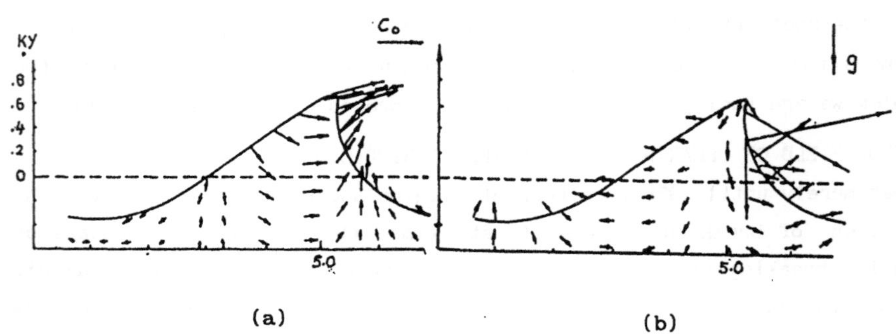

Fig.1 velocity and acceleration distributions in plunging breakers for ak = 0.44, d=L/2, t=0.46 (a) velocity (b) acceleration

trough than predicted by the Stokes higher order theories which are normally used in a deterministic design process." (refer to [6]). the discrepency in Stokes fifth-order wave theory and wave flume experiments direct us to find out the genuine causes why it .s so In contrast to most methods available ,which seem to be somewhat experiential, we try to base ourselves on more theoretical arguments.Starting from the solution by Song & li, we sum up the 15th order series and improve its behaviour nearby the convergence circle by using Pade approximation. The results are proved satisfactory, especially at the crest. The following facts strongly verify the theory:

* They are entirely in accord with Airy theory for small amplitude cases.
* Horizontal velocity component by Pade approximation from the 15th order Stokes wave reduces considerably just as reported by Lloyd Register of Shipping. Sometimes, the calculated ones are slightly smaller than the experimetal ones. It seems that the shallow water effect is responsible for the deviation.
* The observation that the horizontal velocity component under trough is larger than that under the crest at the same depth in absolute value is also consistent with the experiment in Delft Hydraulic Lab.
*The theory predicts the crest is steeper than the trough.

2. INDUCED FORCE BY BREAKING AND NONLINEAR PERIODIC WAVES

With velocity and acceleration given, it is not difficult to evaluate the forces induced either by breaking waves or by nonlinear periodic waves. Here by induced force we mean that imparted on the submarine parts in the wave-induced flow field. However, there still exists some uncertainty about how to apply Morison equation to inclined piles. Although the more conservative method of Ippens' by using velocity module or the less conservative one of Bogma's by using component normal to the surface instead of velocity itself can be selected, we prefer a compromised one derived from the Lighthill principle, i.e. the power done by drag should be proportioal to the energy flux in the wake: $D.U \propto \rho U^2 |U_n| A$ and only the acceleration in the direction normal to the body surface plays its role in the estimation of inertial force. Thus, we have

$$F(t) = \frac{1}{2} C_d \rho U_x |U_x| + \frac{\pi}{4} \rho C_M D \dot{U}_x \tag{3}$$

based on which we have yielded some interesting results as shown in Fig.2 and then led to the conclusions:

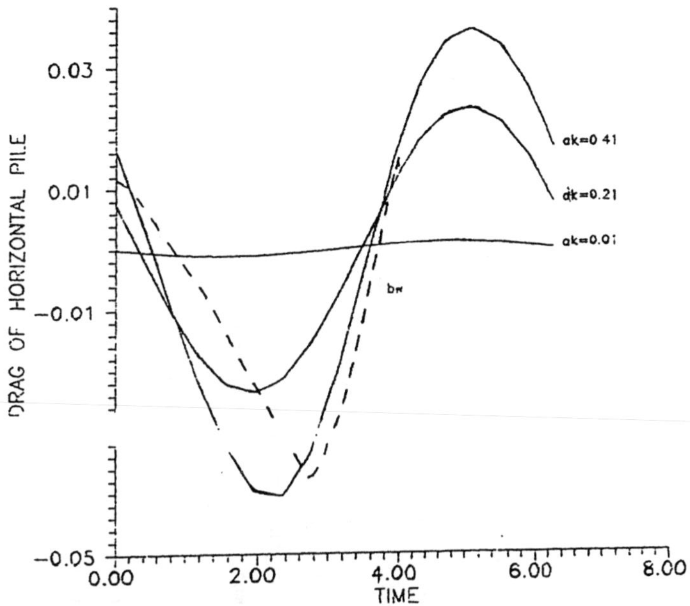

(a)

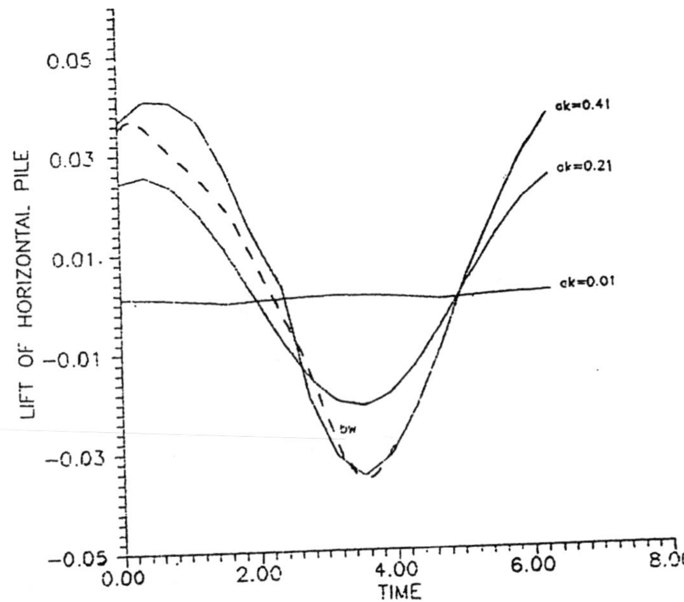

(b)

Fig.2,comparison of forces exerted on horizontal piles by periodic and breaking waves. Solid lines indicate periodic waves for ak = 0.01,0.21,0.41, dash line is for the breaking wave (a) drag,(b) lift

* As the amplitude enhences, the curves of force and moment exerted on both vertical and horizontal piles versus time depart from sinusoidal line due to the square term in drag and the subharmonic components in the inertial term.

* As for breaking waves, the velocity and acceleration in the flow field above MSL are large enough to diminish those in the flow field under MSL from the view point of energy conservation. Hence, It is not surprising that the force imparted on the structures under water by breaking waves sometimes is even smaller than that by periodic waves.

3. MOMENTUM THEOREM FOR IMPACT FORCE BY BREAKING WAVES

Although the impact forces from breaking waves are greatly dominant over the induced one, most methods to determine them are still based on experiences. The earliest one according to the measurements by Ross & Hall turned out modified Morison equation with $C_d = 2.5$. Whereas Japanese Goda applied Von Karman's entry theory to calculating impact forces by breakers, which is basically in accord with those exerted on the piles from breakers to a certain degree with unreasonable pressure distribution. In a word, all the formulae are not so general that they have to be investigated from case to case. Since we are capable of knowing the flow field in breaking waves in advance, it is very likely for us to suggest a method to calculate the impact force stemming from the most fundamental law in mechanics and propose a momentum theory based on the following hypotheses:

H1, Long crest assumption: It means that the breaking waves are assumed infinitely long in the lateral direction so as to reduce the problem to a two-dimentional one. Generally speaking, the hypothesis keeps valid for piles with small diameter in the wave with wavelength around tens meters.

H2, Frozen flow field assumption: It means that once the piles are impacted ,the wave shape and flow field are kept unvaried and a mass of water body moves forward as a whole at the same speed as that of the particle which touches the piles the earliest. It is because what we concern most is the maximum impact force at the beginning. The process appears so short that no considerable changes of flow field are detected during the process due to $g\ t << c$.

H3, Total loss of normal momentum assumption: It means that the existence of the piles hinders the plunging of moving water particles so as to transform all the normal momentum into impact force with viscous force neglected.

Based on the foregoing mentioned hypotheses,we have produced a formula exerted on a verticle cylinder omitting the derivation:

$$F(t) = \int_0^{\eta(t)} dy \int_{-w(t)}^{w(t)} \rho u^2 \cos^2\theta \ dz \tag{4}$$

where u denotes velocity ,ρ the density of water ,θ the angle between vevocity vector and the normal of body surface, z the vertical coordinate,$w(t) = \sqrt{2RU_0t - u_0^2t^2}$ the half width of immerged cylinder ,U_0 that of moving mass of water and η the wave height. With the flow field given,It is not difficult for us to determine the impact force anyway. Fig.3(a) shows the impact force versus time for ak= 0.42 and 0.44. We found that two curves manifest themselves with salient pulse and extremely short rise duration. Furthermore, the more steep the amplitude, the earlier the curve rises and the larger the impact force which can be attributed to the fact that such kind of wave evolves faster,its maximum velocity and acceleration larger and the tongue at the front longer. The ascend and descend duration are proved in good agreement with Y.B.Li's experiment. In Fig.3(b),we study pressure profiles .As compared with Goda's result, triangular profile of ours, which is quantitavely consistent with Reddish's, seems to be more reasonable due to the maximum of mass multipled by velocity in the middle.

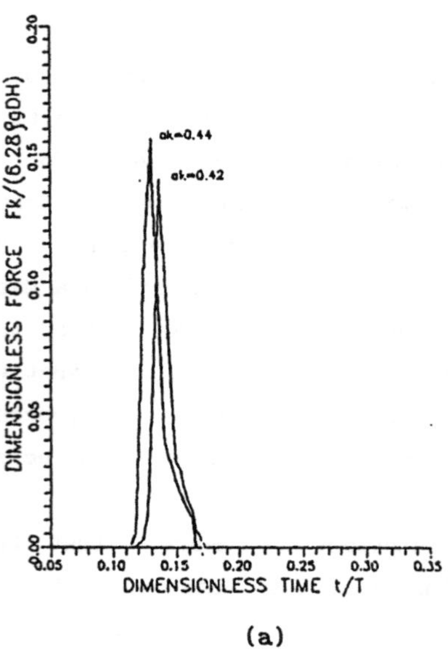

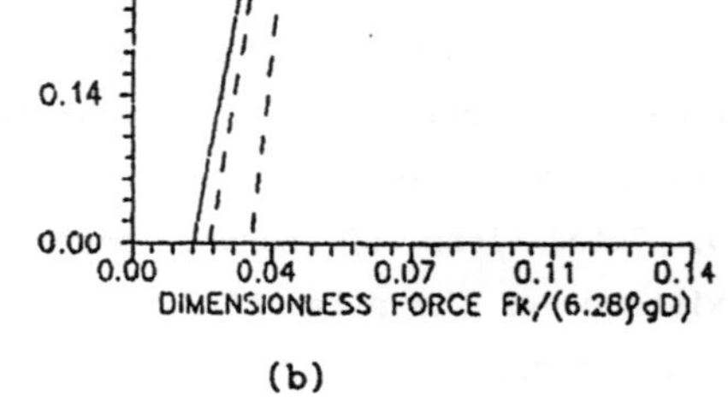

(a) (b)

Fig.3,impact force of breaking waves on the vertical column. (a) comparison oftotal forces for ak = 0.42 and 0.44 (b) comparison of pressure profiles

4. RANDOM FORCE BY BREAKING WAVES IN THE OCEAN

In reality, ocean waves are never so regular as those produced in the lab. As one kind of stochastic phenomena, they are described in terms of ocean wave spectrum. Similarily, the occurence of wave breaking events is also completely random. Up to now, there are only a few studies to statistically predict the magnitude of impact pressure. we would like to deal with them by applying Ochi's theory to the situation more suitable for China sea, making use of the Momentum theorem mentioned above. The approach to the problem is divided into three steps: (a) the probability density function of period associated with breaking waves is derived from the joint probability density function of excursion and time interval considering the wave breaking criterion; (b) the functional relation between pressure and breaking wave period is subsequently established relying on the Momentum theorem; (c) Finally, we yield the probability density function of pressure by using random variable transformation technique. In the present paper, we have analyzed probability density functions in Fig.4(a) for various conventional ocean wave spectra such as Neumann, Bretschnider and HZ spectra for China sea. Moreover, the comparisons of them with different wave significant height are also made in Fig.4(b).

Acknowledgements: The study are sponsored by National Natural Sciences Foundation of China and the Bereau of Resources and Environment, CAS.

REFERENCES

[1] E.D.Cokelet, Nature, *267*(1977) 769-774.
[2] M.S.Longuet-Higgins& E.D.Cokelet, Proc.Roy.Soc.London, *A350*, (1976)1-26.
[3] T.Vinge & P.Brevig, J.Adv.Wat.Res., *4*(1981) 77-82.
[4] J.C.Li & H.Chen, Proc. 7th Cong.APD-IAHR, *3*(1990) 363-368.
[5] O.T.Gudmestad, Appl.Ocean Res.No.2, *8*(1986) 76-88.
[6] L.Engevik, ibid.No.2, *9*(1987) 104-113.
[7] C.A.Brebbia & S.Walker, Dynamic Analysis of Offshore Structures, London(1979)138-139.
[8] J.Lighthill, Proc.Int.Cong.Fluid mech., Beijing (1987) 1-10.
[9] R.Q.Dai, M.Lin & J.C.Li, Acta Mechnica Sinica, No.4, *23*(1991).
[10] Y.Goda et al., Rep. port& Harb.Res.Inst.No.6, *5* (1966)
[11] H.J.Reddish, in Coastal Hydrodynamics, eds by R.A.Dalrymple(1987)184-195
[12] M.K.Ochi & C.Tsai, Appl.Ocean Res.No.3, *6*(1984) 157-165.

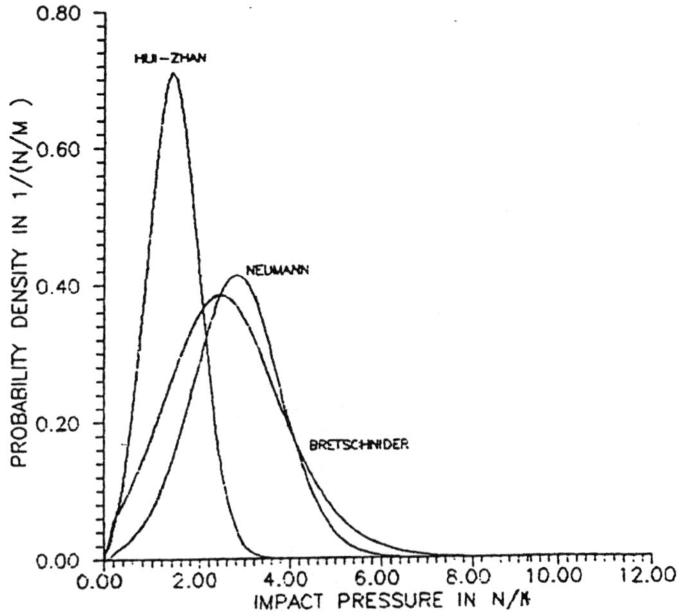

(a)

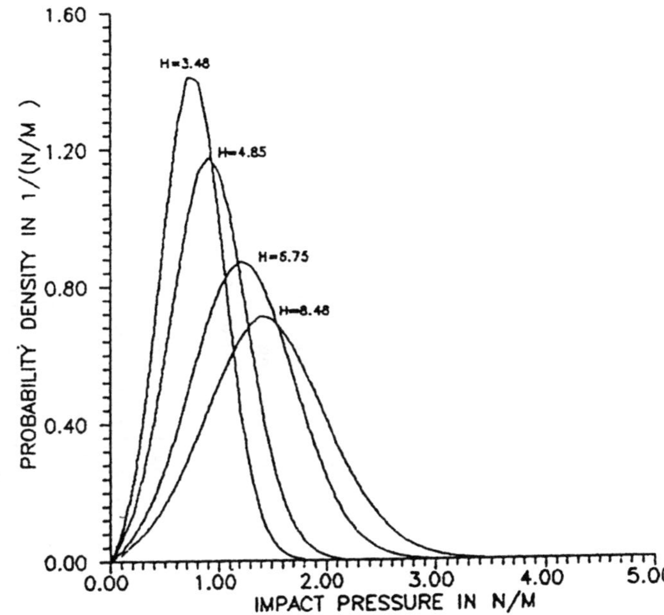

(b)

Fig.4,probability density functions of impact pressure (a)comparison of those for various ocean wave spectra. (b)comparison of those for various wave height.

Inclusion of Wave-Breaking Mechanism in a Modified Mild-Slope Model

S.R. Massel
Australian Institute of Marine Science
PMB No.3 Townsville, M.C., QLD 4810, Australia

Abstract
The energy dissipation mechanism due to wave breaking is modelled. The dissipation term is included in an elliptic, extended mild–slope equation, developed recently by the author [9]. An extension makes it possible to handle relatively rapid undulations in water depth. The resemblance between breaking waves and hydraulic jump is used to model the energy losses due to breaking. Three various types of wave trains are considered. Results of comparison with experimental data are also presented.

Introduction
It has become evident that practical and realistic results on the wave motion in surf zone may be obtained only from models which introduce a dissipation function in the depth averaged equations, i.e. mild–slope equation. This approach was used successfully in a two – dimensional model applying the parabolic approximation of the mild–slope equation [6]. However, because of the parabolic approximation involved, the information on the reflection wave field is totally lost. On the other hand, in the areas with the substantial reflecting boundaries, the elliptic mild–slope equation with the dissipation term should be considered. Originally it has been given by Booij [4] who introduced the energy dissipation in the classical mild–slope equation.

In this paper, the dissipation function is applied to the extended mild–slope equation, developed by the author [9]. Representation of the dissipation mechanism is based on the modified periodic bore approach for breaking [1,13]. Three different types of shoaling waves are discussed and the comparison with experiments is given.

Problem formulation
Consider the coordinate system $0(x, y, z)$ with z-axis positive upwards and equal to zero at the free surface, and the x-axis normal to the shore

M. L. Banner, R. H. J. Grimshaw (Eds.)
Breaking Waves
IUTAM Symposium Sydney/Australia 1991

line, and the y-axis extending along the shore line. For $0 \leq x \leq b$ and $-\infty < y < \infty$ (Region 2), the water depth $h(x, y)$ is a varying function of x and y. For $x \leq 0$ (Region 1) and $x \geq b$ (Region 3), the water depth is constant and equal to h_1 and h_3, respectively.

A plane wave train arrives at an angle θ_1 with respect to the x-axis. The wave height is equal to H and the wave frequency is equal to ω. In Regions 1 and 3, the solutions of the boundary value problem are quite straightforward and are not written here.

To formulate the equation for wave motion in the area of slowly varying bottom contour (Region 2), Berkhoff [3] used the classical perturbation scheme and after averaging the equations of the boundary value problem over depth, he developed a well known mild–slope equation [8]. In the author's solution [9], the Bubnov–Galerkin approach [11] was used, under the assumption that the bottom slope is not necessarily small. Particularly, the velocity potential $\phi_2(x, y, z, t)$ is taken in the form:

$$\phi_2(x, y, z, t) = \frac{-igH}{2\omega} \hat{\varphi}_2(x, y, z)\, e^{-i\omega t}, \tag{1}$$

where:

$$\hat{\varphi}_2(x, y, z) = \sum_{n=0}^{N} {\varphi_2}^{(n)}(x, y)\, \tilde{Z}_n(x, y, z). \tag{2}$$

The functions $\varphi_2^{(n)}(x, y)$ are determined from the requirement that the left–hand side of the resulting equation for function $\hat{\varphi}_2(x, y, z)$, i.e. the Laplace equation:

$$\left(\nabla^2 + \frac{\partial^2}{\partial z^2}\right) \hat{\varphi}_2(x, y, z) = \mathcal{L}\hat{\varphi}_2 = 0, \tag{3}$$

after substituting of (2) for $\hat{\varphi}_2$ herein, is orthogonal to the coordinate functions $\tilde{Z}_n$. Details of the development of function $\hat{\varphi}_2(x, y, z)$ are given in [9]. Here we write only the final result in the form:

$$\hat{\varphi}_2(x, y, z) = \varphi_2(x, y)\, Z(h_2, z) \tag{4}$$

where:

$$Z(h_2, z) = \frac{\cosh k_2(z + h_2)}{\cosh k_2 h_2} \tag{5}$$

and:

$$\mathcal{M}\,\varphi_2 = \left\{ \nabla^2 + \frac{\nabla(CC_g)}{CC_g}\nabla + k_2^2\left[1.0 + R_1(k_2h_2)\frac{\nabla^2 h_2}{\lambda_0} + R_2(k_2h_2)\left(\nabla h_2\right)^2\right]\right\}\varphi_2 = 0 \tag{6}$$

where: k_2–local wave number, R_1, R_2, m are functions of (k_2h_2).

Energy dissipation due to wave breaking

Usually, the energy dissipation due to wave breaking is much stronger than that due to bottom friction. Hence, we consider only the dissipation due to wave breaking. The following representation for function φ_2 is assumed:

$$\{\mathcal{M} + ik_2\gamma\}\,\varphi_2 = 0, \tag{7}$$

where γ – damping factor which should be determined in each particular case.

Determination of damping factor (γ)

A) Single wave train

To estimate the damping factor γ we apply the resemblance between surf zone waves and bores. It is well known that the dissipation in actual measurements is larger than in a hydraulic jump of the same height. However, Svendsen and Madsen [14] have found that the deviation from the hydraulic jump dissipation is less than 20%. Therefore, in order to keep the model as simple as possible, the hydraulic jump expression for energy dissipation, with some modifications given by Svendsen [13], is used. Final result is:

$$\gamma = \left\{\left(1 + 0.65\frac{H}{h_2}\right)\left[1 - 0.35\frac{H}{h_2}\right]\right\}^{-1}\frac{\omega}{\pi C_g}\frac{H}{h_2} \tag{8}$$

B) Harmonic wave train contaminated by random components

The laboratory wave generators usually produce regular wave trains in which the wave heights are not exactly constant. They are varied in the random manner being dependent on the wave maker performance, wave channel condition, human error etc.

We consider the wave maker producing a wave train in which the target wave height $\hat{H}$ is in a certain range $(\hat{H} \pm \Delta\hat{H})$. We denote the probability of that by P. The corresponding probability density function $p(H)$ is:

$$p(H) = \frac{\sqrt{n}}{\sqrt{\pi}\hat{H}A(n)}\cdot\exp\left[-n\left(\frac{H}{\hat{H}} - 1\right)^2\right], \tag{9}$$

in which: n - constant which should be specified and $A(n)$ - normalization function. For $n \Rightarrow \infty$, the probability density $p(H)$ has a limit in the form of δ function [7].

To include the dissipation due to wave breaking, we use a sharp cutoff of the distribution (9) with all waves that are breaking or have already broken, having heights equal to H_m defined by Miche criterion with γ_0 value estimated from the breaking limit calculated by Cokelet [5]. For the damping factor γ we obtain [10]:

$$\gamma = \frac{\omega}{\pi} \frac{\sqrt{gh}}{CC_g} \frac{Q_b}{b_0^2} \Psi \left(\frac{H_m}{h_2} \right) . \tag{10}$$

where Ψ is a function of H_m and h_2, and $Q_b$is the fraction of breaking or broken waves.

C) Totally random wave train

Consider now the random wave train propagated shorewards. We assume that in Region 1, the probability distribution of non – broken waves is given by Rayleigh distribution. The derivation of damping factor γ is similar to that given above and will be not repeated here.

Comparison with experiments

To verify the described method of calculation of breaking waves we use the laboratory and field data [1,2,12]. Particularly in Fig. 1, the theoretical and experimental results for regular waves climbing on the steep, plain beach of the varying slopes are shown [2]. In the next example (Fig. 2), the random waves propagate over schematized bar–trough bottom. The significant wave height $H_s = 0.194m$ and wave period $T_p = 2.22s$ was assumed [1]. Finally in Figure 3, the theory is confronted with the data taken in the shallow water zone of the North Sea. In all cases, the agreement is quite satisfactory.

References

1. Battjes, J.A.; Janssen, J.P.F.M.: Energy loss and set–up due to breaking in random waves. Proc. 16th Coastal Eng. Conf., 1 (1978) 569–587.

2. Bendykowska, G.; Ignerowicz, J.; Massel, S.; Winiecki, J.: Transformation and dissipation of wave over inclined bed. Laboratory experiments. Rep. IBW–PAN, Gdańsk, (1990) 35pp (in Polish).

3. Berkhoff, J.C.W.: Mathematical models for simple harmonic linear water waves. Wave diffraction and refraction. Publ. no. 163, Delft Hydr. Lab. (1976) 103 pp.

4. Booij, N.: Gravity waves on water with non–uniform depth and current. Delft Univ. of Techn. (1981) 130 pp.

5. Cokelet, E.d.: Steep gravity waves in water of arbitrary uniform depth. Phil. Tras. Roy. Soc. London, A286 (1977) 183–230.

6. Kirby, J.T.; Dalrymple, R.A.: A parabolic equation for combined refraction–diffraction of Stokes waves by middle varying topography. Jour. Fluid Mech. 136 (1983) 433-466.

7. Lighthill, M.J.: Fourier analysis and generalised functions. Cambridge University Press (1975) 79pp.

8. Massel, S.R.: Hydrodynamics of coastal zones. Amsterdam, Oxford, New York: Elsevier Science Publ. (1989) 336pp.

9. Massel, S.R.: Extension of mild–slope model for surface waves. (1991) Submitted for publication in Coastal Engineering.

10. Massel, S.R.; Belberova, D.Z.: Parametrization of the dissipation mechanisms in surface waves induced by wind. Proc. 19th Biennal Symposium on Advanced Problems and Methods in Fluid Mechanics, Kozubnik (Poland) (1989).

11. Michlin, S.G.: Variational methods in mathematical physics. New York: Macmillan Comp. (1964) 582pp.

12. Oelerich J.; Dette H.H.: About the energy dissipation over bared beaches. Proc. 21 Coastal Eng. Conf., 1 (1988) 293–306.

13. Svendsen, I.A.: Wave heights and set–up in a surf zone. Coastal Engineering 8 (1984) 303–329.

14. Svendsen, I.A.; Madsen, P.A.: Energy dissipation in hydraulic jumps and breaking waves. Inst. Hydrodyn. Hydraul. Eng. , Rep. 55 (1981) 39–47.

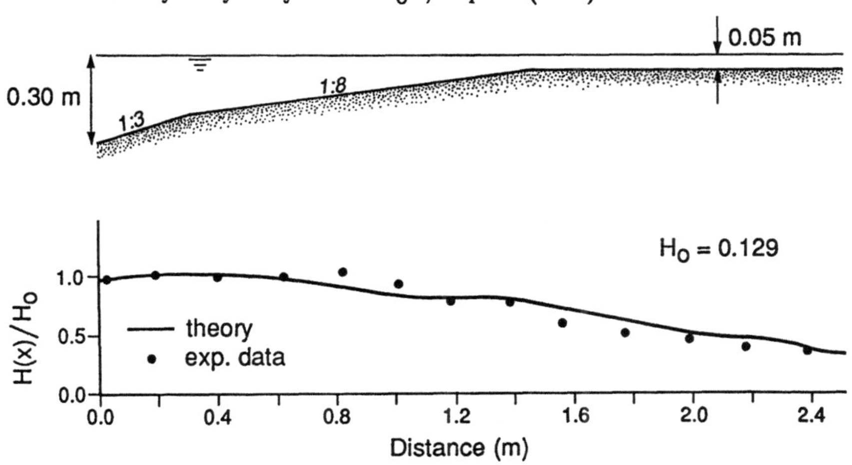

Fig.1 Wave breaking over plain beach with changing slope

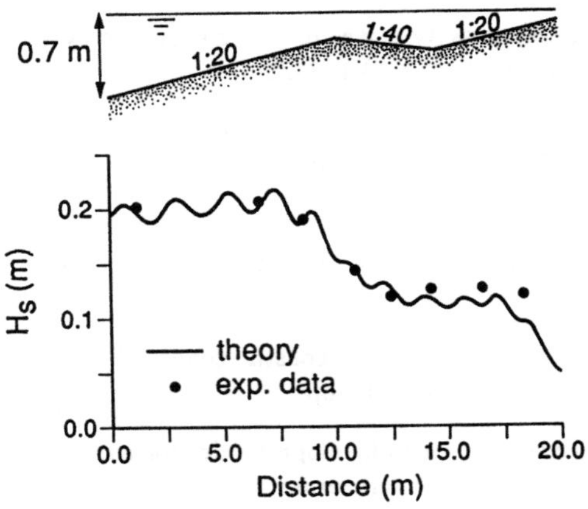

Fig.2 Random waves breaking over schematized bar–trough bottom

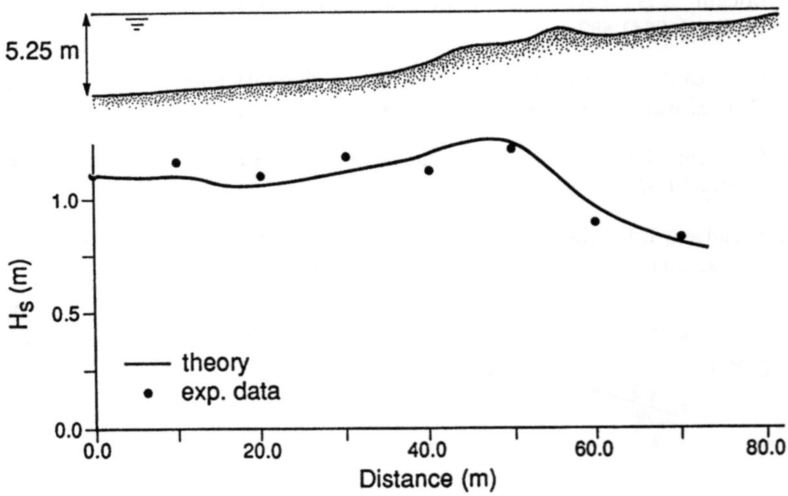

Fig.3 Wind waves breaking in shallow water zone of the North Sea

Imaging of the Ocean Surf Zone Using Interferometric Synthetic Aperture Radar

L. SHEMER

Department of Fluid Mechanics,
Faculty of Engineering,
Tel-Aviv University, ISRAEL

Summary

Microwave remote sensing of the ocean surface using a Synthetic Aperture Radar (SAR) is of great interest due to its potential of nearly instantaneous coverage of large areas. The accumulation in recent decades of experience with ocean imaging using SAR indicates that there are substantial difficulties in the interpretation of the obtained images. The present paper presents some results obtained in imaging of the nearshore regions of the Monterey Bay using a modification of the conventional SAR which employs two spatially separated antennas. This allows us to construct an interferogram of the image which provides direct mapping of the observed surface velocities. The advantages of this novel technique in imaging of the ocean surf zone are demonstrated.

Introduction

Microwave remote sensing of the ocean in general, and of the surf zone in particular, is of great interest due to its potential of nearly instantaneous coverage of large areas. The application of SAR to imaging of the permanently moving ocean surface reveals, however, that interpretation of ocean imaging by SAR represents considerable difficulties.

Recently [1], [2], an interferometric SAR (INSAR) was suggested for imaging of the ocean surface. In this modification of conventional SAR technique, complex SAR images are obtained by two antennas separated along the platform flight path. These images are further combined interferometrically thus providing a remote sensing tool with a well defined mechanism of ocean imaging. Due to the direct imaging mechanism, INSAR can measure quantitatively the actual velocity at the ocean surface. The advantages of INSAR over regular SAR in imaging of the ocean were discussed in [3], [4].

In contrast to conventional SAR, where the image represents the map of the absolute value of the complex reflectivity of the ocean [5], in INSAR image we are foremost interested in the phase component of the interferogram, which contains direct information about the radial velocity component of the ocean surface.

M. L. Banner, R. H. J. Grimshaw (Eds.)
Breaking Waves
IUTAM Symposium Sydney/Australia 1991

Principle of operation of an interferometric SAR

The microwave scattering from a rough surface for a side-looking radar is attributed to the Bragg resonance between the electromagnetic wave and the undulating surface. The crests of the resonant surface waves (Bragg waves) are perpendicular to the radar viewing direction, and there exists the following relation between the wave vector of the surface Bragg waves **k** and the radar wave vector $\boldsymbol{\kappa}$ is

$$|\mathbf{k}| = 2|\boldsymbol{\kappa}| \sin\gamma, \tag{1}$$

γ being the radar beam incidence angle. For an L-band radar and $\gamma = 30^0$, the length of the resonant Bragg waves is 24.5cm. These surface Bragg waves propagate to and/or from the radar platform with the phase velocity $c_p = \sqrt{g/|\mathbf{k}|}$. Bragg wave frequency is given by the linear dispersion relation for surface gravity waves $\omega = \sqrt{g|\mathbf{k}|}$.

Synthetic Aperture Radar (SAR) is a high resolution coherent remote sensor carried either by an aircraft or by an orbiting platform. The high along-track resolution is achieved by using the phase history of the Doppler-shifted return signal over the finite integration time in which moving with the constant velocity V radar forms a large aperture in the flight direction. The high cross-track resolution is obtained by modulating the transmitted signal [6]. The theoretical background for the INSAR imagery of ocean surface is given in [7]. In the present study the modification suggested in [1] and [2] was used. The NASA/JPL DC-8 research aircraft equipped with an interferometric SAR system operating at L-Band with a full digital recording capability and an inertial navigator system for three dimensional measurement of the aircraft attitude was used. The backscattered echoes received by physically separated fore and aft antennas mounted on the left looking side of the aircraft fuselage from the ocean surface are recorded and processed into two separate complex (magnitude and phase) maps, which are then combined interferometrically into a single complex image. The spatial separation between the antennas B = 19.6m. Pulses are transmitted from the rear antenna and are received simultaneously by both antennas. The frequency of the backscattered radar signals at both antennas is Doppler shifted due to non-zero surface velocity vector **U** by

$$\omega_D = 2\mathbf{k}\cdot\mathbf{U} = 2kU_r \tag{2}$$

where U_r is the radial component of **U**. The velocity **U** represents the vector sum of surface currents, orbital velocity of the ocean waves and the phase velocity c_p of the resonant Bragg waves. For a platform moving with the velocity V, the image of the identical surface area by the two antennas is obtained with the time interval $\Delta t = B/2V$. The phase difference α between the return signals of the two INSAR antennas, which is due to the Doppler shift ω_D and the time interval Δt, is related to the measured radial component of the surface velocity U_r by:

$$\alpha = \omega_D \, \Delta t = \frac{kB}{V} U_r \tag{3}$$

Equation (3), first obtained in [2], serves as a basis for extracting the quantitative information from the INSAR images.

Experiment

The experiment was carried out on September 8, 1989 in Monterey Bay. The experiment consisted of four overflights of a NASA/JPL DC-8 airplane carrying an L-band interferometric SAR. The average aircraft altitude was about 8200m. Additional radar system and flight data are reported in [3], [4]. The radar overflights were conducted around 1300 Pacific Daylight Savings Time, close to the time of maximum tidal flood current. The flight pattern and the scene locations of radar data acquired are illustrated in Figure 1.

The environmental conditions during the Marina experiment were typical for late summer in Monterey Bay. Light westerly winds (about 2 m/s), surface air temperature of 14°-16°C and clear skies up to 300 m blocked by a stratus cloud deck, were observed during experiment period. The mild meteorological conditions during this period resulted in almost no local wave generation of sea. The sea surface was predominantly long crested, narrow-band swell waves propagating shoreward. The surface wind had mainly an eastward component with a weaker ambiguous northerly component. Thus in flight legs 360° and 180°, the radar was looking at an upwind or

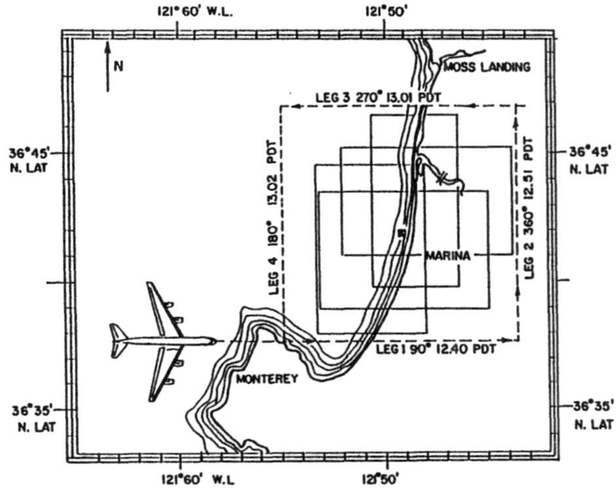

Fig. 1. Flight pattern and scene locations.

downwind surface roughness while in legs 90° and 270° the radar looked at weakly crosswind scenes. The resonant Bragg waves at the ocean surface which are responsible for the radar reflectivity, during flight path 90° and 270° travelled in both directions towards and away from the aircraft due to wave and wind spreading.

Since only the velocity component in the radar viewing direction is sensed by the interferometric SAR, the images of 180° and 360° flight legs make it possible to estimate the on-shore or off-shore current component, while the flight legs of 90° and 270° provide information about the longshore currents. For this reason, in the present study we mainly are interested in the last two images. An interesting feature of the imaged scene, as can be seen in Fig. 1, is the Salinas river, which at that time of the year is a reservoir of standing water blocked from the ocean. The Salinas river is visible in flight legs of 360° and 270°. Due to the absence of currents and long gravity waves in this small reservoir, the measured scatterer velocity in the Salinas river corresponds to the Bragg waves phase velocity. This allows one to compare the measured value of c_p with the linear theoretical result.

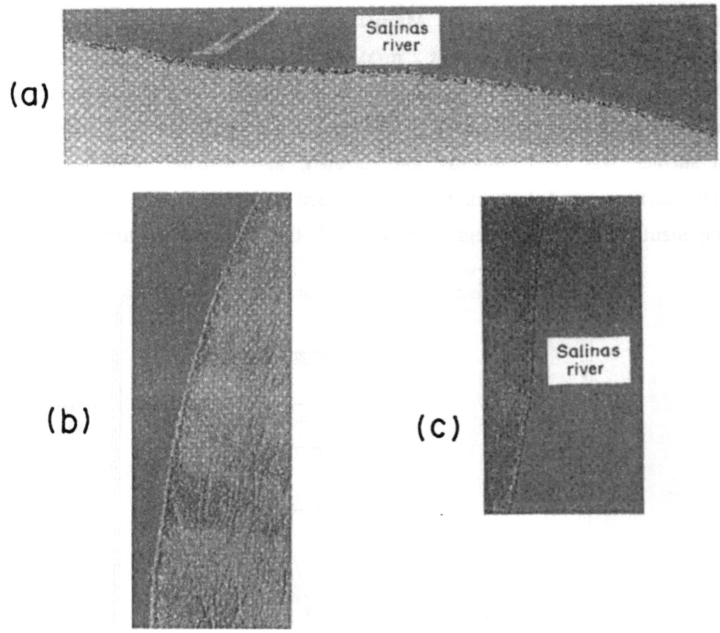

Fig. 2 INSAR images of the nearshore region in a) northbound (360°) flight leg, image size 12.4km by 3km, b) eastbound (90°) flight leg, image size 3km by 8.4km; and c) westbound (270°) flight leg, image size 3km by 6.2km.

Experimental results

The nearshore area of INSAR images obtained in the flight legs 360°, 90° and 270°, are presented in Fig. 2. The Salinas river is clearly seen in images of 360° and 270°. In the 360° case, the radar was looking upwind, the Bragg waves therefore propagated toward the platform, so that the pixels in the Salinas river region have higher values than in the surroundings. The apparent Bragg wave phase velocity estimated using (3) from the Salinas river area of Fig. 2a is 53 cm/s. Substituting the corresponding incidence angle $\gamma = 46°$ into (1) yields the wave number $|k|$ for which the linear theory gives $c_p = 51$ cm/s. In the crosswind 270° flight path the resonant Bragg waves travelled in both directions, to and from the radar, so it is difficult to estimate *a priori* the contribution of each propagation direction. The estimates of Bragg phase velocity for the Salinas river region from Fig. 2c gives the value of -41 cm/s, while application of (1) for $\gamma = 33°$ yields $C_p = 58$ cm/s. It thus appears that Bragg waves in crosswind were mainly southbound, away from the radar in 270° flight path. These measured values of the Bragg phase velocity are assumed to be valid also for the open sea regions in the image, thus making it possible to get absolute estimates of longshore current velocities for both 90° and 270° flight directions.

The image of the surf zone in Fig. 2a (flight leg 360°) indicates that there is a wide range of observable surface velocities in this region. This is in agreement with earlier results which indicate that the bandwidth of the radar Doppler spectrum is drastically increased by breaking waves [8]. The dark pixels in the image may correspond to velocities with the phase shift calculated according to (3) exceeding the value of π [7].

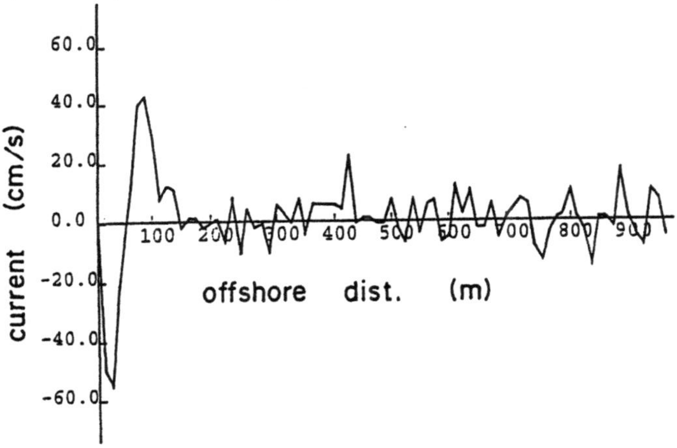

Fig. 3 Current distribution obtained in the westbound (270°) flight leg. Radar beam incidence angle 40°, positive current direction - northbound.

The spectral analysis of the available INSAR images, as well as simultaneous ground-based measurements, indicate that a bimodal wave system existed in Monterey Bay during the experiment [3]. [4]. The two spectral peaks correspond to wave periods of 9.1s and 15.9 s, where longer swell propagates from west-southwest (propagating from direction of about 255°), while the shorter wave comes from west-northwest (direction of about 295°). The longer swell thus generates northbound longshore current, while the shorter wave can result in a southbound current, in particular in the northern part of the imaged area, where the angle of approach of this wave is larger (cf. Fig. 1). Both these currents are clearly visible in the current distribution in the off-shore direction obtained from the 270° flight leg (Fig. 3). As it could be expected, the peak of the longshore current corresponding to a shorter wave is located considerably closer to the waterline (at offshore distance of about 40m) than that of the longer swell ((about 100m). Fig. 4. depicts the current distributions obtained in the overlapping regions of images presented in Figs. 2b and 2c. The agreement between both Figures is reasonably good.

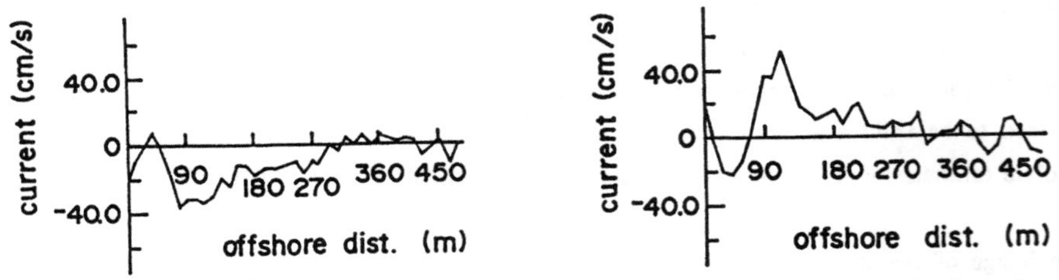

Fig. 4 Comparison of current distributions obtained in the overlapping region of a) eastbound (90°) and b) westbound (270°) flight legs. Positive current direction is southbound in 90° flight leg and northbound in 270° flight leg.

In conclusion, the present preliminary results indicate that the interferometric SAR has a considerable potential in providing information about spatial distribution of ocean currents, which can hardly be acquired by alternative means. In particular, longshore current distribution corresponding to a bimodal wave system is observed. This INSAR capacity to measure complex spatial patterns ocean currents augments its capability to measure directional 2D spectra of ocean surface elevation, as demonstrated in an earlier work [4].

References

1. Goldstein, R.M.; Zebker, H.A.: Interferometric radar measurement of ocean surface currents, *Nature* 328 (1987) 707-709.

2. Goldstein, R.M.; Barnett, T.P.; Zebker, H.A.: Remote sensing of ocean currents. *Science* 246 (1989) 1282-1285.

3. Marom, M.; Goldstein; R.M.; Thornton, E.B.; Shemer, L.: Remote sensing of ocean wave spectra by interferometric synthetic aperture radar. *Nature* 345 (1990) 793-795.

4. Marom, M.; Shemer, L.; Thornton, E.B.: Energy density directional spectra of nearshore wavefield measured by interferometric synthetic aperture radar. Submitted to *J. Geophys. Res.*, 1991.

5. Hasselmann, K; Raney, R.K.; Plant, W.J.; Alpers, W.; Shuchman, R.A.; Lyzenga, D.R.; Rufenach, C.L.; Tucker, M.J.: Theory of sythetic aperture radar ocean imaging: a MARSEN view. *J. Geophys. Res.* 90 (1985) 4659-4686.

6. Tomiyasu, K.: Tutorial review of synthetic-aperture radar (SAR) with applications to imaging of ocean surface. *Proc. IEEE* 66 (1978) 563-583.

7. Shemer, L; Kit, E.: Simulation of an interferometric SAR imagery of an ocean system consisting of a current and a monochromatic wave. To appear in *J. Geophys. Res.*, 1991.

8. Keller, W.C.; Plant, W.J.; Valenzuela, G.R.: Observation of breaking ocean waves with coherent microwave radar. In *Wave dynamics and radio probing of the ocean surface*, ed. by Phillips, O.M. and Hasselmann, K. New York, London: Plenum Press 1986.

Wave-Breaking due to Moving Submerged Obstacles

A. F. Teles da Silva

Inst. Mat. & Lab. Mec. Flui./PEM/ Coppe UFRJ CP68503, CEP21910, RJ Brazil

D. H. Peregrine

School of Maths, Univ. of Bristol, Bristol BS8 1TW, England

SUMMARY

In this work we study some aspects of a 2D free surface flow produced by a submerged cylinder moving close to the bed (see Figure 1). The flow evolution depends on cylinder shape and on the Froude number of the undisturbed flow. According to these we find three main types of evolution which may include steady or transient breaking.

INTRODUCTION

The flow around a moving disturbance on or near a free surface of water allways catches the attention, because of the variety of possible outcomes and of the presence of different régimes in different regions of the same flow. In two dimensions we have two quite different flows one upstream and the other downstream of the disturbance. There may be a bore upstream and a more complicate wave flow downstream.

As the disturbance moves, at a speed smaller than a certain critical value, water piles up in front of it; from this mass of excess water a bore develops and moves away from the disturbance. Bores are produced when water of a certain depth and speed is made to move onto shallower slower water, see Teles da Silva and Peregrine [1]. Bores can be produced in a variety of ways: moving pistons, pushing plates over a surface of water, Binnie and Orkney [2], displacing floating bodies, Ertekin, Webster and Wehausen [3], moving regions of applied pressure, Ertekin, Webster and Wehausen [4] or moving submerged obstacles. Bores fall into one of three types: purely undular, undular breaking and turbulent (which does not develop waves).

Behind the moving disturbance there appears a depression at the free surface; this depression gives rise to a flow which is interesting on its own. Fornberg and Whitham [5] show computations, using K de V's equation, of the evolution of an initial well on a free surface. One of the slopes of the well smoothes away and the other at first steeppens and develops a succession of waves which at a later stage start to get lower and longer. Ultimately, both slopes of the well will be smoothed but one of them go first through steeppening and wave formation. When behind the moving disturbance a well first appears it may, sometimes, be too steep to obey the behaviour predicted by K de V's equation and breaking of the downstream slope is a possible outcome.

The problem of wave generation by moving disturbances has been studied mostly by weakly nonlinear models. Several of the features of the flow have been described. Mei [6] classifies four main types of flow evolution; i)***highly supercritical*** with no bore and downstream waves fastly smoothed and washed away; ii) ***critical*** and ***supercritical*** with a bore and downstream waves smoothed and washed away; iii) and iv) are respectivelly

M. L. Banner, R. H. J. Grimshaw (Eds.)
Breaking Waves
IUTAM Symposium Sydney/Australia 1991

subcritical and *low subcritical* where bores are gentler than in ii) and the downstream waves are not smoothed and trail close to the body. Wu [7], calculates, along with other results, the drag coefficient C_{DW} which is found to depend on the upstream bore; peaks of C_{DW} correspond to the generation of new waves in the bore; amplitudes of C_{DW} oscillations decay with time, faster for smaller Froude numbers. Lee, Yates and Wu [8] model the same problem as Wu, [7], by both K de V and Boussinesq's equations; they show among other results that periods of C_{DW} oscillations increase with Froude number.

For subcritical as well as highly supercritical flows fully nonlinear stationary asymptotic solutions have been calculated by Forbes and Schwartz [8], Wanden Broeck [9], Dias and Vanden Broeck [10]. A steady state with no waves has been calculated by Forbes [11].

MATHEMATICAL MODEL

The problem of a moving submerged obstacle, sketched in Figure 1, as a potential flow problem. The following boundary value problem for the potential Φ, with the reference frame set at the obstacle, is to be satisfied at any time along the flow evolution:

$$\nabla^2\Phi = 0 \tag{1}$$

$$\frac{\partial\Phi}{\partial t} + \frac{1}{2}\nabla\Phi^2 + gY = 0 \tag{2}$$

$$\frac{D\mathbf{R}}{Dt} = \nabla\Phi \tag{3}$$

$$\frac{\partial\Phi}{\partial\mathbf{n}} = 0 \tag{4}$$

$$\lim_{x\to-\infty} Y = \lim_{x\to\infty} Y = 0 \qquad \lim_{x\to-\infty} \nabla\phi = \lim_{x\to\infty} \nabla\phi = -U \tag{5}$$

(1) is to be valid inside the fluid region. At the free surface which is described parametricaly by $\mathbf{R} = (X, Y)$ in order to allow for overturning, (2) and (3) must be satisfied. At the bed, placed at $y = -h$, and on the body (4) is to apply, where $\mathbf{n}$ is a normal to bed and body surfaces. At both extremes $x = \pm\infty$ the free surface is to be flat at $y = 0$ and the speed equal to -U as expressed in (5). The value U must be given as part of the problem as well as the curve $\mathcal{C}$ describing the obstacle shape. As initial condition we set the free surface flat at level zero, $\mathbf{R} \equiv 0$, the potential $\Phi = -Ux$, the cylinder center at $x = 0$ and the flow is started impulsively. Throughout this work the undisturbed depth h and the acceleration of gravity g are, respectively used as units of length and acceleration; the unit of time is thus $\sqrt{h/g}$. Two types of obstacle have been used: half-circles and half-elipses with the horizontal semiaxis a fixed to $a = 1$ The flow depends on two parameters: F and either the radius r or the semi axis b. We compute flows with Froude numbers ranging from 0.6 to 1.6. The free surface evolution is calculated through a boundary integral numerical scheme presented in Teles da Silva and Peregrine, [13]. Computations have been made with IBM's 3090 in the University of Bristol, England, and in CENPES/PETROBRÁS, Brazil.

NUMERICAL RESULTS

Nonbreaking flows behave in the way described by weakly nonlinear calculations. If we take the régimes described by Mei, an interesting feature appears as a result of the finiteness of the obstacle size; for a given obstacle size there exists two Froude numbers, F_{Sb} and F_{Sp}, $F_{Sb} < 1 < F_{Sp}$, which mark the boundaries between *subcritical/low subcrical* and *critical/supercritical* and between *critical/supercritical* and *highly supercritical*. The words subcritical critical and supercritical will be from now on used in this sense. For very small obstacles they both get near one. For greater obstacles they depart from one. For instance, for a half circle of radius 0.1 $F = 1.1$ is already *high supercritical* whilst for a half elipse with $a = 1$ and $b = 0.1$ it is *supercritical*; see Figure 2 a). Figure 2 a) and b) show two long computations for the flow evolution with a half elipse with $a = 1$, $b = 0.1$ both and $F = 1.1$ and $F = 0.95$ respectively. With lower Froude numbers there is more mass accumulation upstream of the obstacle. Mass accumulates in pulses; Figure 3 shows a graph of the rate of mass accumulation with time for the two computations above ; the troughs correspond roughly to the beginning of a new wave at the back of the bore. Figure 4 shows graphs of force components against time for the same computations. The oscillations in Figures 3 and 4 are out of phase and appear to decay with time, more conspicuously for lower Froude numbers.

Breaking occurs both upstream and downstream of the cylinder. Downstream breaking appears to be a more complicated phenomenon. Some computations have been made with a single trough on the free surface and a flat bed. The trough takes the shape of $\epsilon\, sech(\alpha x)$. Figure 5 shows the flow evolution for this initial state with $\epsilon = 0.2$; it can be seen that on the downstream side waves form, get steep and then smooth continually; in this situation breaking may occur during the steepenning stage; depending on the magnitude of ϵ breaking may occur at the very first stages of steepenning as in Figure 6. In all the critical and supercritical régimes downstream waves first get steep and then smooth, that happens in both cases of Figure 2; for these cases the deficit of mass downstream is smaller compared with the subcritical régimes and the well model may approximately apply to them. Figure 7 shows two downstream breakers where breaking occurs at two different stages of the steepenning process. The subcritical flow régimes however present, in the well region, downstream of the obstacle an equilibrium between steepennig and stretching; in the sense that what characterise these flows is the formation of a steady wave train that does not smooth nor gets washed away. We conjecture that this effect may be reproduced with the single well computations if mass, in some way, is taken from the well. Figure 8a) and b) shows a subcritical breaker which is somewhere near two frontiers: between breaking and nonbreaking and between subcritical and critical régimes. If either the obstacle increases in size or the Froude number is reduced breaking happens earlier. In fact for lower subcritical Froude numbers breaking occurs for smaller obstacles. The breakers of Figure 7 will be eventually washed downstream whilst *subcritical*. breakers will probably stay close to the obstacle.

Upstream breakers are bore breakers which are discussed in some detail in [1]. As computations stop after breaking, and breaking downstream is a far more likely event, upstream breakers are computed only for moderately large obstacle sizes and for higher Froude numbers, somewhere above one (one excluded); only for these cases the downstream waves will survive the steepenning stage of their evolution. Moreover if big obstacles cannot be used to produce greater breaking bores smaller ones will do if the period of wave formation is increased; which means to use higher Froude numbers; as in the breaker in Figure 8. Figure 10 shows a breaker that appears to be a steady solitary breaker.

The occurrence of downstream breakers for subcritical flows for any but very moderate obstacle sizes, poses a limitation on the existence of laminar subcritical solutions of the steady problem for obstacles that are not small enough. By another side breaking in the subcritical régime is steady breaking; meaning that there is a turbulent wake trailling at a short distance behind the obstacle. These are the cases discussed by Duncan [14], where he finds incresed resistance due to steady breaking in the wake of a hydrofoil. At these lower Froude numbers the rate of accumulation of mass upstream of the obstacle is greater but the period of wave formation in the bore is shorter meaning that individual waves in the bore are very gentle.

REFERENCES

[1] Teles da Silva, A. F. & Peregrine, D. H. **Nonsteady computations of undular and breaking bores**, Proc. of the 22nd Int. Conf. Coastal Eng., Delft, 1990.

[2] Binnie, & Orkney **Exper. on the flow of water from a reservoir through an open channel. II. The formation of hydraulic jumps.**, Proc. Roy. Soc. **A230**.

[3] Ertekin, R. C., Webster, W. C. & Wehausen, J. V. **Ship-generated solitons**, Proc. of 15th Symp. Naval Hydrodynam., Hamburg, 1984.

[4] Ertekin, R. C., Webster, W. C. & Wehausen, J. V. **Waves caused by a moving disturbance in shallow channel of finite width**, J. Fluid Mech. **169**, 1986.

[5] Fornberg, B. & Witham, G. B. **A numerical and theoretical study of certain nonlinear wave phenomena**, Phil. Trans. R. Soc. Lond. **A289**, 1978.

[6] Mei, C. C. **Radiation of solitons by slender bodies advancing in a shallow water channel**, J. Fluid Mech. **162**, 1986.

[7] Wu, T. Y. **Generation of upstream advancing solitons by moving disturbances**, J.Fluid Mech. **184**, 1987.

[8] Lee, S., Yates G. T. & Wu, T. Y. **Exp. and analysis of upstream advancing solit. waves generated by moving disturbances**, J.Fluid Mech. **199**, 1989.

[9] Forbes, L. K. & Schwartz L. W. **Free surface flow over a semicircular obstruction**, J.Fluid Mech. **114**, 1982.

[10] Vanden-Broeck, J.-M., **Free surface flow over an obstruction in a channel**, Phys. Fluids **30**, 1987.

[11] Dias, F. & Vanden-Broeck, J.-M., **Open channel flows with submerged obstructions**, J. Fluid Mech. **206**, 1989.

[12] Forbes, L. K. **Critical free surface flows over a semi-circular obstruction**, J. Eng. Math. **22**, 1988.

[13] Teles da Silva, A. F. & Peregrine, D. H. **Nonlinear perturbations of on a free surface induced by a submerged body: A Boundary Integral Approach**, special issue of the J. Eng. Anal. Bound. Elem., 1990.

[13] Duncan, J. H. **The breaking and non-breaking wave resistance of a two dimensional hydrofoil**, J. Fluid Mech., **126**, 1983.

FIGURES

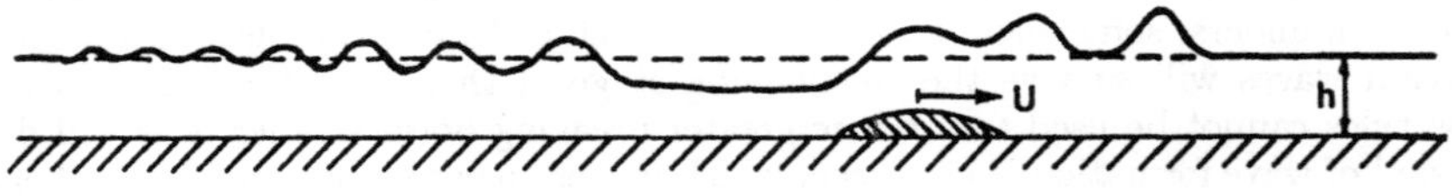

Figure 1. Wave generation problem.

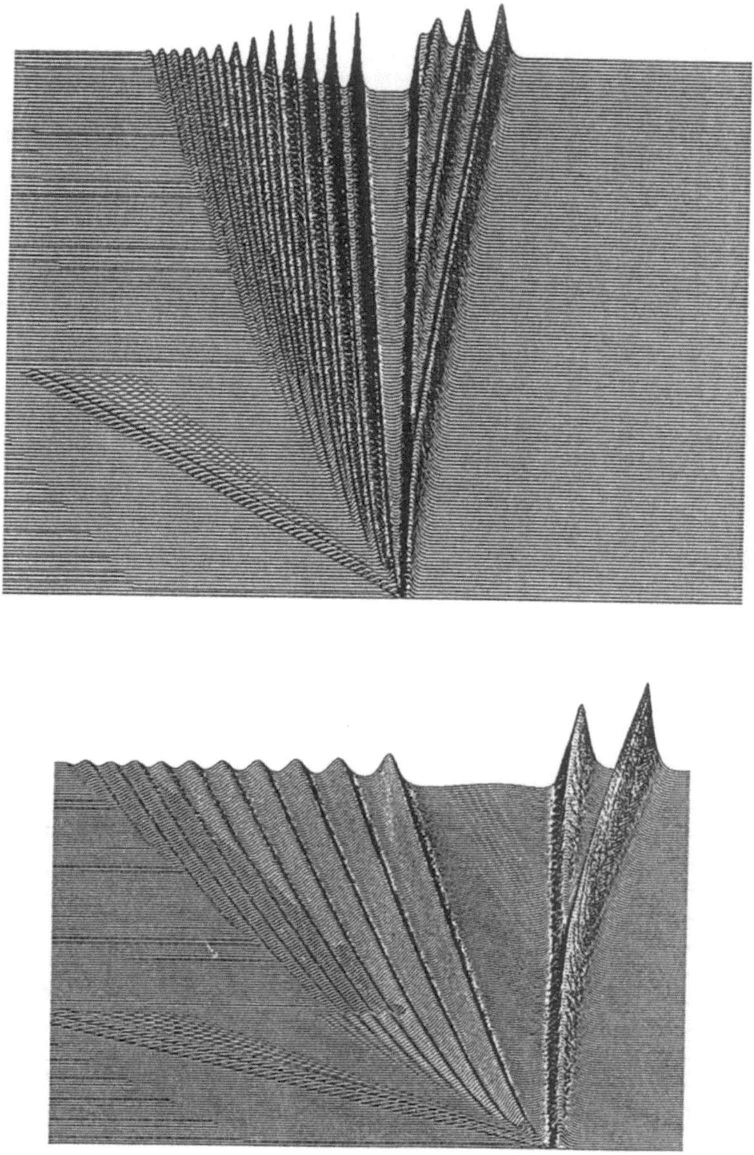

Figure 2. Evolution of free surface flows past half elipses with semi axis $a = 1.0$ and $b = 0.1$ until time 150: a) $F = 0.95$, above b) $F = 1.1$, below. cylinders:

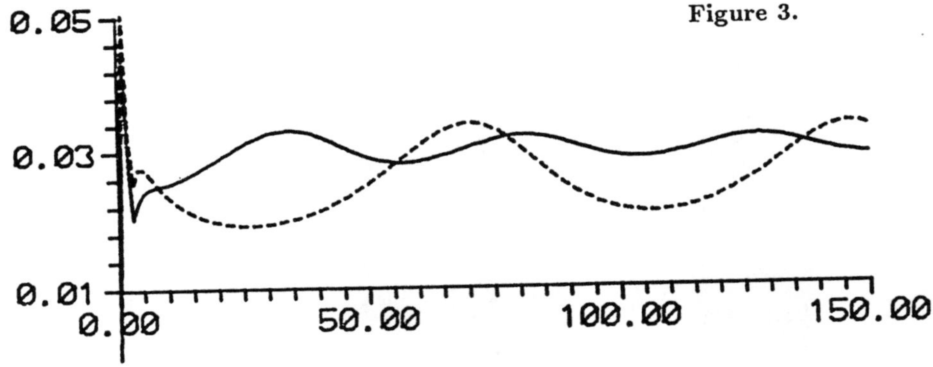

Figure 3.

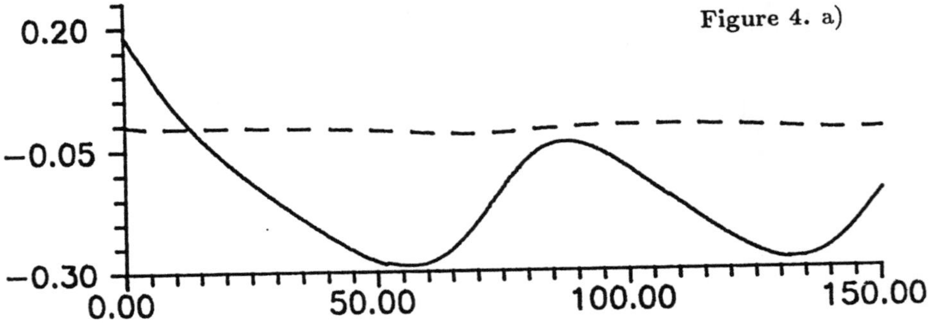

Figure 4. a)

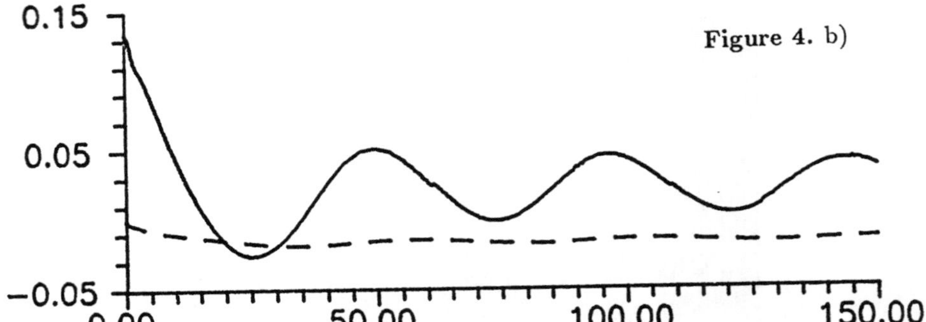

Figure 4. b)

Figure 3. Rate of mass accumulation upstream of half elipses with $a = 1.0$ and $b = 0.1$ against time. Full line: $F = 0.9$; dashed line: $F = 1.1$.
Figure 4. Components of the force on the cylinder, same as in Figure 3, against time. Full line: vertical component; dashed line: horizontal component. a) $F = 0.9$ b) $F = 1.1$.

Figure 5.

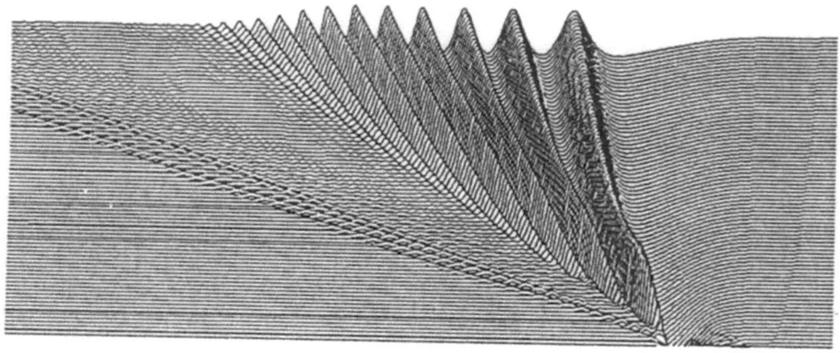

Figure 6.

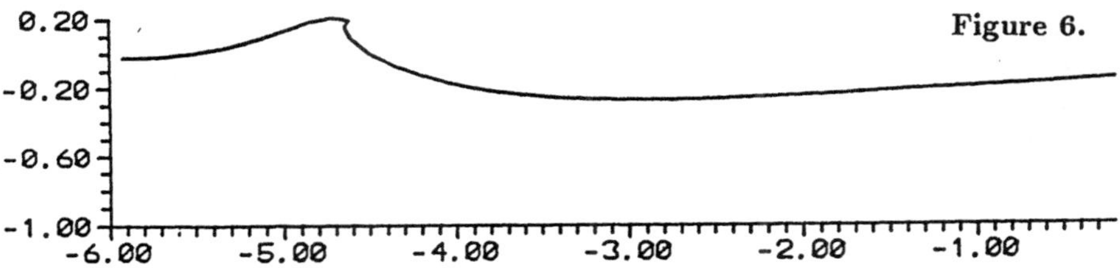

Figure 7. a)

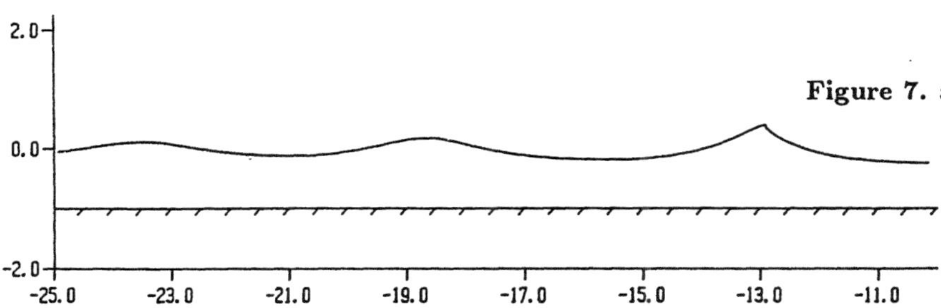

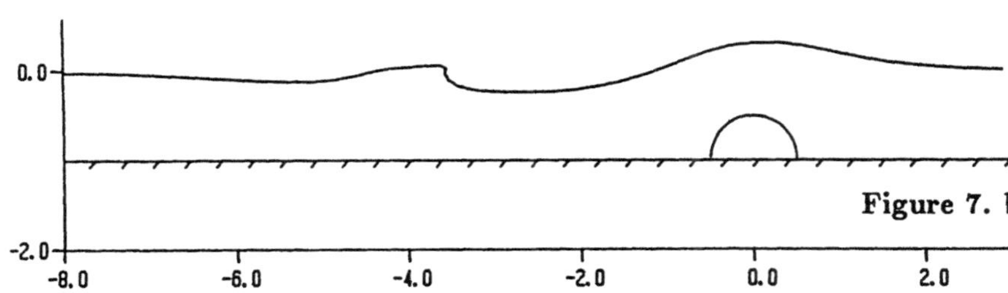

Figure 7. b)

Figure 5. Flow evolution for an initial well, $y = 0.20\, sechx$, until time 80.
Figure 6. Detail of the breaking profile; the flow is started with an initial well, $y = 0.35\, sechx$, at time 6.75. The breaking crest is 0.21 high, the maximum slope is 156°.
Figure 7. Downstream breakers for semi-circular cylinders. a) $r = 0.3$ and $F = 1.0$ time 55.8, the breaking crest is 0.399 high at $x = -13.0$, the maximum slope is 123.78^o. b) $r = 0.5$ and $F = 1.6$ at time 2.50, the breaking crest is 0.070 high at $x = -3.7$, the maximum slope is 145.41^o .

Figure 8. a)

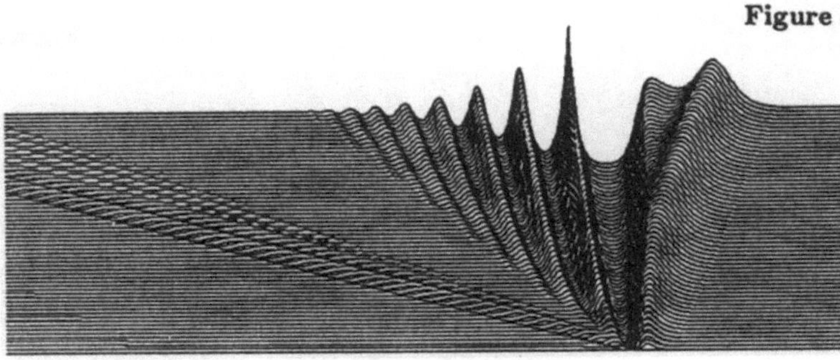

Figure 8. b)

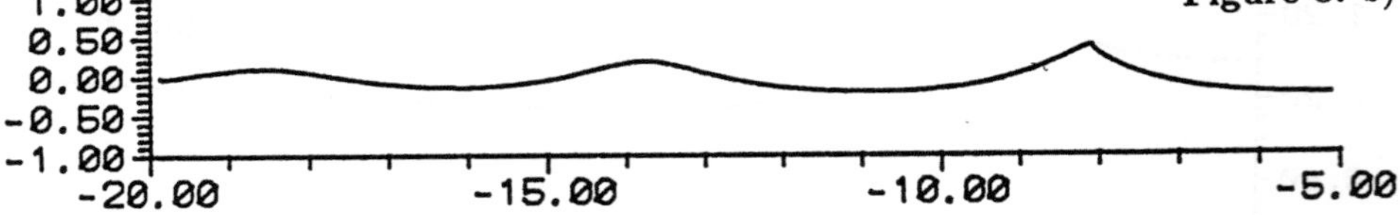

Figure 9.

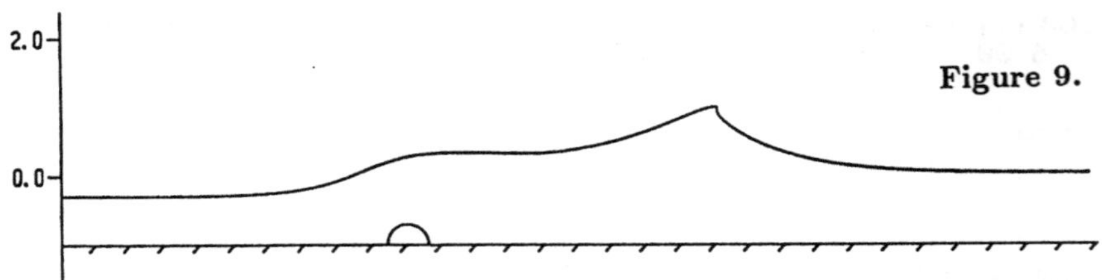

Figure 10.

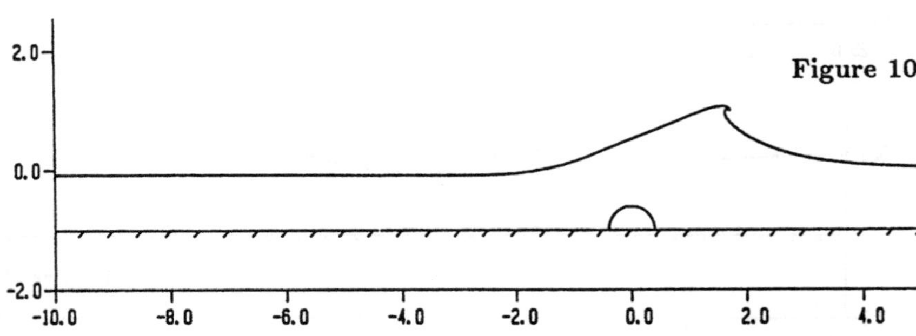

Figure 8. a) Flow evolution until time 60 for a half elipse with $a = 1.0$ and $b = 0.1$ and $F = 0.9$. b) Detail of breaker in a) at time 60.4. The breaking crest is 0.39 high at $x = -8.1$ and the maximum slope is 130°.
Figure 9. Upstream breaker for $r = 0.4$. *a)* $F = 1.1$, time 32, the breaking crest is 0.987 high at $x = 4.5$, the maximum slope is 113°.
Figure 10. Upstream breaker for $r = 0.4$. *a)* $F = 1.3$, time time 53.8, the breaking crest is 1.054 high, at $x = 1.7$, the maximum slope is 184°.

Can Shallow-Water Theory Describe Breaking?

E. O. TUCK

Applied Mathematics Department
University of Adelaide

Summary

The conventional answer is no, except in the sense of allowing approximate discussion of shock-like post-breaking events. But solutions by characteristics of the nonlinear shallow-water equations do give a rather impressive picture of an apparent plunger during the actual breaking itself, and one wonders if there is anything in it, nevertheless. Shallow-water theory should demand only that the fluid velocity vector be nearly horizontal, nothing else. In particular, it should not make any assumption that the free-surface slope is small. The fluid particles in a near-vertical wall of water could be moving nearly horizontally. However, even if this were so, unfortunately there is still a kinematic barrier to acceptance of shallow-water theory beyond the point where the free surface first beomes vertical, since most derivations assume that the free-surface height is a single-valued function of the horizontal co-ordinate. There are alternative derivations, e.g. Lagrangian, that do not demand such an assumption. An apparent singularity occurs when the free surface becomes vertical, but this singularity is not present for pure simple waves. The opportunity is also taken to discuss similar breaking-like problems for lubrication equations, which are the viscous-fluid equivalent of the shallow-water equations, with applications such as to dripping of freshly-painted vertical surfaces.

In water of constant small depth h, the two-dimensional nonlinear shallow-water equations are

$$\begin{aligned} u_t + uu_x + g\eta_x &= 0 \\ \eta_t + (u\eta)_x &= 0. \end{aligned} \tag{1}$$

Here $u(x,t)$ is the horizontal velocity and $y = \eta(x,t)$ is the equation of the free surface. These equations can be solved by characteristics, see [1]. In particular, there exist "simple waves" such that $u - 2\sqrt{g\eta}$ is constant, and these waves satisfy the uni-directional kinematic wave equation

$$\eta_t + c\eta_x = 0, \tag{2}$$

with wave speed

$$c(\eta) = 3\sqrt{g\eta} - 2\sqrt{gh}. \tag{3}$$

The solution of (2) for any initial disturbance $\eta(x,0)$ is given implicitly by

$$\eta = \eta(x - c(\eta)t, 0), \tag{4}$$

or explicitly as a formula for x as a function of η and t, namely

$$x = c(\eta)t + x_0(\eta), \tag{5}$$

M. L. Banner, R. H. J. Grimshaw (Eds.)
Breaking Waves
IUTAM Symposium Sydney/Australia 1991

where $x_0(\eta)$ is the function inverse to $\eta_0(x,0)$. For example, if we start with the simple mound

$$\eta(x,0) = h + \frac{A}{1+x^2/L^2}, \tag{6}$$

so

$$x_0(\eta) = \pm L\sqrt{\frac{A}{\eta - h} - 1}, \tag{7}$$

the resulting explicit solution (5) is plotted in Figure 1 at $A/h = 1$ for various times t (scaled by $L/\sqrt{gh}$), showing plunger-type breaking behaviour.

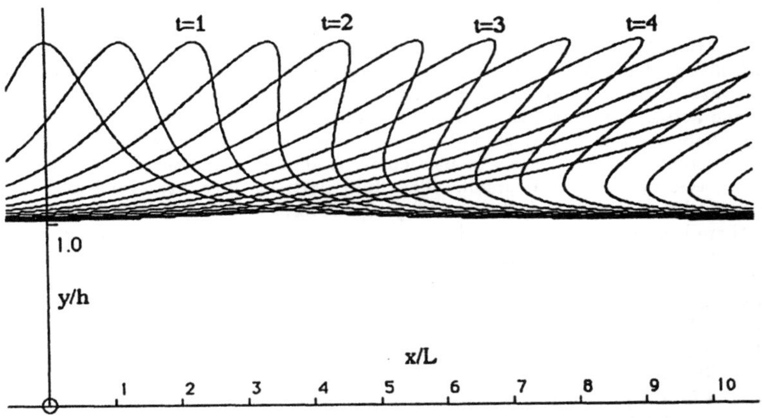

Fig 1. Breaking wave computed by shallow-water theory.

The above illustrates the simplicity of shallow-water theory; a whole discussion, from statement of problem and equations, to plotting of results, fits into one long but straightforward paragraph. However, all is not this satisfactory. It is conventional wisdom that shallow-water theory is not valid for breaking waves, so one must discard these results, however plausible they look.

I do not wish here to make a case for shallow-water theory, but simply to highlight some of the issues. It may be that the conventional wisdom is correct. Some things that look like ducks and quack like ducks and fly like ducks are not ducks.

There is one sense in which the above shallow-water results are widely accepted, and that is as a tool for approximate study of the bore that propagates after breaking. This can be interpreted as accepting validity of portions only of Figure 1. That is, at post-breaking times, when the computed free surface $y = \eta$ is a triple-valued function of x, one discards the "intermediate" free surface, and accepts the "upper" free surface for $x < x_F(t)$ and the "lower" free surface for $x > x_F(t)$. Here $x = x_F(t)$ is the location of the bore or shock front, which is assumed to be a moving step-function discontinuity, determined by conservation of mass. The (false?) nose portion of the computed broken wave lying ahead of the vertical line at $x = x_F(t)$ is discarded, and $x_F(t)$ is chosen so that the area of this discarded portion is identical to the area that is then "filled in" behind the front, see [2], p. 170.

The above interpretation is familiar in other contexts, such as gasdynamic shock waves, and is not controversial. Essentially, the shallow-water approximation holds separately for $x < x_F(t)$ and for $x > x_F(t)$. It breaks down locally near the shock front, and if one needs to know details of the flow in that neighbourhood, one must work a lot harder, and in particular may have to abandon the shallow-water assumption.

As an aside, note that in this and other similar areas, there are possibilities for studying shock-like behaviour by a shallow-water approximation that remains valid even in the neighbourhood of the shock itself. This is justified if the horizontal length scale near the shock is small compared to that elsewhere, but still large compared to h. One way to discuss this phenomenon is via inclusion of surface tension, which plays a role only where curvatures are high.

For example, for the problem of streaming of a thin layer of viscous fluid down a vertical wall, where lubrication theory is entirely analogous to shallow-water theory, a complete asymptotic theory has recently been developed [3]. In this theory, a kinematic wave equation like (2) is solved (without surface tension) everywhere except close to the shock front itself, generating solutions that break. These solutions, truncated as described above to a single-valued form, are then matched near the front to a local flow that does involve surface tension.

The equivalent of avoidance of breaking in this class of viscous flows is avoidance of detaching "drips", a matter of importance in painting and coating technology. On the other hand, there are also generalisations involving "stripped" coatings [4] where the whole purpose of the exercise is to cause the viscous layer to break, i.e. to force excess fluid to drip off the coating. Again, lubrication theory seems able to describe this phenomenon to adequate industrial accuracy, in spite of the fact that it makes approximations equivalent to shallow-water theory.

Returning now to classical inviscid water waves, the question remains of whether there is any validity to the original continuous but multi-valued Figure 1, as distinct from its single-valued but discontinuous shock-like corruption. There are many reasons why one is inclined to doubt the validity of Figure 1, but some of these are illusory, and my aim is to put these to rest. Other serious doubts remain, though.

The first concern with Figure 1, as derived in the first paragraph of this paper, is precisely about its triple-valuedness. Whatever basis one has for believing or not believing the shallow-water equations (1), they themselves cannot justify Figure 1, since they assume that $\eta = \eta(x,t)$ is a well-defined (single-valued) function of x. In some ways, it is a mystery how we were able to obtain a "solution" of these equations which is triple-valued. Note that, whenever there are three values of η for a given x, there are also three values of u at the same x, and this is hard to reconcile with an interpretation of $u(x,t)$ as a velocity field.

However, there are other forms of the shallow-water equations that do not suffer this kinematic paradox. For example, one could formulate the problem in a Lagrangian manner, determining the trajectories of individual particles. A feature of the solution in such a framework is that the independent variable is not a horizontal co-ordinate, but rather a label which is like an arclength on the curves of Figure 1. Hence the solution can be determined in terms of functions that never become multi-valued, even after breaking.

Of course this still begs the question, if one begins with the Eulerian shallow-water equations (1), since these do demand single-valuedness with respect to x. Merely converting them to a Lagrangian form that does not demand such single-valuedness, is not sufficient, if multi-valuedness was ruled out in the first place by the "governing" equations.

So one comes back inevitably to the justification of shallow-water theory, i.e. to its formal asymptotic derivation as an approximation to the full two-dimensional equations of water wave theory. On what basis?

Shallow-water theory is given many justifications, ranging from physical ("assume hydrostatic pressure") through kinematic ("neglect vertical acceleration") through naive ("depth averaged equations") to formal geometrical ("letting $h/L \to 0$ in the exact equations"), even to denial of its true name in favour of too-specialised terminology like "long-wave theory", or denial of the need for any justification at all, as in intrinsically one-dimensional ("Cosserat") flow theories. The essence is however that the vertical velocity component v is much smaller than u, i.e. that the velocity vector is nearly horizontal. This usually occurs because the depth h is small compared to the horizontal length scale L, so a formal development in the small parameter h/L is indeed appropriate. Such a formal view also warns us that there may be trouble near breaking if the effective horizontal length scale shortens dramatically.

One alternative both to a conventional Eulerian derivation and a full Lagrangian derivation is a simple reversion of dependent and independent variables. That is, suppose the full equations of motion, for $u = u(x, y, t)$ and $v = v(x, y, t)$ are re-written as equations for $x = x(u, v, t)$ and $y = y(u, v, t)$. Then (since $v = 0$ on the bottom and $v << u$ everywhere in the flow), the shallow-water approximation is nothing more than a Taylor expansion for small v!

The derivation goes like this, again just about one paragraph's worth. If we define $X(u,t) = x(u, 0, t)$, then these Taylor series must begin $x = X(u,t) + O(v^2), y = -vX_u(u,t) + O(v^3)$. (The second of these equations expresses conservation of mass via the Cauchy-Riemann equation $x_u = -y_v$.) Now if the free surface transforms to $Y(u,t) = \eta(X(u,t),t)$, we have $Y_t = \eta_t + \eta_x X_t$ and $Y_u = \eta_x X_u$. The exact kinematic free-surface condition states that $v = \eta_t + u\eta_x$ on $y = \eta$. To leading order in small v (where the value of v on the free surface is given by $Y = -vX_u$) this reduces to

$$Y_t X_u = -Y + (X_t - u)Y_u \tag{8}$$

In order to revert the dynamic free-surface condition, we need temporarily to use the velocity potential $\phi(x, y, t)$, writing $\Phi(u, t) = \phi(X(u, t), 0, t)$. Then $\Phi_t = \phi_t + uX_t$ and $\Phi_u = uX_u$. Thus the exact dynamic free-surface condition $\phi_t + (u^2 + v^2)/2 + g\eta = 0$ has as its leading-order approximation for small v, $\Phi_t - uX_t + u^2/2 + gY = 0$, or upon u-differentiation to eliminate Φ,

$$X_t = gY_u + u \tag{9}$$

It is convenient to substitute (9) in (8) and (for the first time in this derivation) divide by X_u, giving

$$Y_t = \frac{gY_u^2 - Y}{X_u}. \tag{10}$$

Equations (9) and (10) are written in the form of time-evolution equations for (X, Y), and fully describe nonlinear shallow-water theory. These equations are as simple as the original shallow-water equations (1); indeed they are exactly equivalent to (1), as is seen by identifying Y with η, and noting that $u_t = -X_t/X_u$, $u_x = 1/X_u$, $\eta_t = Y_t - X_tY_u/X_u$ and $\eta_x = Y_u/X_u$. Although (9,10) and (1) are equivalent, (9,10) has the advantage of being derived in a way that has never assumed single-valuedness of flow quantities like u and η as functions of the horizontal co-ordinate x. Rather the reverse, it has assumed that x is a single-valued function of flow quantities. In effect, it has anticipated the form of the forward face of the curves in Figure 1, which have x as a single-valued function of η.

Equations (9) and (10) also have the advantage that the simple-wave solution (5) is immediate. That is, suppose that $Y = Y(u)$ is independent of time (at fixed u). Then (10) requires that $gY_u^2 = Y$, an ODE which solves to show as before that $u - 2\sqrt{gY}$ is constant. The right-hand side of (9) is then just the kinematic-wave speed c given by (3), and that equation then integrates with time immediately to give the explicit solution (5).

The catch is that there is a formal singularity in (10) when $X_u = 0$, and this happens when the free surface is locally vertical. Hence we have jumped out of the frying pan into the fire! A derivation that formally allows multiple-valuedness, nevertheless yields an equation that fails precisely when such multi-valuedness first appears. Notice that this failure is only at the last hurdle, when we obtained (10) from (8) by dividing by X_u, and in one sense the system (8),(9) is preferable.

However, this singularity might not be present for simple waves. That is, the singularity in (10) is caused by a denominator function X_u vanishing. If the numerator also vanishes, this singularity may or may not be significant. But the numerator is $gY_u^2 - Y$, and this is identically zero for simple waves. Obviously there are many interesting issues to be resolved, associated with this 0/0 situation.

References

1. Stoker, J.J. Water Waves. New York: Interscience 1957.

2. Lighthill, M.J. Waves in Fluids. Cambridge University Press 1978.

3. Moriarty J.A., Schwartz, L.W. and Tuck, E.O. Unsteady spreading of liquid films with small surface tension. Physics of Fluids A, 3 (1991) 733-742.

4. Tuck, E.O. Continuous coating with gravity and jet stripping. Physics of Fluids 26 (1983) 2352-2358.

Breaking Criteria of a Solitary Wave Passing Over a Submerged Obstacle

T. YASUDA and Y. SAKAKIBARA
Department of Civil Engineering Gifu University, Yanagido 1-1 Gifu 501-11, Japan

M. HARA
Idemitsu Engineering Co., Ltd., Shinden-Cho 37-24 Chiba 260, Japan

Summary

Numerical simulations using a computational model [1] are made for fully nonlinear two-dimensional irrotational free-surface flows over beds containing arbitrary submerged obstacles, to clarify the criteria of breaking of solitary waves caused by depth changes and formulate them.
Comprehensive discussion based on the simulated results is made on the breaking criteria, so that they are found to be governed by the parameters which commonly mean the rate of the depth change to the wave-height and formulated as functions of the parameters by applying the regression analysis to the simulated results. The emphasis in the following article will be on the presentation of the results.

1 Introduction

A sound knowledge on the breaking criteria of coastal waves propagating over a submerged obstacle is clearly important for planning and designing submerged coastal structures. However, the knowledge is insufficient, although the interaction between a solitary wave and a submerged semi-circular cylinder and its resultant breaking are investigated by Cooker et al. [2].
It has often been remarked that waves on beaches resemble solitary waves. In fact, steep coastal waves are demonstrated to be representable as a random train of solitons [3]. Hence, it may be better to consider each individual wave as a solitary wave and investigate its breaking criteria, rather than to examine directly that of steep coastal waves, as well as Cooker et al.
A computational model [1] to be used here does not utilize a conformal mapping. so that can work in the physical space containing a submerged obstacle with an arbitrary form,

M. L. Banner, R. H. J. Grimshaw (Eds.)
Breaking Waves
IUTAM Symposium Sydney/Australia 1991

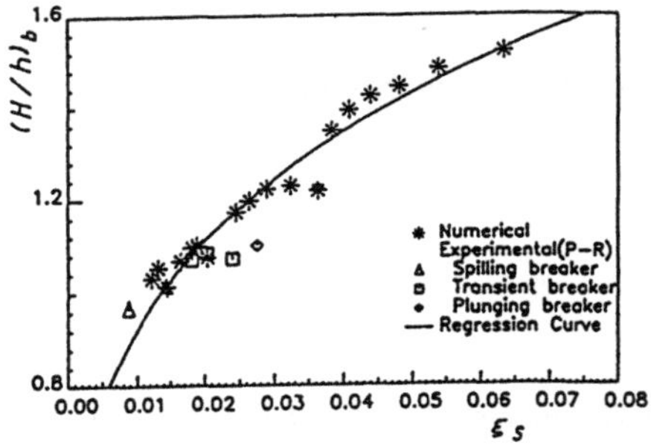

Fig.1 Variation of the breaker height $(H/h)_b$ of a solitary wave over a sloping bottom with the parameter ξ_s .

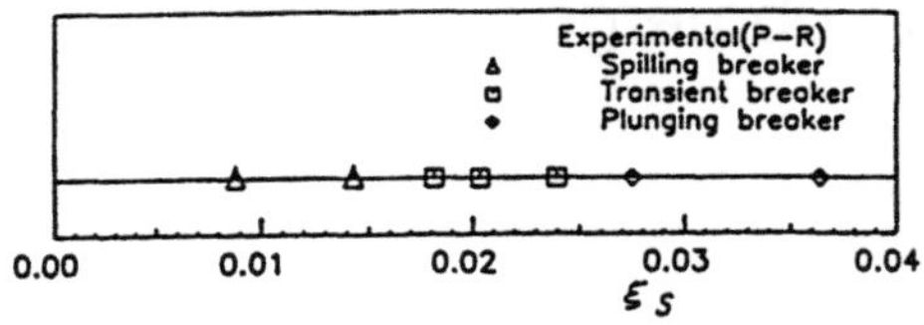

Fig.2 Relation between the breaker type classified by Papanicolaou-Raichlen and parameter ξ_s .

while adopting many of the developments in the formulation of Dold and Peregrine [4]. The validity of the model, that can not only describe accurately the deformation up to breaking but also predict almost exactly the breaking point, has been already verified by comparing the computed results with the experimental measurements of breaking solitary waves [1,5].

2 Breaking of a Solitary Wave over a Sloping Bottom

Many studies have been made on the breaking of a solitary wave over a sloping bottom, but they are yet insufficient to clarify the governing parameter and formulating the breaking criteria as a function of the parameter. Here, a parameter $\xi_s = \tan\theta/(H_1/h_1)^{0.4}$ is defined for a solitary wave having the initial wave-height H_1 on a plane bottom with the still water depth h_1 and travelling over a sloping bottom with a gradient of $\tan\theta$. This parameter is derived on the basis of the rate of the perturbed term by depth change to the nonlinear term of the variable coefficient KdV equation.
Figure 1 shows the relation between the nondimensionalized breaker height $(H/h)_b$ at the onset of breaking of a solitary wave over a sloping bottom and the parameter ξ_s^*. The onset of breaking in the present model is defined as the instant when the front face at the top of the solitary wave becomes just vertical and is experimentally confirmed to agree almost with the time when the formation of a jet at the crest is initiated in the real fluid. The solid line in the figure indicates the regression curve computed by

$$(H/h)_b = 3.24\xi_s{}^{0.275} + 0.005. \tag{1}$$

Further, the asterisk $*$ denotes the numerical results and the experimental (P-R) indicated by the symbols (triangle $\triangle$, square $\square$ and rhombus $\diamond$) points at the experimental results obtained by Papanicolaou and Raichlen [5]. Both the breaker heights $(H/h)_b$ of the numerical and experimental results agree considerably well with the regression curve and are found to be governed by the parameter ξ_s alone.
Figure 2 shows the relation between the breaker type of the measured waves classified by Papanicolaou and Raichlen [5] and the parameter ξ_s. It is found from this figure that the breaker type is also governed by the parameter ξ_s as well as the breaker height. These

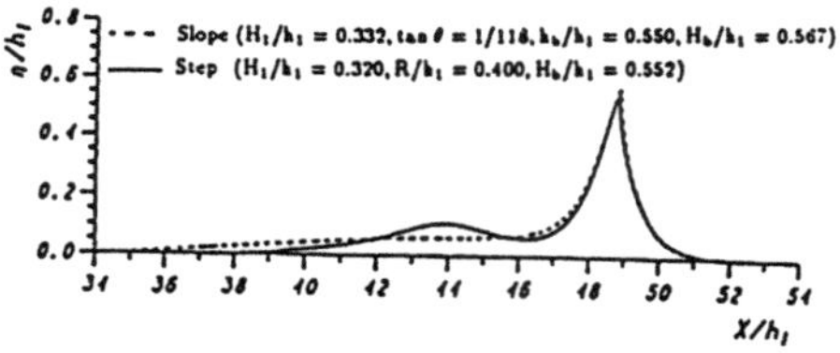

(a) Breaking wave

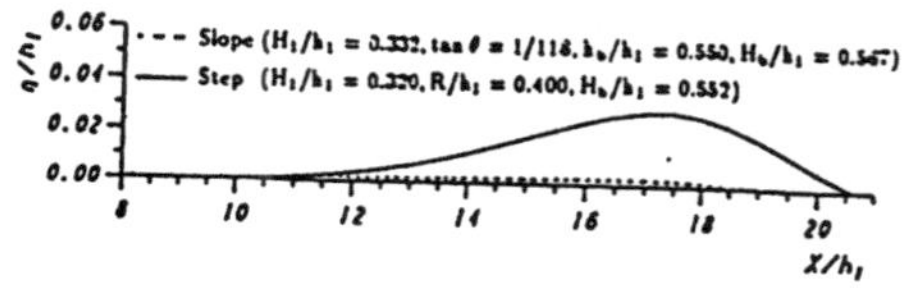

(b) Reflected wave

Fig.3 Comparisons of water surface profiles of breaking wave and reflected wave between a solitary wave over a sloping bottom and that over a rectangular step.

results state that the breaking of a solitary wave over a sloping bottom is governed by a single parameter ξ_s and can be formulated as a function of it.

3 Breaking of a Solitary Wave over a Step

In order to investigate the effect of depth change on the breaking due to a slope, some comparisons of deformation up to breaking are made between the solitary wave with gradient $H_1/h_1 = 0.332$ over a sloping bottom with $\tan\theta = 1/118$ and the one with $H_1/h_1 = 0.320$ over a rectangular step with height $R/h_1 = 0.40$. Figure 3 shows the comparisons with regard to the wave profiles of the breaking wave and the reflected wave. While the breaker depth h_b/h_1 on the sloping bottom is 0.55, that on the step is 0.60. And, the accumulation S of water depth change up to the breaking point is 11.90 over the sloping bottom and is 8.99 over the rectangular step. The accumulation S means the total depth change which an incident wave suffers during its propagation process up to breaking. The values of H_1/h_1, S and h_b/h_1 do not agree closely between both the waves, in spite of efforts to obtain close agreement. However, it is found from the figures that the difference in wave profile between both the waves is almost negligible with regard to a primary portion of the transmitted wave except for the shelf and reflected wave, in spite of the disagreement of H_1/h_1 and S. From this result, the breaking of a solitary wave is supposed to be not so dependent on the qualitative bottom topography, but on the quantitative one, that is, the aforementioned accumulation S. However, it should be noticed that the reflected wave and shelf are remarkably influenced by the qualitative bottom topography.

Furthermore, the effect of a step face slope on the breaking is investigated. Figure 4 shows the comparison with the wave profile at the breaking point over a rectangular step and steps with a face slope of gradient $\tan\theta = 1/5$ and $1/10$. The breaker profile coincides well with each other except on the shelf. Hence, it could be said that the face slope at front of the step is unimportant for breaking.

Figure 5 shows the relation between the breaker height H_b/h_1 of a solitary wave over a rectangular step or a step with an inclined face slope and the similar parameter $\xi_s^* = (R/h_1)^{0.1}/(H_1/h_1)^{0.4}$ with ξ_s . ξ_s^* means the rate of the depth change due to the step with the height of R/h_1 to the incident wave height H_1/h_1 . It is found that all the breaker heights agree well with the regression curve given by following equation, independent of the gradient of the face slope ($\tan\theta = 1/10 \sim \infty$).

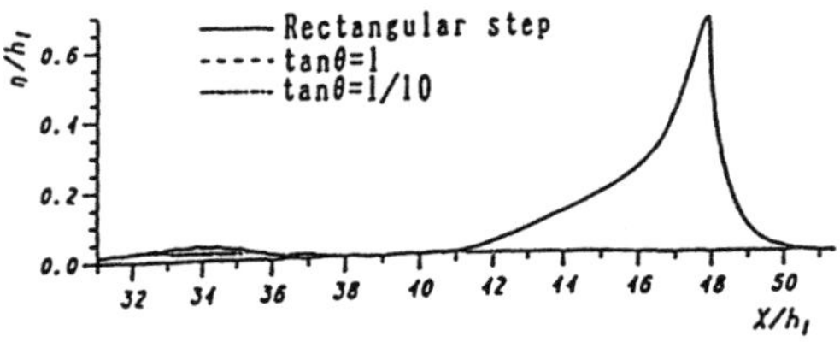

Fig.4 Comparisons of breaking wave profiles over a rectangular step and steps with face slopes of gradient $\tan\theta = 1/5$ and $1/10$.

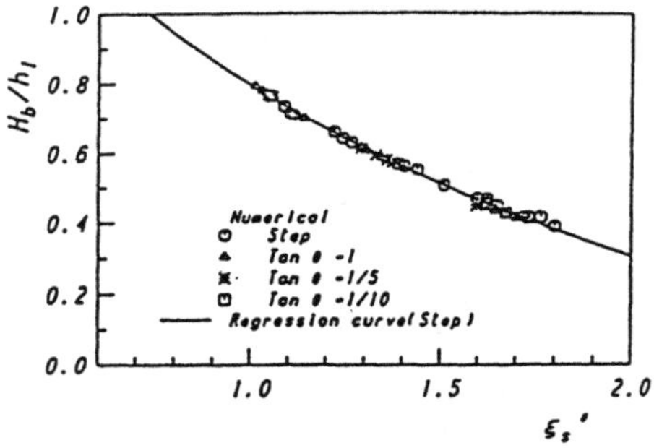

Fig.5 Relation of the breaker height of a solitary wave over a submerged step having a vertical face slope or inclined one of $\tan\theta = 1, 1/5$ and $1/10$ to the parameter ξ_s^* .

$$H_b/h_1 = 5.885 - 5.090\xi_s^{*0.133}, \tag{2}$$

4 Breaking of a Solitary Wave Passing over a Submerged Dike

Intensive investigations are made on the breaking of a solitary wave caused by a submerged rectangular or trapezoidal dike.

Figure 6 shows examples of the computed shape of the free surface at some evolution times up to breaking. While the breaker shape shown in Fig.6 (a) seems to be a spilling breaker, the wave profile shown in Fig.6 (b) could be regarded clearly as a plunging breaker because its horizontal asymmetry is large. Although both of them break forwards, the surface of the wave on the shelf shown in Fig.6 (c) breaks backward onto the dike. This breaking found firstly by Cooker et al. [2] is characterized by backward breaking occurring before any other breaking event and is called a backward breaker. Further, it is noticed from Fig.6 (d) and (e) that the complicated deformations up to breaking occur in the case of a solitary wave passing over a trapezoidal dike. Especially the breaker profile shown in Fig.6 (e) is remarkably different from usual breaker types and requires more intensive investigations including experimental work.

Figure 7 indicates the regions of forward breakers (o, circle), backward breakers (///, hatching) and nonbreaking (×, cross) for solitary waves incident on rectangular dikes having arbitary heights R/h_1 and the widths of $B/h_1 = 1.0$, 3.4 and 5.0 . From these figures, the conditions where forward or backward breaker occur are easily found with regard to $H_1/h_1, R/h_1$ and B/h_1 .

Figure 8 shows the limiting initial wave height H_c/h_1 of a solitary wave incident on rectangular and trapezoidal dikes. The limiting height H_c/h_1 means that the solitary wave with initial wave height H_1/h_1 exceeds H_c/h_1 can not be transmitted without breaking forwards. The solid line in the figure denotes a regression curve computed by

$$H_c/h_1 = 0.952 - 0.591\gamma^{0.76}, \quad \gamma = [(B/h_1) + (R/h_1)/2\tan\theta]^{0.4}(R/h_1), \tag{3}$$

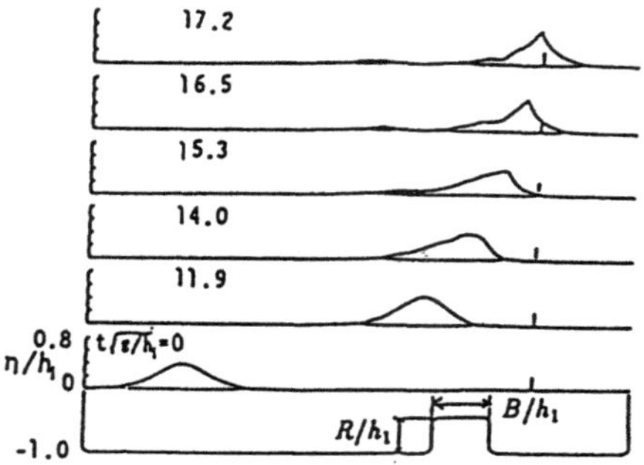

(a) $H_1/h_1 = 0.40, R/h_1 = 0.6,$
$B/h_1 = 3.4$

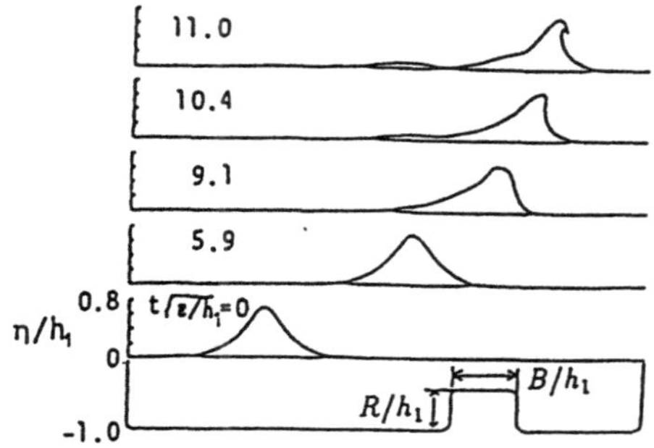

(b) $H_1/h_1 = 0.70, R/h_1 = 0.6,$
$B/h_1 = 3.4$

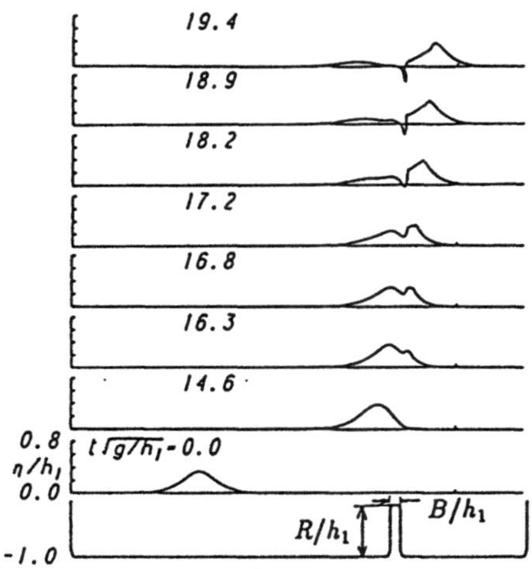

(c) $H_1/h_1 = 0.34, R/h_1 = 0.8,$
$B/h_1 = 1.0$

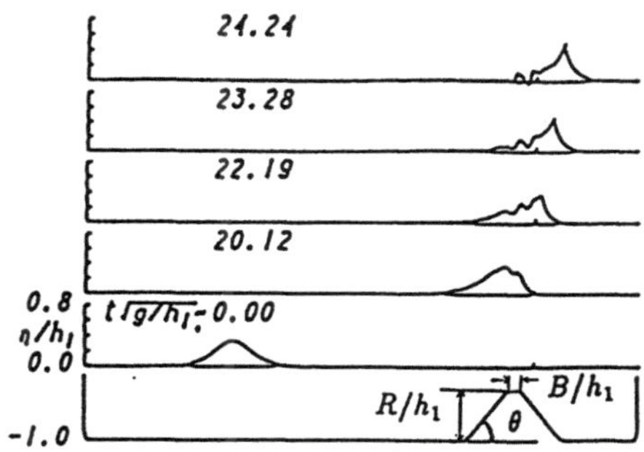

(d) $H_1/h_1 = 0.34, R/h_1 = 0.7,$
$B/h_1 = 1.0, \tan\theta = 1/5$

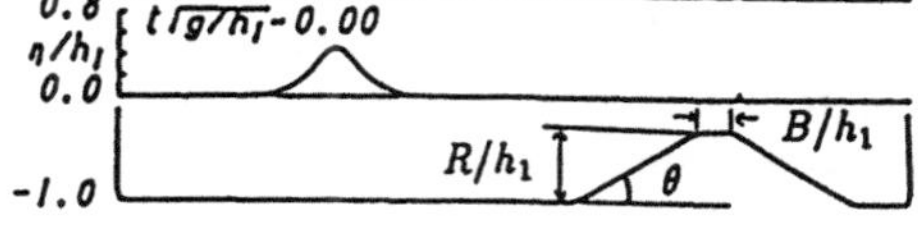

(e) $H_1/h_1 = 0.46, R/h_1 = 0.7,$
$B/h_1 = 2.0, \tan\theta = 1/10$

Fig.6 Temporal changes of water surface profile of a solitary wave passing over a submerged dike.

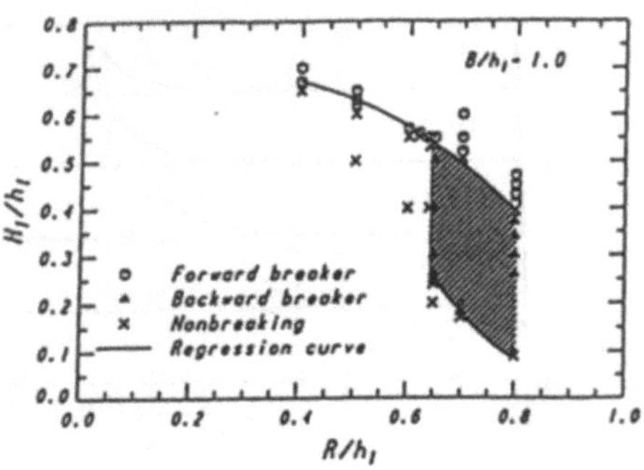

(a) $B/h_1 = 1$

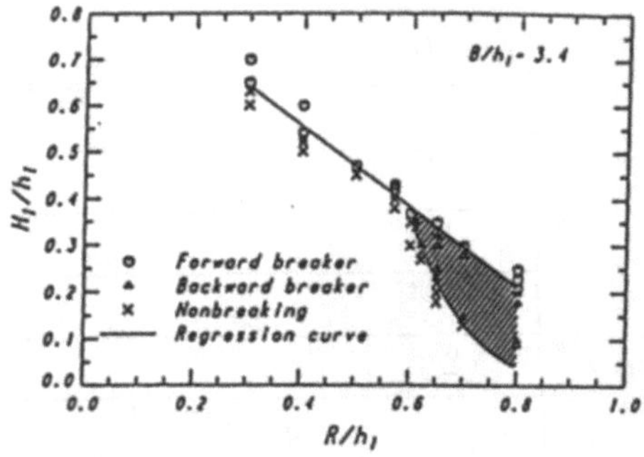

(b) $B/h_1 = 3.4$

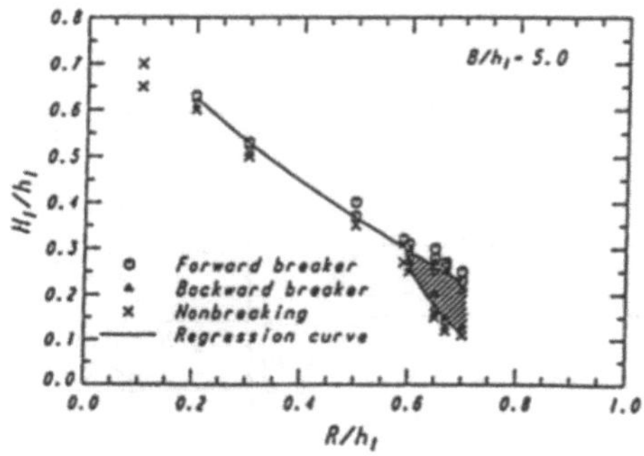

(c) $B/h_1 = 5.0$

Fig.7 Relation of the presence of breaking caused by a submerged rectangular dike to the wave-height H_1/h_1 of an incident solitary wave and the relative step-height R/h_1 .

where the symbols ($\triangle, *, \square$) are the numerical results for trapezoidal dikes and the circles are those for rectangular dikes. It is found that the limiting heights H_c/h_1 for the submerged dikes are almost governed by a single parameter γ .
Figure 9 shows the relation between the breaker heights H_b/h_1 of solitary wave passing over a submerged rectangular or trapezoidal dike and the parameter $\xi_s'' = [B/h_1 + (R/h_1)/3.5\tan\theta]^{0.2}(R/h_1)\big/(H_1/h_1)^{0.4}$. It is found that all the breaker heights almost agree with the regression curve given by

$$H_b/h_1 = -0.463\xi_s''^{0.721} + 1.039. \tag{4}$$

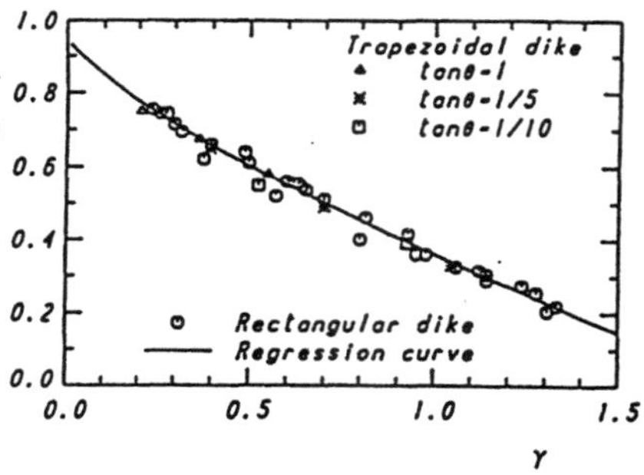

Fig.8 Relation of the limiting wave height H_c/h_1 of a solitary wave incident to a submerged rectangular and trapezoidal dike to a parameter γ .

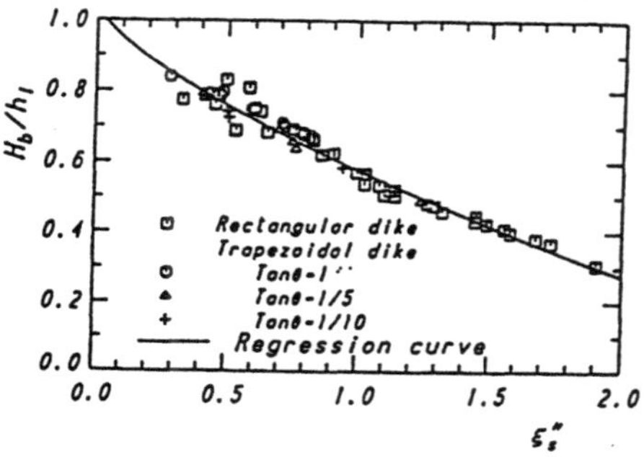

Fig.9 Relation of the breaker height of a solitary wave passing over a rectangular submerged dike or trapezoidal one with face slopes of $\tan\theta = 1, 1/5$ and $1/10$ to the parameter ξ_s'' .

Nevertheless the deformations of wave profile are remarkably different between the wave over a rectangular dike and that over a trapezoidal one. The authors wish to express their thanks to Dr. Tanaka of Gifu Univ. for his valuable advice.

References

[1] Yasuda, T., M. Hara and M. Tanaka (1990), Numerical Methods in Fluid Dynamics II (Eds. M. Yasuhara, H. Daiguji & K. Oshima, JSCFD), pp.919–924.

[2] Cooker, M.J., D.H.Peregrine, C.Vidal and J.W.Dold (1990), J.Fluid Mech., pp.1–22.

[3] Tsuchiya,Y. and T.Yasuda(1984), Proc. 19Int. Conf. on Coastal Eng., pp.435–451.

[4] Dold,J.W. and D.H.Peregrine (1984), Proc. 19Int. Conf. on Coastal Eng., pp.955–967.

[5] Yasuda, T., and M. Hara (1990), Proc. 22Int. Conf. on Coastal Eng., pp.923–934.

[6] Papanicolaou, P. and F. Raichlen (1988), Sea Surface Sound (Ed. B. Kerman, Kluwer Academic Pub.), pp.97–109.

Wave-Breaking in other Contexts

Processes Leading to Filamentation of a Potential Vorticity Interface over a Topographic Slope

R. GRIMSHAW[1] **and Z. YI**[2]

[1] School of Mathematics, University of New South Wales
Kensington, N.S.W. 2033, Australia
[2] NRCMEF, No. 8 Da Hui Si, Hai Dian Division
Beijing, 100081 P.R. China

Summary

Using the barotropic, nondivergent model for flow over a topographic slope adjoined by a coastal barrier, a contour dynamics algorithm is developed to describe wave evolution on an interface separating two regions of uniform potential vorticity. We find that wave steepening due to advective nonlinearity is partially controlled by wave dispersion; the usual outcome for even small–amplitude waves is filamentation. A kinematic mechanism proposed by Pullin et al. [1] is used to identify criteria for when filamentation will occur. Of the various parameters in this system, the most important is found to be the ratio of the potential vorticity jump across the interface to the background potential vorticity.

Introduction and formulation

The tendency for wave–like perturbations to a vorticity interface to develop into filaments is now well–documented (e.g. Dritschel, [2]). Filamentation can be interpreted as a form of wave–breaking, which is quite different from that associated with water waves or interfacial waves. The essential difference is that at a vorticity interface, the velocity field is continuous but the vorticity is discontinuous, whereas for water waves or interfacial waves the tangential velocity is discontinuous. In a previous paper (Grimshaw and Yi, [3]) we used the methodology of contour dynamics to study the evolution of an interface separating two regions of constant potential vorticity. The context was that of barotropic, nondivergent flow over a topographic slope. The motivation was to study a nonlinear mechanism for the modeling of the current meanders and squirts often observed in coastal regions. Related work in the oceanographic context are the contour dynamics studies by Pratt and Stern [4] and Send [5], the main differences being that, respectively, we do not make the quasi-geostrophic approximation of a weak topographic slope, and we consider only linearly stable basic flows. In our previous paper we

M. L. Banner, R. H. J. Grimshaw (Eds.)
Breaking Waves
IUTAM Symposium Sydney/Australia 1991

adapted a suggestion of Pullin et al. [1], and proposed that filamentation occurs when a hyperbolic stagnation point (relative to the wave crest) almost coincides with the wave crest; that is, there is a critical layer near the wave crest. Here, we complement our previous study by examining this mechanism more closely.

First, however, it is necessary to recapitulate briefly the contour dynamics algorithm and our previous results; for further details see Grimshaw and Yi [3]. For barotropic, non-divergent flow the $x-$ and $y-$components of velocity, u and v respectively, are given in terms of the transport streamfunction ψ by

$$hu = \psi_y, \qquad hv = -\psi_x. \tag{1}$$

The equation for conservation of potential vorticity q is then

$$q_t + uq_x + vq_y = 0, \quad \text{where} \quad q = \frac{f+\zeta}{h}, \tag{2}$$

and
$$\zeta = v_x - u_y = -\left(\frac{\psi_x}{h}\right)_x - \frac{\psi_{yy}}{h}. \tag{3}$$

Here f is the constant Coriolis parameter, and $h(x)$ is the topographic slope, which we shall assume is given by $h = h_1 \exp\{s(x-\ell)\}$. The coastal and offshore boundary conditions are that

$$\psi = 0, \text{ at } x = 0, \quad \text{and} \quad \psi_y = 0, \text{ as } x \to \infty. \tag{4}$$

Next we suppose that $x = L(y,t)$ denotes the interface between two regions of constant potential vorticity, where $L > 0$ and $L \to \ell$ as $y \to -\infty$. Hence $q = Q_1$ for $0 < x < L$ and $q = Q_0$ for $x > L$. To allow for the development of folds and filaments, we assume that $L(y,t)$ can be a multi-valued function of y. The boundary conditions on the interface are the kinematic condition

$$L_t + vL_y = u, \text{ on } x = L \tag{5}$$

and the continuity of u, v. For the purposes of contour dynamics we can replace these conditions with the equivalent Lagrangian conditions for each point (L, y) on the interface,

$$\frac{dL}{dt} = u(L,y,t), \quad \frac{dy}{dt} = v(L,y,t). \tag{6}$$

Next, we introduce the basic flow, which consists of a longshore flow with streamfunction $\psi_0(x)$ and current $v_0(x)(hv_0 = -\psi_{0x})$, where

$$v_0(x) = -\Delta Q h_1 \{ \frac{H(\ell - x)}{s}[1 - \exp(-s(\ell - x))] + \gamma_1(x - \ell)\}, \tag{7}$$

and $\gamma_1 = f/\Delta Q h_1, \Delta Q = Q_1 - Q_0$. Here $x = \ell$ is the undisturbed position of the interface, and we have put $v_0(\ell) = 0$. Note that this basic flow has a discontinuity in potential vorticity of $-\Delta Q$ across $x = \ell$. Then we put $\psi = \psi_0(x) + \phi$, and deduce from (3) that

$$\left(\frac{\phi_x}{h}\right)_x + \frac{\phi_{yy}}{h} = \begin{cases} -\Delta Q h, & \text{for } \ell < x < L, \\ \Delta Q h, & \text{for } L < x < \ell, \\ 0, & \text{elsewhere.} \end{cases} \tag{8}$$

Also $\phi = 0$ at $x = 0$ and as $x^2 + y^2 \to \infty$. To solve (8) we introduce the Green's function $G(x, x', y - y')$ by

$$\left(\frac{G_x}{h}\right)_x + \frac{G_{yy}}{h} = \delta(x - x')\delta(y - y'), \tag{9}$$

$$G = 0 \text{ at } x = 0, \quad \text{and} \quad G \to 0 \text{ as } x^2 + y^2 \to \infty. \tag{10}$$

With h given by

$$G = \{h(x)h(x')\}^{\frac{1}{2}}\{H(R) - H(R_I)\}, \tag{11}$$

where $$H(R) = -\frac{1}{2\pi}K_0(\tfrac{1}{2}sR), \text{ and } R, R_I = \{(x \mp x')^2 + (y - y')^2\}^{\frac{1}{2}}. \tag{12}$$

Note that $H(R)$ has a logarithmic singularity as $R \to 0$. The solution of (8) is now given by

$$\phi = -\Delta Q \int_{-\infty}^{\infty} dy' \int_{\ell}^{L(y',t)} G(x', x; y' - y)h(x')dx' \tag{13}$$

Finally we obtain the velocity components u, v from (1) and (13). After various integration by parts we obtain for u

$$u = -\frac{\Delta Q}{h^{\frac{1}{2}}} \int_{-\infty}^{\infty} \{h(L')\}^{\frac{3}{2}} \{H(R') - H(R'_I)\} \, dL', \tag{14}$$

where $$R', R'_I = \{(x \mp L')^2 + (y - y')^2\}^{\frac{1}{2}}, \quad \text{and} \quad L' = L(y', t).$$

The corresponding expression for v is

$$\begin{aligned} v - v_0 = &-\frac{\Delta Q}{h^{\frac{1}{2}}} \int_{-\infty}^{\infty} \{h(L')\}^{\frac{3}{2}}\{H(R') + H(R'_I)\}dy' \\ &- \frac{\Delta Q h_1^{\frac{3}{2}}}{s h^{\frac{1}{2}}}\{\exp(-\tfrac{1}{2}s|x - \ell|) + \exp(-\tfrac{1}{2}s|x + \ell|)\} \\ &+ \frac{s\Delta Q}{h^{\frac{1}{2}}} \int_{-\infty}^{\infty} dy' \int_{\ell}^{L'} \{h(x')\}^{\frac{3}{2}}\{2H(R) + H(R_I)\}dx'. \end{aligned} \tag{15}$$

Combined with (6) we now see that (14) and (15) form an integro–differential equation for the evolution of the interface.

Contour dynamics algorithm and results

The essence of the contour dynamics method is to represent the interface by N points $(L_i(t), y_i(t))$ for $i = 1, \ldots, N$ in a discretized formulation

$$\frac{dL_i}{dt} = u(L_i, y_i), \quad \frac{dy_i}{dt} = v(L_i, y_i), \tag{16}$$

where u, v are given by discretized versions of (14), (15) respectively. Here we use a trapezoidal rule so that (14) is approximated by

$$\begin{aligned} u(L_i, y_i) = &-\frac{\Delta Q}{h_i^{\frac{1}{2}}} \sum_{j=1}^{N+1}{}' (L_j - L_{j-1})\overline{h_j^{\frac{3}{2}} H(R_{ij})} \\ &+ \frac{\Delta Q}{h_i^{\frac{1}{2}}} \sum_{j=1}^{N+1} (L_j - L_{j-1})\overline{h_j^{\frac{3}{2}} H(R_{Iij})} - \frac{\Delta Q}{h_i^{\frac{1}{2}}} \int_{L_{i-1}}^{L_{i+1}} \{h(L')\}^{\frac{3}{2}} H(r')dL', \end{aligned} \tag{17}$$

where $\qquad R_{ij}, R_{Iij} = \{(L_i \mp L_j)^2 + (y_i - y_j)^2\}^{\frac{1}{2}},$

and $\qquad h_i = h(L_i), \quad \overline{f}_j = \frac{1}{2}(f_j + f_{j-1}). \qquad (18)$

Here $\sum'$ denotes the sum over j with the terms $j = i, i+1$ omitted; for notational convenience we introduce the fixed end–points (L_0, y_0) and (L_{N+1}, y_{N+1}) where $L_0 = L_{N+1} = \ell$, and we assume that the contour is not displaced for $y \geq y_0$ and $y \leq y_{N+1}$. The last term in (17) is a singular integral, and is evaluated by first extracting the singular part (for details, see Grimshaw and Yi, [3]). The corresponding approximation for (15) is

$$\begin{aligned} v(L_i, y_i) - v_0(L_i) = &-\frac{\Delta Q}{h_i^{\frac{1}{2}}} \sum_{j=1}^{N+1}{}' (y_j - y_{j-1})\overline{h_j H(R_{ij})} \\ &- \frac{\Delta Q}{h_i^{\frac{1}{2}}} \sum_{j=1}^{N+1} (y_j - y_{j-1})\overline{h_j H(R_{Iij})} - \frac{\Delta Q}{h_i^{\frac{1}{2}}} \int_{y_{i-1}}^{y_{i+1}} \{h(L')\}^{\frac{3}{2}} H(R')dy' \\ &- \frac{\Delta Q h_1^{\frac{3}{2}}}{h_i^{\frac{1}{2}}} \left\{ \int_{-\infty}^{y_0} + \int_{y_{N+1}}^{\infty} [H(R_i) + H(R_{Ii})]dy' \right\} - \frac{\Delta Q h_1^{\frac{3}{2}}}{s h_i^{\frac{1}{2}}} \{\exp(-\tfrac{1}{2}s|L_i - \ell|) \\ &+ \exp(-\tfrac{1}{2}s|L_i + \ell|)\} + \frac{s\Delta Q}{h_i^{\frac{1}{2}}} \sum_{j=1}^{N} (y_j - y_{j-1})\{2\overline{K}_{ij} + \overline{K}_{Iij}\}, \end{aligned} \tag{19}$$

where $$K_{ij}, K_{Iij} = \int_{\ell}^{L_j} \{h(x')\}^{\frac{3}{2}} H(R'_{ij}, R'_{Iij}) dx',$$

$$R_i, R_{Ii} = \{(L_i \mp \ell)^2 + (y_i - y')^2\}^{\frac{1}{2}}, \text{ and } R'_{ij}, R'_{Iij} = \{(L_i \mp x')^2 + (y_i - y_j)^2\}^{\frac{1}{2}}. \quad (20)$$

The third term is a singular integral and is evaluated in an analogous way to the singular integral in (18). The fourth term, comprising two "tail" integrals, are evaluated by Simpson's rule over a range large enough to ensure that the Bessel function integrands have achieved exponential smallness. Finally the integral terms K_{ij}, K_{Iij} are also evaluated by Simpson's rule. Altogether, these discretizations convert (16) into $2N$ ordinary differential equations for the nodal points. They are stepped forward in time using a standard fourth-order Runge-Kutta method. As time progresses the nodal points tend to clump together in some places and to diverge in other places. Hence we also use a node insertion/deletion scheme to maintain adequate resolution (for details, see Grimshaw and Yi, [3]). In a typical calculation we begin with about 100 nodal points and allow this to increase to a maximum of 200 to 300 nodal points.

In the numerical calculations we choose the length and time scales so that $\ell = 1$ and $\Delta Q h_1 = -1$, where we note that a change of sign of $\Delta Q h_1$ is equivalent to a change of sign of y. The available parameters are then $\gamma_1 = f/\Delta Q h_1$ which is an inverse measure of the strength of the potential vorticity front, and the topographic slope s, together with parameters associated with the initial condition. In Grimshaw and Yi [3] we considered as the initial condition small-amplitude isolated disturbances of the form

$$L(y, 0) = 1 + a_0 \operatorname{sech}^2 \beta y. \quad (21)$$

A typical result is shown in Figure 1 where $\gamma_1 = 2, s = 1, a_0 = 0.1$ and $\beta = 5$. We see that the wave steepens in the forward direction and evolves into a thin filament. In Grimshaw and Yi [3] the tendency for filamentation from the initial condition (21) was examined as a function of the available parameters. Summarizing it was found that for $a_0 > 0$, filamentation was enhanced as either γ_1 increases, s increases or β increases when $a_0 > 0$. For $a_0 < 0$ there are analogous results due to the approximate resemblance between the cases $a_0 < 0$ and $\gamma_1 > 0 (< 0)$ with $a_0 > 0$ and $\gamma_1 < 0 (> 0)$.

In Grimshaw and Yi [3] we proposed that a possible explanation of these tendencies could be provided by a kinematic mechanism put forward by Pullin et al. [1].

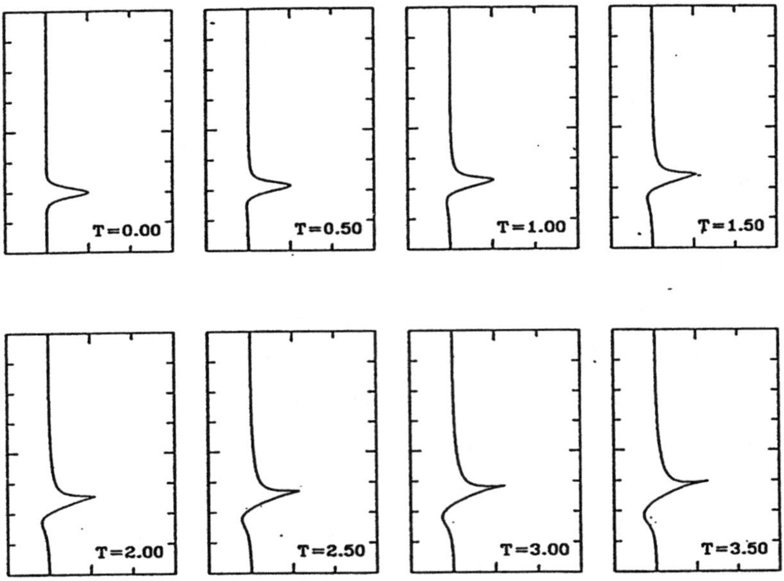

Figure 1: A typical result for the initial condition (21). The x–axis is horizontal over the range (0.9, 1.3) and the y–axis is vertical over the range $(-2, 6)$.

They suggested that filamentation will occur when a hyperbolic stagnation point (in a suitable frame of reference) almost coincides with an interface extremum. To show this concept here we display in Figure 2 the velocity field at $t = 0$ for the case shown in Figure 1, viewed from a frame of reference in which the crest is at rest. We see that the resulting stagnation point is hyperbolic, and note that the crest speed is 0.29. Further Pullin et al. [1] conjecture that criterion for filamentation can be approximately determined by considering waves on the interface of phase speed c and identifying critical layer sites where $v_0(x_c) = c$ with interface extremum. In Pullin et al. [1] the waves considered were modulationally unstable disturbances on a basic periodic wave. Here, since we are dealing with small– amplitude initial disturbances, it is sufficient to consider the critical layer determined by the basic wave itself. Thus we consider infinitesimal waves with wavenumber k propagating on the otherwise undisturbed interface. These have phase speed c given by

$$c = -\frac{\Delta Q h_1}{2(k^2 + \frac{1}{4}s^2)^{\frac{1}{2}}}\{1 - \exp[-2\ell(k^2 + \frac{1}{4}s^2)^{\frac{1}{2}}]\}. \tag{22}$$

Our hypothesis, then, is that a wave will filament when its initial amplitude a_0 is such that $a_0 > x_c - \ell > 0$ (or $a_0 < x_c - \ell < 0$) where x_c is the critical layer position given by $v_0(x_c) = c$. However, the expression c is for a periodic wave, while our initial state (21) is isolated. Hence there is no definitive mechanism for determining the wavenumber k. Nevertheless we choose $k = 3$ as a representative value when $\beta = 5$, and find that (with $\Delta Q h_1 = -1, s = 1$ and $\ell = 1$) $c = 0.164$ so that $x_c - \ell = 0.033, 0.097, -0.024$ and -0.033 for $\gamma_1 = 5, 1.7, 0.2$ and -5 respectively. These values of γ_1 correspond to cases considered by Grimshaw and Yi [3]. We conclude that with $a_0 = 0.1$, we would expect filamentation to occur when $\gamma_1 > 1.7$ and to be inhibited when $\gamma_1 < 1.7$. This is indeed what was found. Further c increases as either γ_1 increases, s increases, or k decreases, and hence $x_c - \ell$ also increases. With $a_0 > 0$ the critical layer hypothesis implies that these same trends should enhance filamentation and this is indeed what we found.

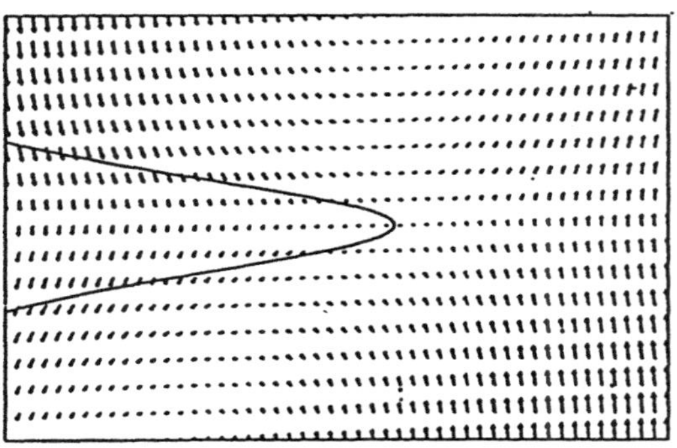

Figure 2: A close–up view of the velocity field near the crest for the case of Figure 1 at $t = 0$, for a frame of reference where the crest is at rest.

To obtain a more definitive estimation of k we consider two further initial conditions,

$$L(y,0) = 1 + \frac{1}{2} 3^{\frac{3}{2}} a_0 \operatorname{sech}^2 \beta y \tanh \beta y, \tag{23}$$

or

$$L(y,0) = 1 + a_0 \cos ky \exp(-\beta^2 y^2). \tag{24}$$

The first of these (23) is an antisymmetric critical disturbance, and a typical result is shown in Figure 3 for $\gamma_1 = 5, s = 1$, $a_0 = 0.1$ and $\beta = 5$. The positive displacement, whose initial maximum is 0.1, evolves into a filament but not so the negative displacement. This is consistent with the critical layer hypothesis since the critical layer is such that $x_c - \ell > 0$. Here k can be determined from the distance between the two initial extremum which gives $k \approx 3.85$, and then $x_c - \ell = 0.026$. Next, for the initial condition (24) which is a wave packet, a typical result is shown in Figure 4 for $\gamma_1 = 5, s = 1, a_0 = 0.1, k = 3$ and $\beta = 0.5$. For these parameters the main peak has an amplitude of 0.1, and the second positive peak an amplitude of 0.033. With $k = 3$, $x_c - \ell = 0.033$ and indeed both these positive peaks produce filaments, whereas there are no filaments produced from the negative peaks.

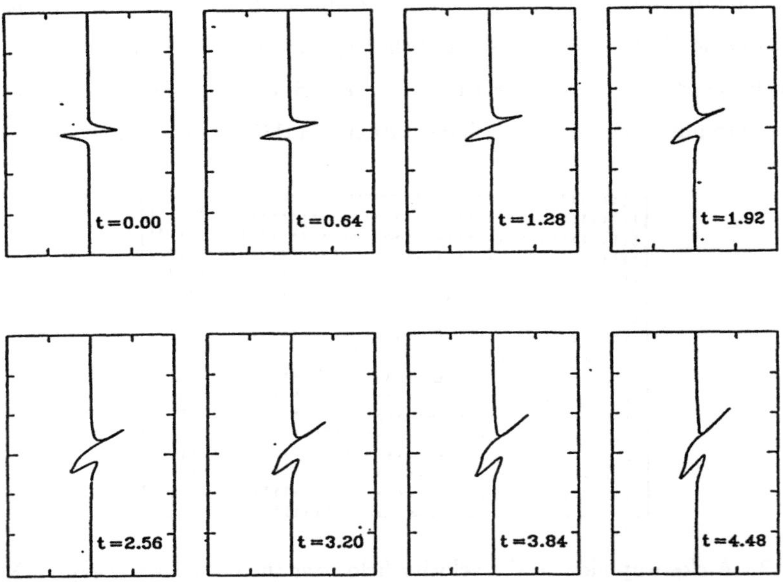

Figure 3: A typical result for the initial condition (23). The x-axis is now $(0.7, 1.3)$ and the y-axis is $(-6, 6)$.

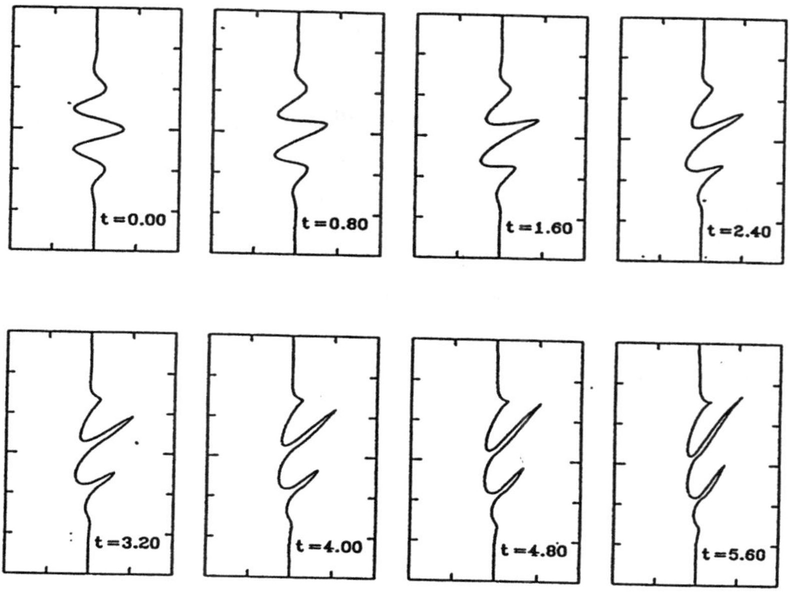

Figure 4: A typical result for the initial condition (24). The x–axis is now $(0.7, 1.3)$ and the y–axis $(-6, 6)$.

To conclude, the critical layer hypothesis seems to provide a robust criterion for the onset of filamentation. Here we have tested the hypothesis against the initial wave amplitude, which provides a criterion for immediate filamentation. However, if this criterion is not met, the filamentation may still eventually occur if there is a mechanism for wave growth so that the wave extremum can reach the critical level. Possible mechanisms are modulational instability of the basic wave (Pullin et al. [1]), or linear instability of the basic flow (Send, [5]). But note that here our basic flow (7) is linearly stable.

References

[1] Pullin, D.I., Jacobs, P.A., Grimshaw, R.H.J. and Saffman, P.G., "Instability and filamentation of finite–amplitude waves on vortex layers of finite thickness", *J. Fluid Mech.*, **209**, (1989) 359–384.

[2] Dritschel, D.G., "Contour dynamics and contour surgery: numerical algorithms for extended, high–resolution modeling of vortex dynamics in two–dimensional, inviscid, incompressible flows," *Comput. Phys. Rep.*, **10**, (1989) 77–146.

[3] Grimshaw, R. and Yi, Z., "Evolution of a potential vorticity front over a topographic slope", *J. Phys. Ocean.*, **21**, (1991) 1240–1255.

[4] Pratt, L.J. and Stern, M.E., "Dynamics of potential vorticity fronts and eddy detachment", *J. Phys. Ocean.*, **16**, (1986) 1101–1120.

[5] Send, U., "Vorticity and instability during flow reversals on the continental shelf", *J. Phys. Ocean.*, **19**, (1989) 1620–1633.

Observations of Shock and Undular Bore Formation in Internal at a Shelf Break

P.E. Holloway

Department of Geography and Oceanography
University College, University of New South Wales
Australian Defence Force Academy
Canberra, ACT, Australia, 2600

Summary
Large amplitude internal waves at semi-diurnal tidal frequency (internal tides) are observed to be generated over the slope region of the Australian North West Shelf and propagate shoreward over shoaling bathymetry. In deep water the waves are nearly linear in shape and are seen to steepen and become strongly non-linear as they propagate. The waves steepen at the leading edge to the point where an internal hydraulic jump or shock forms. The shocks are generally downwards and are weak shocks seen as undular bores. At times a second, and weaker, backward shock forms on the trailing edge of the wave. The undular bores are followed by a train of oscillations that can be reasonably described as solitons. An analytical model using a perturbed extended Korteweg-de Vries equation is used to describe the evolution of an initially sinusoidal wave as it progagates over an exponential depth profile, in the presence of a background current of constant shear, and breaks to form a shock. The model predictions, of the distance the wave propagates before forming a shock, and of the height of the shock as it propagates, compare well with observations.

Introduction

Barotropic tidal currents oscillating stratified water over the steeply sloping bathymetry of a continental slope region can generate a long internal wave (known as an internal tide). Most favourable conditions for the generation seem to be when the slopes of the bathymetry and internal wave characteristics are of similar magnitude. For sub-critical slopes (sea floor is steeper than the characteristics) the internal wave propagates forward i.e. towards the coast, whereas for super-critical slopes, the waves are reflected off the bathymetry and propagate seawards. A number of studies have modelled these processes e.g. Baines [1] and Craig [2].

M. L. Banner, R. H. J. Grimshaw (Eds.)
Breaking Waves
IUTAM Symposium Sydney/Australia 1991

Observations from the Australian North West Shelf (NWS) show that as the long semi-diurnal internal tide (typical wavelength of 30km) progagates shoreward, the initially sinusoidal shape is distorted and eventually the leading face of the wave breaks to form an undular bore, or shock, as discussed by Smyth and Holloway [3]. In most occurrences the bore further breaks up into a packet of short period waves that can be reasonably described as solitons, Holloway [4]. As the solitons propagate shorewards they are gradually dissipated, presumably by turbulent and viscous processes, see Holloway [5].

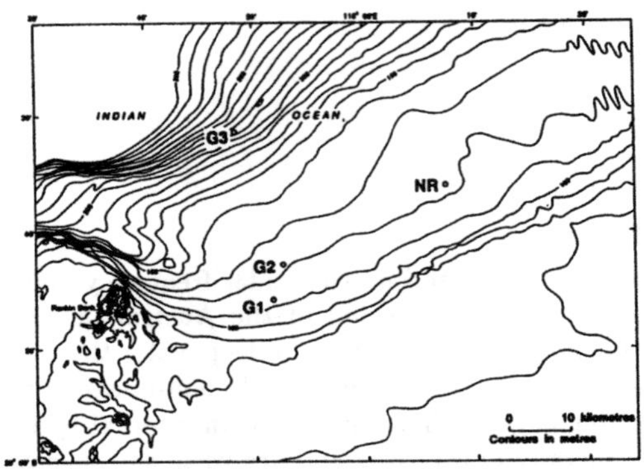

Figure 1. Bathymetry and mooring locations on the Australian North West Shelf

Observations

Current meter observations from 4 moorings and various depths from around the shelf break region on the NWS (Figure 1) are plotted, in Figure 2, as time series of currents in the cross-shelf direction, positive in the direction 135^0 east of north, which is the approximeate direction of wave propagation. All the records show an approximately 12 hour peroid oscillation which is a combination of both the barotropic and internal tides. At the deeper mooring (G3) the velocity has a relatively smooth sinusoidal shape with few shocks and short period waves. However, as the internal waves propagate into shallower water they are seen to steepen and form shocks or bores on the leading face. These records are less sinusoidal than from the deeper mooring and in some cases a backwards shock is seen on the trailing edge of the waves as well as the forward shock e.g. around 1000 19 February at G1 3m above the sea bed. Following many of the shocks at the shallower moorings, packets of short period waves are evident with typically 5 waves per packet, with the leading wave of largest amplitude. The variation of these short waves with depth, is defined as an initial depth of 400m with an exponential slope giving $K = 5.5 \times 10^{-3}$. Buoyancy frequency is $0.014s^{-1}$ and wave frequency is $1.41 \times 10^{-4}s^{-1}$. Shock strength is plotted as a fuction of distance from the origin of the wave for an initial wave amplitude of 20m and for three values of non-dimensional shear. The distance to breaking can be seen to be a strong function of the shear, decreasing from approximately 60km to 12km for an increase in the shear from $k = 0.1$ to 0.5. The initial shock strength also decreases for increasing shear. It can be seen that some shear is essential for producing a shock in a physically reasonable distance i.e. around 40km in this example. The corresponding solutions for negative shear (and negative mode number) give large distances to breaking (approximately 80km) with the distance only weakly dependant on the magnitude of the shear.

The observations presented in this paper have background shears that vary through the water column and over the measurement period from positive to negative giving values of k ranging from 0.35 to -0.12. The values are not constant with depth as assumed by the model, but are of the required strength to give reasonable agreement with the model.

Short Period Waves

Figure 4 shows an expanded view of a packet of waves from the NR mooring at different depths. Waves have periods of approximately 25 minutes with 3 or 4 well defined waves followed by weaker and poorly defined oscillations. The picture is very similar to theoretical solutions for the breakup of an initial step, e.g. Gurevich and Pitaevskii [7], where the step from one level to another is constructed from an initial shock, of strength twice the step, followed by a train of waves of decreasing amplitude. The first waves are solitons with trailing cnoidal waves gradually become sinusoidal waves. Certainly the observations show the initial wave of at least twice the step strength. The observed initial waves are also clearly strongly distorted from a sinusoidal form and show a soliton like shape.

This paper has shown the evolution of long internal waves as they steepen to the point of breaking. The mode of breaking is the formation of a shock or undular bore which is, at times, followed by a packet of short period cnoidal waves. These waves are dissipated, as they propagate, through turbulent and viscous processes. No large scale overturning of waves is observed.

e.g. at NR around 2300 21 February, are consistent with the modal solutions for internal waves, in particular showing the 180^0 phase change from surface to sea bed.

Model

Smyth and Holloway [3] used a perturbed extended Korteweg-de Vries equation to describe the evolution of an initially sinusoidal internal wave of amplitude a progagating shoreward over an exponential depth profile, $h = h_0 e^{-KX}$ where X is distance, in an ocean of constant buoyancy frequency N. In non-dimensional variables, the shock height is then given as,

$$\Delta\eta = 2ae^{KX} \sin\psi \; \phi(z) \tag{1}$$

where

$$X = \frac{1}{(2+\alpha\beta)K} \ln \left[1 + \frac{(2+\alpha\beta)Kh_0^2 k e^{-j\pi/r}(r^2+\frac{9}{4})\psi}{2Mra\omega(e^{j\pi/r}-1)(1-(-1)^j e^{-3j\pi/2r})\sin\psi} \right] \tag{2}$$

and the modal function, in the vertical coordinate z, is given by,

$$\phi(z) = M\left(1+\frac{k(z+h)}{c_0}\right)^{-\frac{1}{2}} \sin\theta \tag{3}$$

where

$$M = \left(1+\frac{1}{4r^2}\right)^{\frac{1}{2}} \exp\left[\frac{\tan^{-1}(2r)}{2r} + \frac{(j-|j|)\pi}{4r}\right]$$

and

$$\theta = r\ln\left(1+\frac{k(z+h)}{c_0}\right) \quad .$$

The distance the initial wave propagates to the point of breaking, and formation of the shock, is given by (2) with $\psi = 0$. The shock strength in terms of velocity can also be defined and is given as,

$$\Delta v = 2ae^{KX} \sin\psi \; c_0 \; \phi_z(z) \tag{4}$$

where the modal function for velocity is given by,

$$\phi_z(z) = \frac{kM}{c_0}\left(1+\frac{k(z+h)}{c_0}\right)^{-\frac{3}{2}} \left[r\cos\theta - \frac{1}{2}\sin\theta\right] \quad . \tag{5}$$

The phase speed of the waves is given as $c_0 = kh\left(e^{j\pi/r}-1\right)^{-1}$ where $r = \left(\frac{N^2}{k^2}-\frac{1}{4}\right)^{\frac{1}{2}}$. Other non-dimensional variables are the initial wave amplitude a_0, wave frequency ω, and k is the background shear defined from the background velocity profile of $U_0 = -k(z+h)$. All variables are re-dimensionalised by a depth scale H_0 and a buoyancy frequency N_0. We define $\alpha = a_0/H_0$ and $\beta = H_0/L$ where dimensional quantities are wave amplitude a_0, and wavelength L. Typical values from the continental slope region make α and β small,

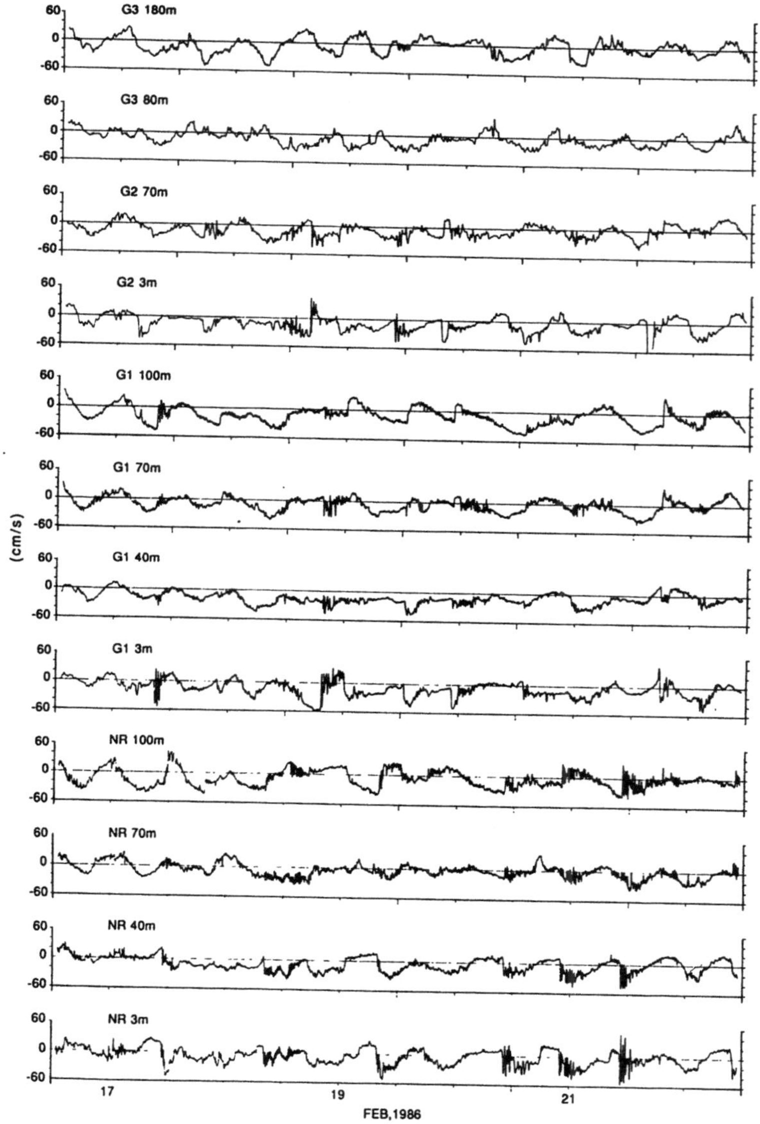

Figure 2. Time series of currents, positive in the direction 135° east of north, from four moorings at heights above the sea bed as indicated. Water depths at moorings NR,G1,G2 and G3 are 123,115,122 and 202m respectively

justifying the weakly non-linear assumption for the use of the extended KdV equation, see Gear and Grimshaw [6].

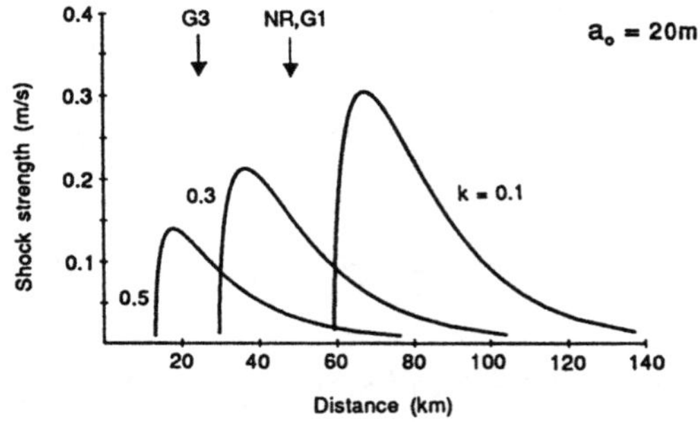

Figure 3. Plot of shock strength against distance from the origin of the wave for an initial wave amplitude of 20m and for three values of non-dimensional shear (k). Approximate mooring locations of the observations presented in this paper are marked

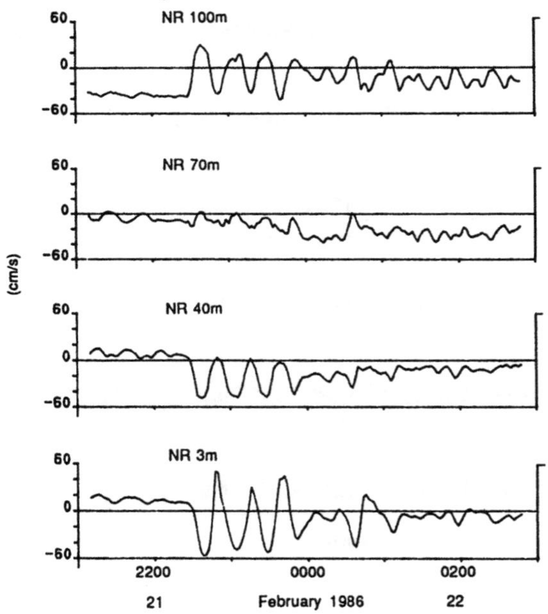

Figure 4. Onshore current components, from four heights above the sea bed at the NR mooring, showing detail of short period oscillations. Water depth is 123m

Using parameter values that are representitive of the conditions on the NWS, shock strength in terms of velocity is calculated from (4) and plotted in Figure 3. The bathymetry

References

1. Baines, P.G. (1974) The generation of internal tides over steep continental slopes. *Phil. Trans. Roy. Soc. London* **A227**, 27-58.

2. Craig, P.D. (1987) Solutions for internal tidal generation over coastal topography *J. Mar. Res.* **45**, 83-105.

3. Smyth, N.F. and P.E. Holloway (1988) Hydraulic jump and undular bore formation on a shelf break. *J. Phys. Oceanogr.* **18**, 947-962.

4. Holloway, P.E. (1987) Internal hydraulic jumps and solitons at a shelf break region on the Australian North West Shelf *J. Geophys. Res.* **92, C5**, 5405-5416.

5. Holloway, P.E. (1991) On the dissipation of internal tides. In, Tidal Hydrodynamics, ed B. Parker, John Wiley, New York.(In Press)

6. Gear, J.A. and R. Grimshaw (1983) A second order theory for solitary waves in shallow fluids. *Phy. Fluids* **26**, 14-29.

7. Gurevich A.V. and L.P. Pitaevskii (1974) Nonstationary structure of a collisionless shock wave *Sov. Phys. JETP* **38**, 291-297.

Internal Waves and Turbulence on a Gravity Current Under Ice

BERNARD C. KENNEY

National Hydrology Research Institute
11 Innovation Boulevard
Saskatoon, Saskatchewan, Canada S7N 3H5

Summary

Ice covered lakes are insulated from the usual turbulence sources such as near-surface shear and breaking gravity waves. Insolation that penetrated the ice and snow produced weak stratification and a weak gravity current between the main body of a lake and a shallow confined bay. At the mouth of the bay, the gravity current formed a two-layer flow with outflow from the bay at the bottom and inflow near surface. The inflow layer was well mixed. The largest temperature gradients occurred in a 2.5 m thick layer at the bottom that was dominated by internal waves. The amplitude of the internal waves decreased overnight and with increasing cloud cover. The temperature structure change rapidly with the arrival of front-like events. Nonlinear effects were also evident in the asymmetry of the internal waves and in a train of two solitary waves with 6 m wavelengths. Internal waves were an order of magnitude shorter in wavelength and period than previously reported in open water conditions.

Introduction

Kenney [1] reported weak gravity currents in early spring that were produced by solar radiation penetrating the ice and snow cover of Southern Indian Lake. The heating generated density differences (and horizontal pressure gradients) between the lake and a confined bay with significantly different mean depths. At the narrow mouth of the bay, the gravity current took the form of a two-layer flow with outflow from the bay near bottom and return flow near surface. The present paper reports on the under-ice temperature structure and the short internal waves observed in the narrow strait that separates Waskesiu Lake from an elongated bay.

Experimental Site and Methodology

The main body of Waskesiu Lake is 14 km long and 4.5 km wide. A bay that is typically 1 km wide extends 12 km north-west from the major axis of the lake. The bay is approximately half as

M. L. Banner, R. H. J. Grimshaw (Eds.)
Breaking Waves
IUTAM Symposium Sydney/Australia 1991

deep as the lake to which it is connected by the Narrows, a channel about 400 m long, 130 m wide and 6.65 m deep.

The temporal variation of the vertical temperature structure in the Narrows was measured with a chain of ten thermistors, spaced 0.5 m apart. Temperatures were measured at 3 second intervals from 14:37 to 21:14 Central Standard Time (CST) on March 14 and at 1 second intervals from 9:00 to 10:40 on March 15, 1991. Horizontal gradients in the vertical temperature structure were determined from measurements at 6 stations along the major axes of the lake and bay.

Horizontal wavelengths and phase speeds were subsequently measured with the same thermistor chain suspended horizontally 5 m below the water surface (1.65 m above the bottom) and aligned with the axis of the Narrows. The variation of the ten thermistors from a horizontal line was estimated to be less than ± 2 mm. Sampling (from 12:36 on March 27 to 10:37 on March 28, 1991) was at intervals of 3 seconds during the day and 6 seconds overnight.

The 10 thermistors were calibrated simultaneously in a well-mixed, constant-temperature bath against two platinum resistance thermometers with an absolute accuracy of ± 0.01 C.

Vertical Temperature Structure

The entire water column was below the temperature of maximum density of fresh water (4 C). The mean temperature structure was similar at all 6 stations with layers of weak stability at the surface and at the bottom, separated by an isothermal region. The mean temperature gradient measured in the Narrows was largest in the bottom layer (minimum Brunt-Vaisala period ≃ 8.3 minutes). The temporal variation of these layers shows that the bottom layer was dominated by short-period internal waves (Fig. 1).

The depth of the mean isotherms in the bottom layer sloped upwards from the lake towards the head of the bay with the most

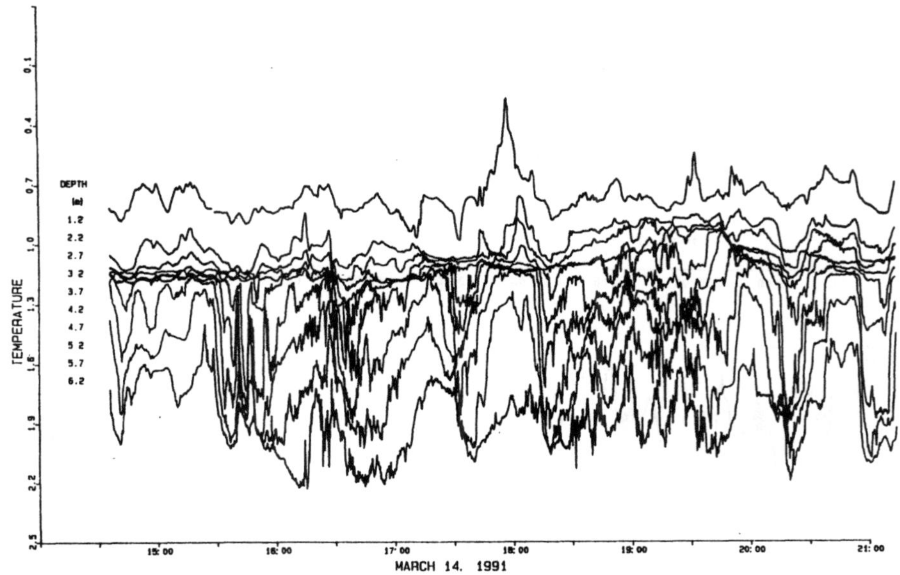

Fig. 1. Temporal variation of the vertical temperature structure in Waskesiu Narrows showing a central isothermal layer between stable layers at the surface and bottom. The bottom layer was dominated by short-period internal waves. Ice 0.39 m thick covered water that was 6.7 m deep.

rapid change occurring in the Narrows. The concomitant gravity current was directed out of the bay near bottom with return flow near the surface. Kenney [1] estimated that penetration of 1% of the incident radiation through the ice was necessary to drive the circulation observed in Southern Indian Lake. This estimate is supported by measurements on Waskesiu Lake where 3 watts/m^2 of the 425 watts/m^2 incident flux penetrated the ice and snow cover at noon on March 15th (Chew, personal communication). Diurnal and longer term variation in the amplitude of the internal waves that dominate the bottom layer also demonstrates the importance of the incident radiation to under-ice dynamics (Fig. 2). The largest amplitude internal

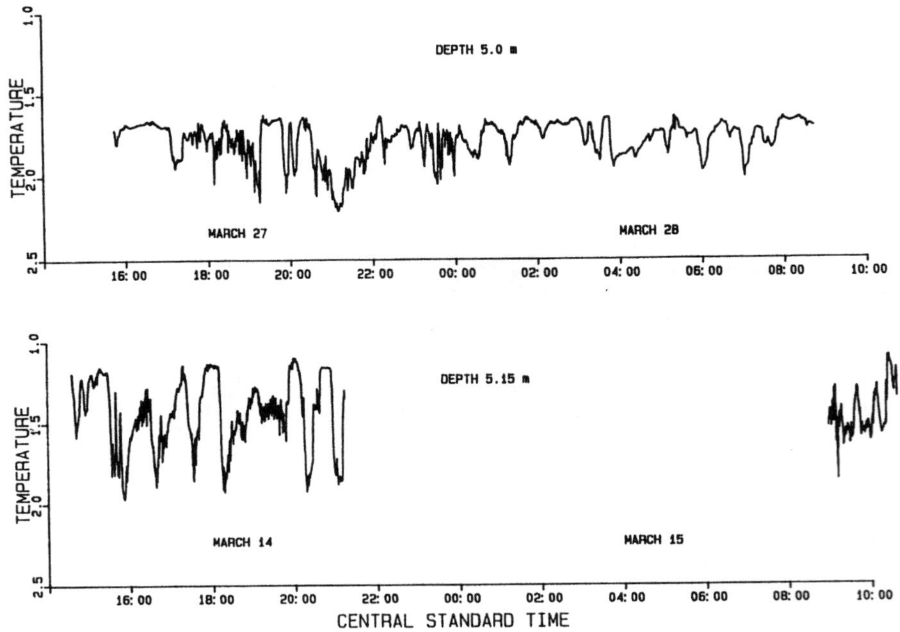

Fig. 2. Time history of temperature in the Waskesiu Narrows March 14-15 and March 27-28, 1991. The amplitude of the internal wave activity varied from day to night and from week to week with changes in the incident radiation.

waves were observed during bright sunshine on March 14th (7.8 hours of bright sunshine). The amplitudes were notably lower during the overcast conditions of March 27th (2.3 hours of bright sunshine). A diurnal variation in amplitude occurred on both days with the largest waves occurring in the late afternoon.

The vertical temperature structure can be explained by three processes: 1) gravity flow of warmer water downslope, 2) convective instability and turbulence caused by heating water (< 4 C) from above, and 3) near-surface fluxes of sensible and latent heat upwards toward the ice due to low air temperatures and melting at the bottom of the ice sheet. The first process produces stable bottom layers by downslope advection of warm water from the shallow littoral zones. The warm bottom layer in the bay flows as a gravity current through the Narrows down into the lake with return flow near surface. The second process is mainly responsible for the generation of isothermal water in the central part of the lake (i.e. wherever the water is deep relative to the light extinction coefficient). The isothermal water is then advected into the Narrows by the return flow. The third process generates the stable layer immediately below the ice.

Internal Waves

Most of the internal waves in the bottom layer propagated from the bay to the lake. On one occasion, a wave propagating from the bay appeared to stall. When it collapsed a train of waves was seen to propagate back into the bay. On other occasions, several thermistors at only one end of the horizontal array would fluctuate suggesting that not all of the internal wave modes were propagating along the axis of the Narrows.

The internal waves were generally asymmetrical with a very sharp increase in temperature followed by a much slower return to initial values (Fig. 2). Several of the waves had a double step structure on the front (Fig. 3). Many waves were similar in shape to internal surges previously reported in the ocean [2], [3] and in stratified lakes [4], [5], [6] but with much shorter wavelengths and periods. The internal waves on the afternoon of March 14 had a dominant period of 52 minutes (Fig. 2) compared with repetition rates of open-water fronts that varied from several hours to several days depending upon the nature of the tidal or wind-driven flows that provided the forcing.

A train of two solitary waves were also observed with symmetical front and back sides (Fig. 4). The first wave had a phase speed of 4.1 cm/s, a wavelength of 6 m and a height of 0.75 m (calculated using the mean temperature gradient). Solitary waves previously reported in the literature were typically 200-500 m in length and travelled at speeds of 0.20 - 50 cm/s.

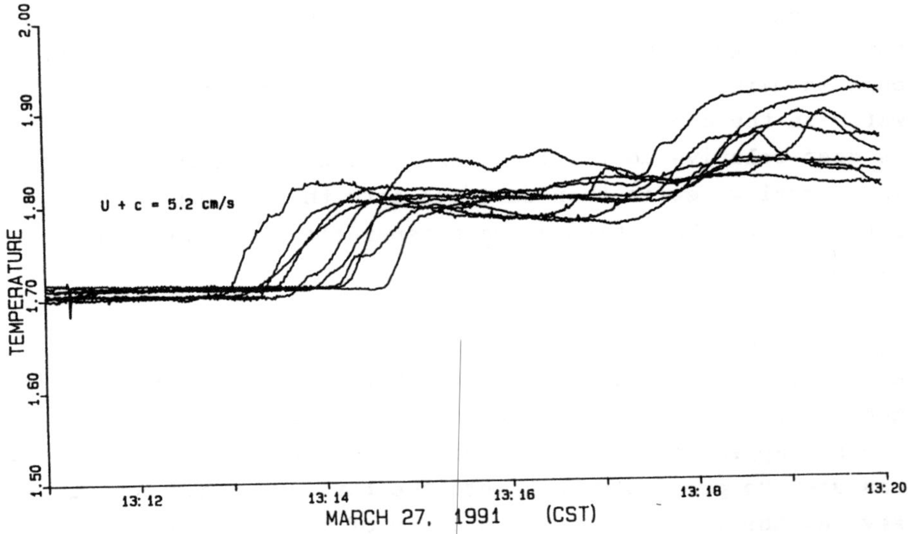

Fig. 3. Evolution of an internal front past a horizontal array of 10 thermistors spaced 0.5 m apart.

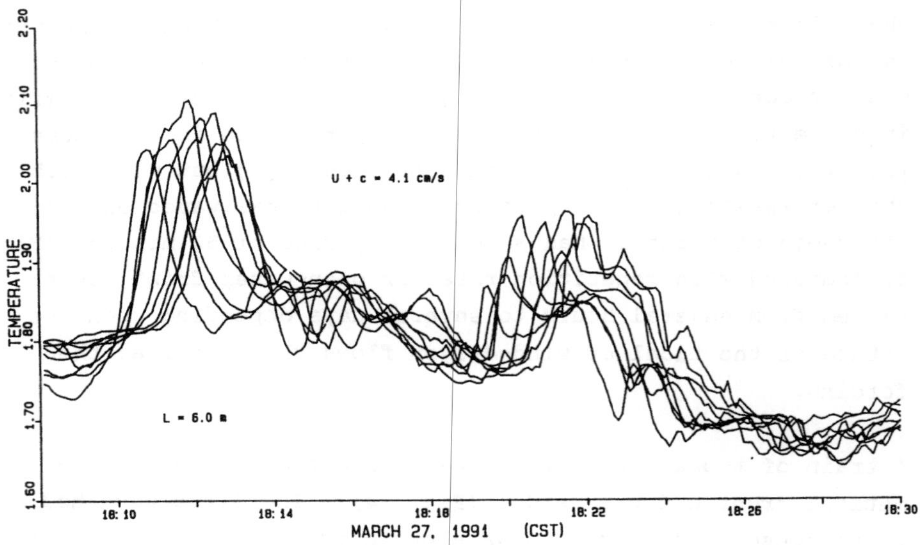

Fig. 4. Evolution of two solitons past a horizontal array of 10 thermistors spaced 0.5 m apart. The length and height of the first wave were 6m and 0.75m respectively.

There was no obvious forcing mechanism for either the solitons or the longer internal waves. The wave speeds were low because the stability of the water column was weak. For example, using the measured temperature gradient, the reduced gravity in a 2.5 m bottom layer was 0.00039 m/s^2. This is two orders of magnitude less than is generally found in stratified regions of lakes and oceans. The theoretical phase speed for long waves is 3.1 cm/s. The observed (fixed frame) phase speeds varied from 1.9 cm/s to 5.1 cm/s.

No velocity measurements were made during this experiment that would permit separating the phase speed of the waves relative to the speed of the current. A detailed study of the importance of the gravity current to the generation of finite-amplitude internal waves (as well as to the hazardous thinning of ice at the mouths of bays and to under-ice nutrient cycling) awaits the development of a current meter that is accurate at speeds < 3 cm/s.

Acknowledgements
The author is endebted to H. Chew for providing the radiation data, D. Bucilla for field support and D. Peters for computer software. D. Anions and S. Oakes provided the lake bathymetry.

References

1. Kenney, B.C.: Under-ice circulation and the residence time of a shallow bay. Can. J. Fish. Aquatic Sci. 48 (1991) 152-162.

2. Ziegenbein, J.: Short internal waves in the Strait of Gibraltar. Deep-Sea Res. 16 (1969) 479-487.

3. Winant, C.: Internal surges in coastal waters. J. Geophys. Res. 79 (1974) 4523-4526.

4. Thorpe, S.: Asymmetry of the internal seiche in Loch Ness. Nature 231 (1971) 306-308.

5. Hunkins, K. and Fliegel, M.: Internal undular surges in Seneca Lake: A natural occurrence of solitons. J. Geophys. Res. 78 (1973) 539-548.

6. Farmer, D.M.: Observations of long nonlinear internal waves in a lake. J. Phys. Oceanogr. 8 (1978) 63-73.

Extreme Standing Waves and Their Stability

G. N. MERCER & A. J. ROBERTS

Applied Mathematics Department
University of Adelaide, Australia

Summary

A stable and accurate numerical method to calculate the motion of an interface between two fluids is used to calculate two-dimensional standing water waves. The general method calculates arbitrary time-dependent motion of an interface, possibly including interfacial tension and different density ratios between the fluids. Extremely steep standing waves are determined, significantly steeper than has been previously reported. The peak crest acceleration is used as the determining parameter rather than the waveheight as the wave steepness is found to have a maximum short of the most extreme wave. Profiles with crest accelerations up to 98% of gravity are calculated and the shape of these extreme standing wave profiles are discussed. The stability of the standing waves is examined and growth rates of the unstable modes are calculated. It is found that all but very steep standing waves are generally stable to harmonic perturbations. However, standing waves are typically unstable to subharmonic perturbations via a side-band type instability.

Introduction

Two dimensional standing waves in deep water were first studied by Rayleigh [1] who calculated a perturbation series to third order using the wave amplitude as the small parameter. Later Penny and Price [2] improved this perturbation series to fifth order and noted that at no time is the interface ever flat. More recently Schwartz and Whitney [3] have used a conformal mapping method to simplify the perturbation series which they then calculated to 25th order.

We choose not to use a perturbation series method to calculate standing waves but instead use a generalisation of the method given in Roberts [4] allowing for a standing wave configuration. This method is based on the semi-Lagrangian approach which has been used for periodic waves by Longuet-Higgins and Cokelet [5]. Others [6,7] have also developed similar algorithms to numerically calculate free surface flows and their stability. The full details of the method used here are described in Mercer and Roberts [8]. In summary, it consists of distributing vortices along the unknown interface and from this determining the motion of the interface. By integrating this profile forward in time and using this to calculate a profile that replicates itself after an unknown period standing wave profiles can be found. It can be shown that this is equivalent to solving a system on N nonlinear equations in N unknowns.

Extreme standing waves

Previous methods of calculating standing waves on deep water have been unable to accurately calculate standing waves with a waveheight (h), defined to be half the crest to trough height, greater than $h = 0.62$. Here we choose not to use the waveheight as the determining parameter but instead use the initial crest acceleration, A_c, which is scaled such that the acceleration due to gravity corresponds to $A_c = 1$. The reason

M. L. Banner, R. H. J. Grimshaw (Eds.)
Breaking Waves
IUTAM Symposium Sydney/Australia 1991

for choosing this parameter to describe the standing waves is that there is a maximum in the waveheight of $h = 0.6202$ at $A_c = 0.9264$ (see Figure 1). This is why methods using perturbation series based on the waveheight [3] could not obtain convergence past $h = 0.62$. Previous estimates of the maximum waveheight are therefore not a reliable estimator of the most extreme standing wave due to the non–monotonic nature of the waveheight. The limiting standing wave is better referred to in terms of the limiting crest acceleration, which in the absence of surface tension is limited by the acceleration due to gravity ($A_c = 1$).

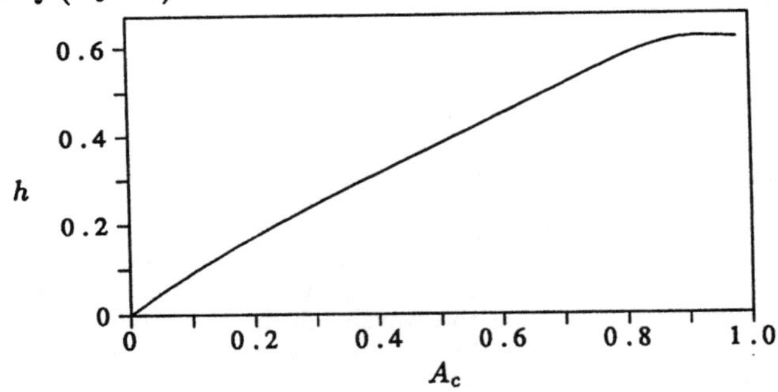

Figure 1. *Waveheight (h) versus the crest acceleration (A_c) showing the maximum waveheight* $h = 0.6202$ *at* $A_c = 0.9264$.

We were able to accurately calculate standing waves with a crest acceleration up to $A_c = 0.98$. Figure 2 is a plot of 4 profiles two of which have the same waveheight ($h = 0.6166$) but different crest accelerations. These profiles were checked using different integration stepsizes and different meshpoint distributions and were found to be consistent. Certain integral invariants, defined in [8], were also calculated and these varied by less than 3 in 10^{10} which compares well with the integration tolerance used. In all the profiles calculated the wave replicated itself after one period with a relative error of less than 1×10^{-8} and thus we are confident they are indeed standing waves.

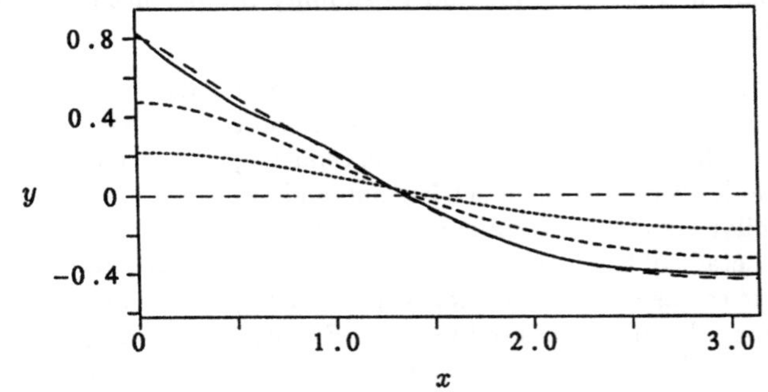

Figure 2. *Standing wave profile for 4 different crest accelerations with the corresponding waveheight.* $A_c = 0.9800$ [$h = 0.6166$] (——), $A_c = 0.8885$ [$h = 0.6166$] (– – –), $A_c = 0.5390$ [$h = 0.4000$] (- - - -), $A_c = 0.2384$ [$h = 0.2000$] (······).

The most extreme profile calculated has some interesting features to note. The profile has a sharp crest and a long broad trough with a pronounced hump in the profile near $x = 1$. There is more than one inflexion point, a fact first noted by [3], and in fact as the crest acceleration is increased the number of inflexion points increases (see Figure 3).

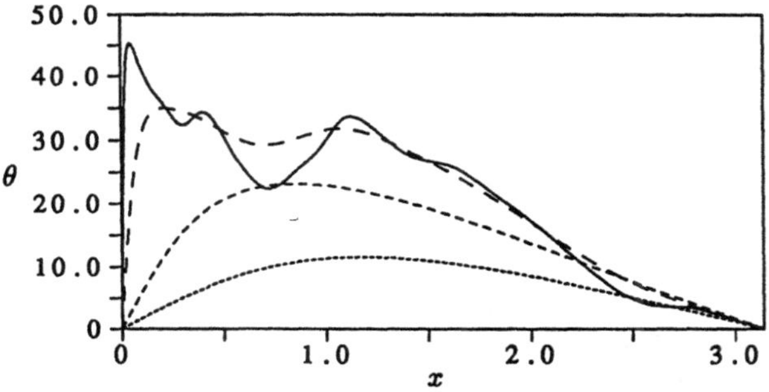

Figure 3. *Surface inclination (θ) of the profiles given in Figure 2.*

The shape of the limiting profile has been of interest for many years. It has been conjectured [3] that the limiting profile has a cusped crest with an enclosed crest angle of 90°, although this result was extrapolated from below the waveheight maximum and hence is in some doubt. Saffman and Yuen [9] argued that the limiting angle was less than 90° but their conclusions were unclear as they were not considering pure standing waves. The current work suggests that the enclosed crest angle is considerably less than 90° (see Figure 4). A surface inclination of 45° corresponds to an enclosed crest angle of 90°. For the highest crest acceleration we calculate the maximum surface inclination is 45.3°. Also shown in Figure 4 are some possible extrapolations to the most extreme profile (the dashed lines). The most likely, but my no means definite, extrapolation gives an enclosed crest angle of between 60° and 70° for the limiting wave ($A_c = 1$). This is clearly an area still open to considerable research and discussion. The shape of the limiting profile is as yet undetermined.

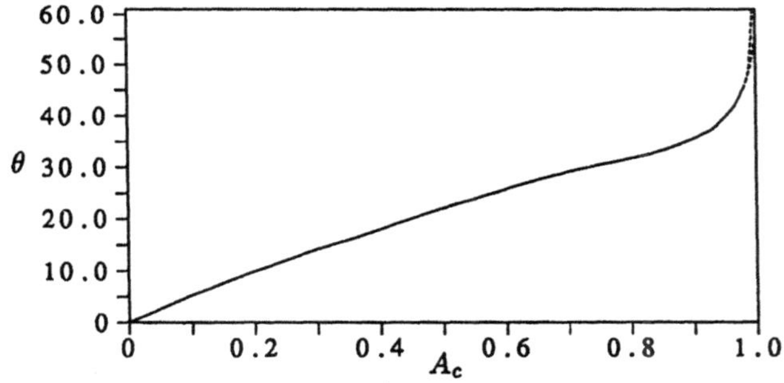

Figure 4. *Maximum surface inclination versus the crest acceleration showing some possible extrapolations to the most extreme profile*

Stability

The stability of steady two dimensional gravity waves on deep water has been extensively studied [10, 11, 12, 13] in contrast to this the stability of two dimensional standing waves has had scant attention. Here we will study both the subharmonic and superharmonic instabilities of standing waves. The method used is described in full in [8]. To summarize, the analysis is based on Floquet theory which reduces the problem to an eigenvalue problem for the complex frequency of the perturbation.

We first consider the superharmonic perturbations, that is perturbations having the same horizontal length scale as the fundamental wave or less. For all standing wave profiles with $A_c < 0.893$ ($h = 0.6175$) the fundamental wave was found to be stable to superharmonic perturbations. For profiles with $A_c > 0.983$ there was some evidence of a physical instability although the exact nature of this instability could not be determined. Calculation with more mesh points and a more detailed analysis of the region is needed to fully ascertain the stability criterion and type. This is an area of current research.

Secondly, we investigate the stability of the subharmonic perturbations, those perturbations having a horizontal length scale greater than the fundamental wave. Using a rescaling similar to that used by Longuet–Higgins [11] we investigate the subharmonic perturbations with length scales up to 8 times that of the fundamental wave. Shown in Figure 5 is the eigenfrequency of each mode versus the waveheight (h). Using a notation similar to [11] we recognize the modes (7,9), (6,10) and (5,11) as becoming unstable at $h = 0.382$, 0.199 and 0.335 respectively. These 3 instabilities are all side–band instabilities resulting from the interaction of the base wave, mode $m = 8$, with modes $m \pm \ell$ where ℓ is integral. The shaded area is the region of instability. Figure 6 shows the average growth rates of the unstable modes. As the waveheight increases the maximum growth rate also increases. It is expected that for higher waveheights more of these side–band instabilities will occur, for example (4,12) and (3,13).

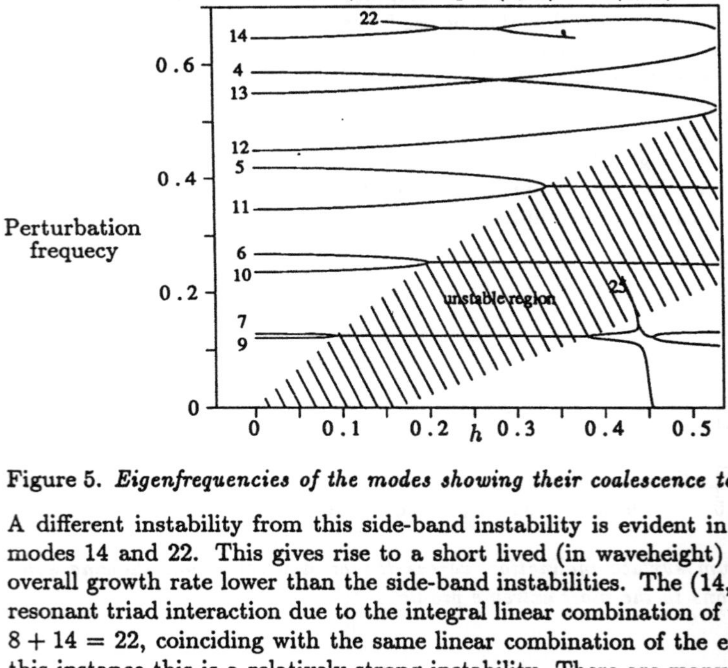

Figure 5. *Eigenfrequencies of the modes showing their coalescence to form instabilities.*

A different instability from this side-band instability is evident in the coalescence of modes 14 and 22. This gives rise to a short lived (in waveheight) instability with an overall growth rate lower than the side-band instabilities. The (14,22) instability is a resonant triad interaction due to the integral linear combination of the mode numbers, $8 + 14 = 22$, coinciding with the same linear combination of the eigenfrequencies. In this instance this is a relatively strong instability. There are many apparent crossings of eigenfrequencies not all of which lead to instability. These crossings fall into the three categories as defined by MacKay and Saffman [13]: avoided crossings where the eigenfrequencies do not actually cross but avoid each other; bubbles of instability where the eigenvalues coalesce and move off the imaginary axis resulting in an instability; and crossings that do not interact. The exact nature of all the crossings here could be determined using the signature method as described by MacKay and Saffman (1986). Most of the crossings appear to be the ones that do not interact. Some of the crossings may well be bubbles of instability but the range of waveheights that they are unstable over must be very narrow. The side–band instability is the dominant instability for all waveheights we have considered.

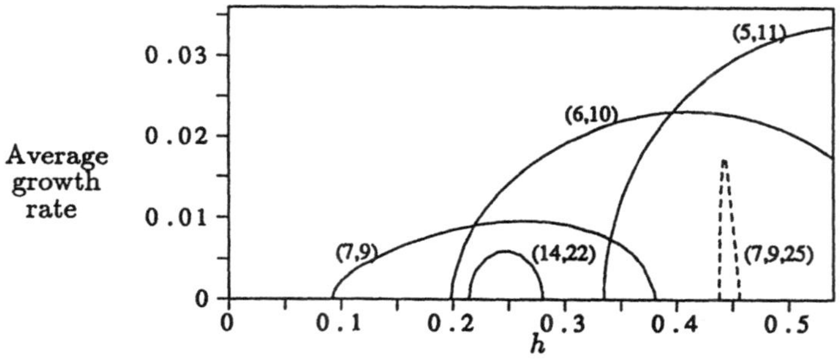

Figure 6. *Average growth rates of the unstable modes*

The instability denoted by (7,9,25) is a non-physical instability due to the fact that as the waveheight is increased the high modes become unstable due to computational error. In this case mode 25 is interacting with modes 7 and 9 to form a non-physical instability. These modes restabilise once the eigenfrequency of mode 25 has reached zero.

References

1. Rayleigh, J.W.S. Deep water waves, progressive or stationary, to the third order of approximation. Proc. Roy. Soc. Lond. A 91 (1915) 345–353.
2. Penny W.G. & Price A.T. Finite periodic stationary gravity waves in a perfect fluid, Part 2. Phil. Trans. Roy. Soc. London A 244 (1952) 254–284.
3. Schwartz L.W. & Whitney A.K. A semi-analytic solution for non-linear standing waves in deep water. J. Fluid Mech. 107 (1981) 147–171.
4. Roberts A.J. A stable and accurate numerical method to calculate the motion of a sharp interface between fluids. IMA Journal of Applied Mathematics 31 (1983) 13–35.
5. Longuet-Higgins M.S. & Cokelet E.D. The deformation of steep surface waves on water. I. A numerical method of computation. Proc. Roy. Soc. Lond. A 350 (1976) 1–26.
6. Pullin D.I. Numerical studies of surface-tension effects in nonlinear Kelvin-Helmholtz and Rayleigh-Taylor instability. J. Fluid Mech. 119 (1982) 507–532.
7. Baker, G.R., Merion, D.I. & Orszag, S.A. Generalised vortex methods for free surface flow problems. J. Fluid Mech. 123 (1982) 477–501.
8. Mercer G.N. & Roberts A.J. Standing waves in deep water: their stability and extreme form. Physics of Fluids A (1991) submitted.
9. Saffman P.G. & Yuen H.C. A note on numerical computations of large amplitude standing waves J. Fluid Mech. 95 (1979) 707–715.
10. Longuet-Higgins M.S. The instabilities of gravity waves of finite amplitude in deep water. I. Superharmonics. Proc. Roy. Soc. Lond. A 360 (1978) 471–488.
11. Longuet-Higgins M.S. The instabilities of gravity waves of finite amplitude in deep water. II. Subharmonics. Proc. Roy. Soc. Lond. A 360 (1978) 489–505.
12. McLean J.W. Instabilities of finite-amplitude water waves. J. Fluid Mech. 114 (1982) 315–330.
13. MacKay R.S. & Saffman P.G. Stability of water waves. Proc. Roy. Soc. Lond. A 406 (1986) 115–125.

MIX
Papier aus verantwortungsvollen Quellen
Paper from responsible sources
FSC® C105338

If you have any concerns about our products,
you can contact us on
ProductSafety@springernature.com

In case Publisher is established outside the EU,
the EU authorized representative is:
Springer Nature Customer Service Center GmbH
Europaplatz 3, 69115 Heidelberg, Germany

Printed by Libri Plureos GmbH
in Hamburg, Germany